Systems Analysis for Effective Planning

Wiley Series on Systems Engineering and Analysis

HAROLD CHESTNUT, Editor

Systems Analysis
for Effective Planning:
Principles and Cases

Bernard H. Rudwick

The MITRE *Corporation, Bedford, Massachusetts*
and Member of the Faculty
Northeastern University Center for
Continuing Education

John Wiley & Sons, Inc.

New York · London · Sydney · Toronto

Library of Congress Catalog Card Number: 69-19098
SBN 471 744867
Printed in the United States of America

to Regina

SYSTEMS ENGINEERING AND ANALYSIS SERIES

In a society which is producing more people, more materials, more things, and more information than ever before, systems engineering is indispensable in meeting the challenge of complexity. This series of books is an attempt to bring together in a complementary as well as unified fashion the many specialties of the subject, such as modeling and simulation, computing, control, probability and statistics, optimization, reliability, and economics, and to emphasize the interrelationship between them.

The aim is to make the series as comprehensive as possible without dwelling on the myriad details of each specialty and at the same time to provide a broad basic framework on which to build these details. The design of these books will be fundamental in nature to meet the needs of students and engineers and to insure they remain of lasting interest and importance.

Foreword

An increasing emphasis in today's world is being placed on change and it is likely that attention to change will continue to grow in the future. The opportunity for change and the presence of uncertainty makes planning desirable and essential. Systems analysis for effective planning has been needed, particularly in the military, and much work has been done over the past decade or more to develop principles for carrying out effectively such analysis.

Experience gained in systems analysis of military systems has been valuable but costly. It is desirable that this sort of experience be presented to a broader readership and that the ideas gained be made available to others for use wherever such systems analysis for effective planning is not so well documented. Bernard Rudwick's book describes systems analysis principles and cases principally in terms of military examples where the major expenditures of time and effort have been placed; however, non-military uses are also indicated.

Rudwick has had many years of experience in developing and teaching the material presented in this book. He has illustrated many of the principles with practical cases which initially have much of the poorly-defined flavor of the real world. His important contributions are to recast the problem as given to make it the problem as understood, and to indicate methods for its solution.

Although many of the cases illustrated here have a military setting, the fundamental principles and methods are frequently applicable to non-military situations as well. Readers of this book will find much to benefit them from a study of the methods described. The principles are useful both in their original context as well as in a number of broadly related fields.

It has been a pleasure for me to work with Rudwick as he has developed and clarified his exposition of the ideas on systems analysis for effective planning.

HAROLD CHESTNUT

Preface

This book is about the application of systematic, quantitative methods and techniques to the task of planning. However, the principles covered here have such broad applications to the more generalized topic of problem solving that other descriptions of the topic might be helpful. These other descriptives might include the following:

1. How to plan for change, particularly technological change, in an organization.
2. A creative, systematic approach to the task of problem solving.
3. An efficient way of looking at problems in an attempt to obtain the preferred solution.
4. A method that can help a planner to convince others of the rationality of his proposal.
5. A method that can help a decision-maker to evaluate proposals effectively for change.

OBJECTIVE OF BOOK

This book closely documents a formal two-semester course entitled "Systems Analysis for Effective Planning," which has been presented a number of times at The MITRE Corporation and at Northeastern University's Center for Continuing Education. In general, the students were technical systems planners, systems engineers, or technical managers of planning groups. In addition, portions of the material have been presented at a number of university and defense courses on systems engineering and resource management.

The course, hence the book, was developed to meet the following observed needs:

Although much has been written on the philosophical or intellectual level regarding the need for systems analysis or cost-effectiveness analysis as part of the systems planning process, there has been a lack of unclassified, methodological material which showed explicitly how to attack complex, unstructured problems involving choice among system alternatives.

It was felt that there is a lack of appreciation of the real problems which higher-level decision makers face in confidently making choices among alternate systems and the type of structured information they require.

Thus the objective of this book is to show the following:

1. There is a describable approach for dealing with complex problems involving high uncertainties.

2. This approach can and does provide much of the key information that decision makers need to choose rationally among alternatives.

3. That systems analysis can serve to sharpen the intuition of experienced system planners.

4. That despite its limitations systems analysis can be far superior to any alternative approach for making policy decisions.

AUDIENCE TO WHOM THIS BOOK IS DIRECTED

By bringing together the key principles and methods of systems analysis and showing how they are applied in representative cases of generalized interest, this book will be of interest to the following audiences: First, to those individuals currently engaged in systems planning who have not been exposed to a methodological presentation of how this work can be accomplished efficiently and who wish to improve their understanding of the formal process involved. This book should provide these planners with more systematic methods for creating more and better system alternatives, as well as a better understanding of how to relate the value of systems to the higher-level organization objectives.

Second, the material presented here should be of interest to managers or decision-makers who wish to have a better understanding of the quantitative methods and techniques of systems analysis so that they may see what information could and should be provided them by a trained systems analysis team located in their organization.

Third, the book should be of interest to systems engineering and other students who are interested in learning more about systems analysis or the "systems approach" before entering the systems planning field in either the defense or nondefense sector of the economy; it should also be of interest to university faculty who want to develop courses for providing such knowledge.

PHILOSOPHY OF PRESENTATION

The principles of systems analysis are relatively straightforward and, perhaps, deceptively simple. A listing of such principles might include the following:

1. Be explicit in all aspects of the analysis.
2. Determine the objectives of the job to be done.
3. Identify and attempt to quantify all of the key factors involved in the system and the job to be done.
4. Determine the complete economic implications of each alternative under consideration.
5. Conduct the analysis in an iterative fashion.

What may be difficult is the task of applying such principles to real problems. I believe that one can truly learn these principles only by grappling with real problems. Hence the teaching approach used in the course on which this book was based consisted of a presentation of key principles of systems analysis followed by a series of complex, lifelike problems used as a vehicle for reinforcing and showing how to apply these principles. Various techniques, such as probability theory, decision theory, and economic theory, are deliberately introduced on an "as needed" basis so that the need for the theory is established, the theory presented, and its application shown. This book has not been written as a customary reference book but more as a self-teaching document. Moreover, since principles are developed in a cumulative fashion, the book should be read completely for maximum reader utility.

Before presenting the overview of the approach to be followed, let me say something about the type of problem used to illustrate the principles discussed. For several reasons many of them are defense oriented. First, they are the problems with which I have been associated; second, they are complex and indicate the need for thorough analysis; third, in my opinion there is a need for a book that documents the detailed approaches used in analyzing defense oriented problems. I have attempted to cover a wide range of planning problems in the defense industry. For these reasons this book provides a double benefit to the defense oriented systems planner. As will be seen, however, these problems can be readily understandable to the nondefense planner. Moreover, as described in the overview, many of the same *classes* of problem also occur in the nondefense field and the same principles of systems analysis also apply. These classes include support systems such as information systems, maintenance systems, logistics, and spare-parts inventory systems as well as the general class of "flow systems" encountered in the analysis of manufacturing or distribution systems. Hence each chapter contains a description of the general principle(s) involved, the specific application of each, and a discussion of the generalized class of problem in which the same principle(s) can be applied.

There are many acknowledgments I wish to make for all the help received in preparing this book. First, I wish to thank the management of The MITRE Corporation, particularly Ken McVicar, for the encouragement and support he has given me. It should be noted that although I have drawn heavily on my experience at The MITRE Corporation in writing the book the contents reflect my own views and do not necessarily reflect the official views or policy of the Corporation or its employees.

Many others have been of help to me in this effort. In particular, I would like to thank my colleagues Jack Porter, Charles Godwin, Lee Morris, Martin Jones, Joseph Ye, William Marcuse, and Harold Glazer of The MITRE Corporation and Lieutenant Colonel James Blilie of the Electronic Systems Division of the United States Air Force for their review of all or parts of the book and the thoughtful comments they provided. I am also indebted to David Votaw for all of his comments on the treatment of the probability and statistics portions of the book and to Clare Farr, Nelson Briggs, Louis Perica, Jr., Pat Chatta, and Rosemary DeFusco, all from MITRE, for their help. My thanks also to Harold Chestnut and Donald Heany of the General Electric Company for my discussions with each which proved so helpful. A special thanks to Mrs. Joan Blanchard and Mrs. Barbara Olson for the excellent typing and all the painstaking efforts that were involved in assembling this manuscript with such skill, patience, and good humor and to Ted Cutting for his meticulous editing of proof.

Last I wish to acknowledge Robert McNamara, Charles Hitch, and Alain Enthoven for their foresightedness in conceiving the need for such analytical approaches to management and their abilities in being able to implement them to the extent that they did; for without these approaches this book could not have existed.

BERNARD H. RUDWICK

Lexington, Massachusetts
June, 1968

Contents

I

SYSTEMS PLANNING AND THE DECISION-MAKING PROCESS

1

Introduction and Overview

This chapter indicates how the book is organized and summarizes the philosophy of presentation. This is accomplished by discussing the term "systems analysis," describing the role which it has played in the acquisition of systems, (particularly defense systems) and showing that it is actually a form of systems planning performed at the decision-making level. The role which President Johnson has directed systems analysis to play in the nondefense area of federal programs is also discussed. Finally, a summary of the key topics of each chapter is presented in order to give an organizational "road map" to the reader.

The field of systems analysis, (sometimes called cost-effectiveness analysis, cost-benefit analysis or cost-utility analysis), as applied to the decision-making process, is relatively new. It has achieved its greatest development in the area of defense systems under the direction of Robert S. McNamara, the eighth Secretary of Defense. However, since the objective of systems analysis is to aid a manager in decision-making, it can be applied beneficially to many other fields when choosing a preferred alternative on some rational basis is the key problem. Application of systems analysis to the nondefense sector of the national government was advanced by President Johnson when he directed * that every department and agency of the government "will set up a very special staff of experts who, using the most modern methods of program analysis, will define the goals of their department for the coming year. And once these goals are established this system will permit us to find the most effective and least costly alternative to achieving American goals." †

Following the presidential announcement, Charles L. Schultze, Director, Bureau of the Budget, made these points about the new approach: ‡

---

* Presidential Directive, August 25, 1965.
† As discussed in Chapter 4, the President no doubt meant "the most effective alternative at given cost, or the least costly alternative to achieve a given level of effectiveness," but not an unachievable mini-max solution.
‡ *The New York Times,* August 26, 1965.

"It is not designed to make decisions but to enable us to ask the right questions. There would be no computerized decision-making.

"It is designed to make the government work all year long on its future programs instead of crowding all the work into the last few hectic weeks of the year as the budget is being drawn up.

"It is also designed to give, where possible, the cost of a given program over several years ahead instead of just for the immediate budget year."

These management techniques are also being applied to large problems of the state and local governments. For example, California has utilized trained systems analysts from the aerospace industry to cope with problems of transportation, waste management, crime, and smog in a state pilot project.

Much has been written about the so-called management revolution which has taken place in the approach to managing what has been called the world's largest business, the Department of Defense. Many dollars have been expended in this area to develop better management techniques for coping with the difficult problems of decision-making with which the Defense Department is faced. Many of these techniques have a direct carry-over to the problems of optimal resource allocations (i.e., systems planning) in any organization.

WHAT IS SYSTEMS ANALYSIS?

The term "systems analysis" has been used by different practitioners to describe different classes of work. To emphasize the differences, I will describe these classes in exaggerated form. At one end of the work spectrum are the mathematically oriented analysts who wish to apply a set of optimization techniques to highly structured problems. Thus at the extreme, if a decision-maker provides them with a well defined structure to his problem, including the objective he wishes to optimize (e.g., company profit or targets destroyed), the analyst will compose a set of mathematical or logical equations containing a set of relevant variables of interest and boundary condition constraints, and will find some way of determining the mathematically optimum operating point of the system. Such work might be called "the mathematics of systems analysis."

On the other end of the work spectrum (e.g., as exemplified by the systems analysts of the RAND Corporation) are those analysts whose starting point is the unstructured problem of the decision-maker. Their major objective is to build a proper structure to the problem, including uncovering the true goals of the decision-maker. Their emphasis is on what might be called "the logic of systems analysis." Only then do they continue to find the preferred solution using the mathematical techniques of optimization already mentioned. Both types of analysis are required for system planning; this book concentrates on the logic of systems analysis needed to structure a problem, and shows how a preferred solution may be reached. While elementary concepts of probability and statistics are touched upon when necessary, the book as a whole is nonmathematical.

One of the best discussions on systems analysis is contained in a brief article by Dr. Alain Enthoven, Assistant Secretary of Defense (Systems Analysis) who states: *

"What is systems analysis? I have not been able to produce a good brief definition. I would describe the art as it has evolved in the Department of Defense as a reasoned approach to problems of decision. Some have defined it as quantitative common sense. Alternatively, it is the application of methods of quantitative economical analysis and scientific method in the broadest sense to the problems of choice of weapon systems and strategy. It is a systematic attempt to provide decision-makers with a full, accurate, and meaningful summary of the information relevant to clearly defined issues and alternatives."

Another definition of systems analysis is given by the RAND Corporation, perhaps the leading developer of this field: †

"Systems analysis is an inquiry to aid a decision-maker choose a course of action by systematically investigating his proper objectives, comparing quantitatively where possible the cost, effectiveness, and risks associated with alternative policies or strategies for achieving them, and formulating additional alternatives if those examined are found wanting."

This definition of systems analysis emphasizes certain key words and their relationships which can be illustrated in Figure 1-1.

Since systems planning may be defined as the process of generating and evaluating alternative ways of changing a system in some beneficial manner, Figure 1-1 may be used as a means of discussing the relationship between systems analysis (as defined by RAND) and systems planning. This figure also shows that:

1. The problem on which both the systems planner and analyst work comes from some identifiable decision-maker or organization manager having resources (such as manpower or dollars) which can be applied to the solution of the problem.

* Navy Review, 1965.
† E. S. Quade, Ed. (1964), p. 4.

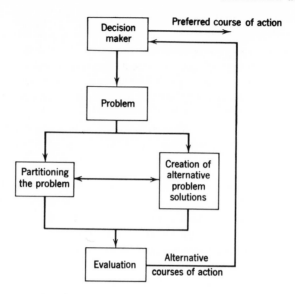

Figure 1-1.   The systems analysis and planning process.

2. Analysis is used in the solution of the problem in two ways: first, in dissecting or partitioning the problem so that alternative solutions can be configured, and second, to evaluate these solutions to determine how well each alternative solves the problem (the benefits), as well as the costs and risks of implementing the alternative. The evaluation procedure must also take into account the set of values of the decision-maker. During this analysis phase, determining proper objectives of "what ought to be done" is many times the biggest problem.

3. If none of the proposed problem solutions meets the objective adequately, (or within acceptable cost restrictions), the results of the analysis accomplished thus far may be used to create additional alternatives which are then also evaluated.

4. The results of the evaluation appear in the form of relevant information structured for the decision-maker. This information includes those alternative courses of action examined, the results which each solution would be expected to produce, the resources required by each, and the alternative recommended for implementation. If the decision-maker approves the recommended course of action or some other alternative examined (including the option of doing nothing at the present time), the systems planning phase is completed. On the other hand, if the decision-maker finds that no solution is acceptable because of certain deficiencies, he may ask that the planning effort be continued, and a reiteration of the process described oc-

curs, based on the reactions obtained from the decision-maker. Such iterations continue until the decision-maker is satisfied with the proposed solution, taking into account the additional costs involved in another iteration of the dynamic process.

Thus, in summary, the process as depicted in Figure 1-1 starts with a problem, finds a number of alternative solutions, evaluates these, and arrives at a preferred course of action which can be recommended to a decision-maker. Systems analysis, as defined by the RAND Corporation, and as used in this book, encompasses the evaluation function of systems planning since the alternatives initially considered were created by the systems planners. However, the RAND definition of systems analysis also involves the "formulation of additional alternatives if those examined are found wanting." Hence, the major difference between the work elements of systems planning and systems analysis appears to be one of emphasis. The systems planner starts, generally, with a need and develops a preferred solution among several developed, whereas the systems analyst starts with the preferred solution and the other alternatives developed and investigates whether the recommended preferred solution does indeed offer the best approach, including the possibility of uncovering other approaches during the evaluation process. How systems analysis is used as part of the systems planning process is described in further detail in Chapter 2.

One additional comment should be made to differentiate systems analysis from decision-making. Recall that the RAND definition stated that systems analysis is an inquiry to *aid* a decision-maker. Invariably there will be factors which have been omitted from the analysis for a variety of reasons: the analyst may not be able to quantitatively include a factor (such as morale), or adding certain factors properly would not permit the analysis to be completed within the constraints of the allocated time or available resources. The decision-maker himself may not wish to have certain "political" factors included explicitly. To operate under such constraints, the analyst is forced to make appropriate compromises in the analysis but should explicitly indicate these other important considerations in qualitative form to the decision-maker, permitting him to factor this into his interpretation of the analysis when making his decision.

OVERVIEW

Part I establishes the context for systems analysis by further detailing its relationship to systems planning and the decision-making process. While systems analysis applies to all phases of the systems acquisition process (especially to that portion known to the defense community as the "concept formulation phase" and to the industrial community as the "systems planning phase"), Chapter 2 examines the activity of systems planning and explores the roles which systems analysis plays within this activity. In doing so, the following questions are explored:

1. Why is systems planning performed?
2. What is the work process involved and what skills are needed for its implementation?
3. What is the role of systems analysis to the systems planning phase?

The need for a systematic method and the use of quantitative techniques in performing the systems planning phase is discussed.

Chapter 2 shows that the objective of systems analysis is to provide meaningful information to a decision-maker faced with the problem of allocating resources among a number of system proposals which have been generated, and whose total costs exceed the amount of funds available, while Chapter 3 develops a set of information needs in the form of questions which should be answered if the decision-maker is to rationally select among proposed alternatives and properly allocate resources available to the organization. Some of the management principles involved in this decision-making process are described in examining the management information system developed by one large organization, the Department of Defense (DOD), in coping with its problems of medium and long range planning and acquisition of large complex systems. As the reader will see when following the problems which led to the development of this management information system, many of the same systems planning problems faced by DOD are also faced to some degree by managers of all organizations and many of the approaches to solutions which were developed by DOD will also be applicable to decision-makers involved in capital budgeting problems within these other types of organizations.

The objective of Part II, Chapter 4, is to develop an over-all procedure for evaluating a system alternative in terms of its cost and its effectiveness in accomplishing an objective. In particular, the role that models (particularly analytical models) play in systems planning is discussed. The various types of models are covered, including those which describe a system, the job to be done, and the environment in which the system operates. The two models used for evaluation are described: (a) the effectiveness model which enables the analyst to predict how well the system will perform a job in a given environment, and (b) the cost model which predicts total system cost requirements. Finally, the interrelationship of these models in performing a systems analysis, using the steps in systems planning as discussed in Chapter 2, are developed.

Part III which contains the first case situation has a number of objectives.

The primary objective is to illustrate how to perform a systems analysis at the mission level, following the procedure previously described in Chapter 4. Chapter 5 consists of a verbal description of the problem, culminating in the development of an over-all qualitative structure to the problem, while Chapter 6 focuses on the development of models used to quantify the system performance characteristics, and the environment, and to determine system effectiveness. Heavy emphasis is given to operations analysis methods of gathering operational data to generate probabilistic models of the many random processes involved in this problem, and to the use of applied probability theory and techniques for combining the series of models representing the random processes involved.

Chapter 7 describes the two primary methods of model exercise (i.e., combining the numerical equations of the model used to predict system effectiveness). These are the Monte Carlo simulation method, and various analytical methods which could be used. Since the output of an analysis is only as good as the logic of the models used and the accuracy of the data employed, Chapter 8 explores the various uncertainties associated with the analysis—technological, statistical, and competitive.

Concluding topics are then presented, including how the models developed in the analysis can be used to generate new system concepts which have not been previously identified. Lastly, methods for presenting the results of the analysis to the decision-maker are also discussed.

While the analyses in Part III focus on the problem of determining how well different system alternatives would accomplish an operational objective as measured by some effectiveness measure, the cost aspects of this problem require that the systems planner be able to estimate the amount of scarce resources required for each alternative. Thus, Chapter 9 focuses on that task which will involve such problems as ways of translating scarce resources into some common denominator(s) of cost, determining which costs should be included in the analysis, and how to combine costs which occur at different times over the system life. Methods of coping with cost uncertainties are also discussed, including techniques for translating performance uncertainties into cost uncertainties.

Part V is concerned with the application of systems analysis to the problems of subsystem planning. By a subsystem we mean those systems which perform some function of the operational mission, rather than the total mission. Hence this part deals with systems planning at the functional level. The objectives of Part V are:

1. To show how to plan or design a system, taking into account the subsystem considerations;
2. To show how to obtain the proper balance of these subsystem ele-

ments so that the mission objectives can be achieved at lowest total system cost;

3. To show how to properly suboptimize (i.e., to plan or design a functional subsystem where other parts of the system cannot be changed or varied as they can in the optimization of a total systems design).

The first subsystem planning case of Chapter 10 was chosen primarily to illustrate one of the most important principles in systems analysis: how to select between two functional subsystems, each of which provides different levels of performance and each having a different cost. We show that even when evaluators and decision-makers can agree that one system is superior to another, if the costs of the first system are higher it may still be difficult to select the preferred system on some rational, defensible basis (i.e., to determine and be able to show if the increase in benefits is worth the increase in cost). It will be shown that the only way to make such a selection is to analyze the higher mission level systems problem.

Chapter 11 addresses the problem of how to properly design a system to meet a future demand in an acceptable fashion. Two specific applications are analyzed to illustrate the system design principles involved. The first is the design of an inventory/supply system where the key question concerns what quantities of spare parts should be stored at different locations to efficiently meet the future fluctuating demands for these parts. The second problem deals with determining the proper capacity of a bridge to satisfactorily meet an uncertain longer term future demand.

There are three other objectives to Chapter 11. The first is to demonstrate methods of quantitatively predicting a future demand based on past data and any other information available. The second is to show how to design a "balanced subsystem" (i.e., one which provides sufficient performance but where adding performance compared to the resources required is inefficient) without having to perform a total mission level systems analysis. The third objective is to illustrate how to build flexibility into a systems design so that the system objective can be met at lowest cost in spite of the uncertainty in the future demand.

Chapter 12 examines the problems of planning automated information systems, such as military command, control, and communications systems, or management information systems, production control systems, inventory control systems, and distribution systems, as used in a business. Many times the problem of comparing the worth of information to the cost of the information system is not obvious, and in Chapter 12 is shown how to determine the proper balance between the degree of sophistication of an information system and other system elements of a higher level system, without having to perform a total mission level analysis.

Finally Chapter 13 deals with the analysis of a subsystem whose performance characteristics are highly interrelated with the characteristics of other subsystems, which is generally the case in subsystems planning. Thus the primary objective of this chapter is to show how systems analysis may be used properly to set the requirements or specifications of a system by taking into account the various intrasystem tradeoffs involved. Here we shall use the same case example to focus on two separate but related contexts which occur in systems design. First, the problem that a subsystem designer has in optimizing his design and his reliance on being provided a higher level systems evaluation model by the higher level system designer. Second, the usefulness of a mission-oriented cost-effectiveness analysis, as performed during the systems planning or concept formulation phase, as a necessary first step in obtaining the higher-level systems evaluation model used in setting system and subsystem specifications.

In Part VI is shown how the same systems analysis methods and techniques previously described can be used to assist the systems engineer in the entire systems planning and procurement cycle, which includes contractor source selection. The problems associated with the examined tasks include:

1. How analysis can be used to set detailed system specifications, particularly when there may be high uncertainties in the job(s) to be performed by the system.

2. How analysis can be used to evaluate the proposals received from each of the responding vendors and a contractor source selected when the specifications of a proposed system may not be identical to the desired specifications, or if development or other vendor uncertainties are involved.

Two case situations are discussed; Chapter 14 involves the use of systems analysis in setting system specifications in a "request for proposal" (RFP) and in evaluating alternative vendor proposals for "off-the-shelf" electronic data processing (EDP) equipment. Chapter 15 involves the same problem of Chapter 13 but introduces methods of evaluating systems where development uncertainties are also involved.

Chapter 16 focuses on some of the practical problems which the analyst runs into, particularly the constraint of insufficient resources in time, manpower, and funds for performing the analysis in a thorough fashion. Methods are described for accommodating this constraint, and still performing an analysis which is acceptable to both the analyst and his client. These methods include ways of initially planning the effort, performing the work, as well as managing the work of others, assuring the standards of quality of the work, and presenting the final results to others. Finally, the limitations of systems analysis are presented, showing that all other methods of evaluation also share the same limitations.

2

The Work Process of Systems Planning

Systems analysis applies to all phases of the systems acquisition process, but especially to that portion known to the industrial community as the *systems planning phase* (or long range planning or preliminary design) and to the defense community as the *concept formulation phase*. This chapter examines the activity of systems planning and explores the roles which systems analysis plays within this activity. In doing so, the following questions are explored:

1. Why is systems planning performed? The need for systems change and the various forces which initiate this phase of activity are discussed.

2. What is the work process involved and what skills are needed for its implementation? The systems planning process is modeled as a flow of information whose required inputs consist of a series of operational, technological, cost and decision-making information, and whose output is a detailed proposal for some recommended course of action, supported by sufficient, substantiating information on which the decision-maker feels he needs to base his decision regarding the proposal. The steps involved, the data required, and the personnel skills needed to perform properly the work process are then described. The need for a systematic method and the use of quantitative techniques in performing the systems planning phase are discussed.

BACKGROUND

Since systems planning is the process of generating and evaluating alternative ways of changing a system in some beneficial manner, we shall first explore the why and when of initiating a systems planning project. In an industrial profit-making organization the primary consideration which leads to this initiation is the continuing desire to increase the profits of the firm. In a nonprofit organization, such as the defense establishment, there are two other prime considerations (which would also be included under the profit motive of the industrial firm): (a) a continuing desire to reduce operating costs, and (b) remaining competitive in accomplishing its objectives.

Increasing Profitability

An industrial firm offers many opportunities for internal change. These opportunities include expansion of production facilities of certain existing product lines, expansion of the product lines, investment in research and development to create new products, the acquisition of other companies, and the increase of expenditures in marketing. Some of these changes may provide relatively short term benefits (e.g., hiring more salesmen) while others may provide longer range benefits (e.g., increasing research and development).

Cost Reduction

All organizations are under continuous pressure to reduce their normal operating costs. In the case of an industrial firm, such cost reductions can directly contribute to increased short term profits. Sometimes, however, a so-called cost reduction (such as radically reducing research and development funds) may also reduce long term profits by much larger amounts than the short term profit increase achieved. Similar pressures exist for reducing the operating costs of nonprofit organizations such as the Defense Department, particularly since, up to the Vietnamese conflict, this country was experiencing the highest costs for maintaining a "peace time" standing military force in its history.

Competitive Forces

System improvements are continually being made by one's competitors, who may be either members of the industrial community or potential enemies of one's country. Where there is room for technological change, even a well entrenched organization may find its position of superiority reduced rapidly unless it can act sufficiently fast in response to a competitive move. For example, the American aircraft industry, with its dominant position in transport aircraft, has had to respond to the supersonic Concorde threat to its position with its own government subsidized supersonic transport. Similarly, a well established computer firm such as IBM must still invest heavily in the development of new systems as new technologically superior components become available to them and to their competitors.

The same competitive pressure exists within the defense establishment, since improvement in the capability of a potential enemy may reduce the

American defense capability to an unacceptable level, causing pressures for a responsive move. This was the case in the 1950's when intelligence estimates indicated an increase in Soviet missile development, resulting in an accelerated effort in the then dormant Atlas ICBM program. A similar situation has occurred in the development by both sides of the antiballistic missile defense system.

Technological Advances as Aids in Improving the System

New technological advances, such as the use of nylon in the fabric industry, the laser for communications systems, or large, rapid-access memories for computer applications, are continually being generated, and such technological developments provide new tools to assist the systems designer to meet the previously mentioned problems. Sometimes the new tool permits implementation of a new function while at other times, it permits the same functions to be performed at higher levels of performance or lower costs. (The latter is exemplified by the use of solid state devices to replace vacuum tubes in electronic devices.)

While technological advances have led to reduced operational costs and improved system performance, they have also increased the number of approaches available for achievement of the system's goal. When these improvements are translated into new proposals, they present the decision-maker with the difficult task of selecting the most promising approach.

Planning for Change

The sum total of the above forces (cost, competition, technological advances) produces both the need and the means for changing an organization. To satisfy this need, special system planning groups are often formed to determine ways of improving the organization. These groups are concerned with understanding the implications inherent in these forces; recommended proposed changes are then sent to a higher-level decision-maker for approval. Each proposal generally requires funds for its implementation. These funds may be used for further research and development, capital investment, and operating and maintaining the proposed systems. Invariably, each proposal promises some increased benefit to be obtained in exchange for the funds requested. For example, a proposal indicates that approving the capital investment of a new computer-driven machine will result in its "paying for itself" in three years; but, will this really occur? Can one really believe the data presented to substantiate this claim? What about the risks and uncertainties which are always present? These uncertainties may be particularly large in the analysis of defense problems because of the long time required for the entire system acquisition process (i.e., research, development, test, system engineering and production). This time lag, which now approaches five to ten years, produces high risks and uncertainties of two types. The first is technological uncertainty, due to the constant uncertainty in the entire process of successfully converting a new technological concept into a workable piece of hardware. Technological uncertainty results in uncertainties in the performance characteristics finally obtained in the system design, coupled with the total cost to achieve such characteristics. Another aspect of technological uncertainty is the problem of technological obsolescence. Will the new system remain in operation for at least the predicted number of years needed to amortize the costs of development and production? The second risk is the competitive uncertainty when an anticipated enemy (or business) threat vanishes and reappears in some other form. Thus, the new system may no longer have the high effectiveness which was predicted, but the resources have still been expended. This is not to say that the expenditures were necessarily wasted; in the design of a defense system, for example, the greater the system effectiveness achieved, the less likely it is to be used in combat. Examples of this are the ICBM missile systems and the Ballistic Missile Early Warning System which accomplish their primary function of deterrence.

Even if the proposed advantages of the new system would really materialize, is this proposal superior to another being made (or which could be made) for the same cost (e.g., hiring four new salesmen for the marketing department)? The problem for the decision-maker is one of allocating the resources among alternative means in order to achieve the best solution. He must determine, from the total proposals submitted, those which promise the highest amount of return for the given amount of resources available, taking into account the amount of risks and uncertainties involved.

Further complexity develops when the systems planning proposal requires funds of any magnitude; then there is, in general, a hierarchy of decision-makers who must approve the system proposal. Since each decision-maker has his own set of values, some means of enabling the system planner to take all of these into account is required when constructing an improved system.

How the Decision-Maker Evaluates Proposals

Some of the proposals of the decision-maker may be rejected, after a relatively small amount of consideration, as being of lesser value than others. If the resource requirements of the remainder of the proposals do not exceed the limited resources available, there is no longer a decision-making problem, since all of these proposals can be approved. On the other hand, if the total resources required still exceed the resources available, further study must be made of each proposal by comparing the projected benefits (called "effectiveness" by the military or "utility" by economists) of each proposal with the estimated costs and the risks or uncertainties involved.

Two factors increase the complexity of the problem:

1. All proposed systems may not be operationally available at the same time.

2. The data describing system proposals, and the environment in which the system will operate are never complete, or completely accurate.

An example of the first factor is as follows: Assume that the Air Force wishes approval of a proposal to procure quantities of new fighter-bombers applicable to limited war. The decision-maker may know that the Army is currently working on a new surface-to-surface missile system which also has applicability to the same limited war objective. The Army's proposal, however, may indicate that its missile system is not ready for procurement at the present time since there are many uncertainties in its performance characteristics. The decision-maker must now determine whether he can afford to wait until further research and development is performed by the Army, making additional information available. The decision-maker does have, as additional information, the Army's prediction of when its development program will be completed. What cannot be predicted, however, is whether the Army will have to revise its schedule because of problems such as technological (developmental) difficulties.

If the problem is sufficiently critical so that some improved capability must be added in the near future, the decision-maker may be forced to choose from among those system alternatives which will be available at the required time. However, the knowledge that other improved approaches for meeting the same objective will eventually be available provides the decision-maker with the opportunity of preplanning periodic re-evaluations of the problem area as improvements become available.

In this way it is possible to introduce later system improvements on an evolutionary basis so that maximum use can be made of past resources expended. Thus, the equipments that need to be shelved for total obsolescence when new equipments become available can be minimized.

THE WORK PROCESS IN SYSTEMS PLANNING

As mentioned previously, the systems planning phase is initiated because of such factors as:

1. The threat of impending competitive action (the competitor in this case can be either a potential political enemy or an industrial competitor);

2. The desire to modify current systems in order to take advantage of a newly developed technology;

3. The decision maker's instructions.

Initiating the Systems Planning Study

A simplified model of the systems planning work process is shown in Figure 2-1. The various elements which make up a systems planning effort and the various ways in which the effort is initiated are described in the following. The first type of input needed for systems planning are the data describing the mission which the organization is to perform and the environment which will interact with the mission at the future time period under study. This mission, in general, is independent of people, equipments, or technology.* An example of a military mission is the strategic mission dealing with general war. This mission has two objectives: (a) to deter a potential enemy from attacking the United States or its allies, and (b) to limit the damage to

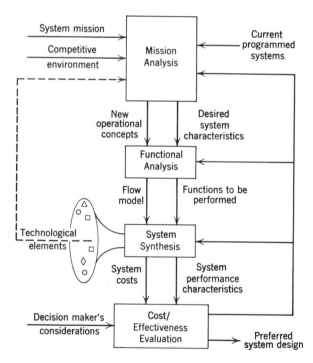

Figure 2-1. The Work Process in Systems Planning

* How the mission is to be conducted and implemented by the organization is, of course, highly dependent on the people, equipment, and technology to be used.

the United States and its allies if war comes. One of NASA's missions, for example, is space exploration. A mission of an industrial contractor is to operate and maintain a profitable business in a particular product area.

The environment consists of a physical environment, a threat or competitive environment, and a market (for the industrial organization) which is expected to exist at the time when the system is to operate. One way in which a systems planning effort may start is by reacting to a newly perceived view of the environment. For example, new military intelligence inputs may indicate that the threat on which the current and programmed systems were based now appears to be increasing. By comparing the current and programmed systems which affect the pertinent mission area, it may appear that the American defense effectiveness in coping with a new projected threat at the future time period may fall below what is felt to be an acceptable level if nothing is done to improve the system. The industrial counterpart to this example occurs when a company learns that a competitor is planning to modify a product or his production facilities. Uncovering a new apparent market need also illustrates a situation in which the present system (i.e., products offered) is inadequate and requires (or would accommodate) change. Thus, one of the functions of the systems planning effort is to determine in an analytical fashion the validity of these intuitive feelings, and, if they are valid, to take appropriate action.

It is possible for the systems planning phase to be initiated in a second way. The system operators (or operations analysts or requirements personnel at the particular military command involved, or advanced planners in an industrial firm) are periodically apprised of the projected improvements in the state of technology which may affect the user's system. This is accomplished by visits of technologists from the service (or industrial) laboratories. This information transfer is represented by the dotted line of Figure 2-1; it has sometimes been called "the solution seeking the problem." An example of this is the development of the laser which caused a number of systems analyses to be conducted with systems planners seeking ways of applying this technology to replace standard microwave power sources in the areas of radar, communications, etc. Another example is the case of new high-speed computers which not only can perform functions which previously could not be automated but can do them at much less cost because of technological improvements. This is why the two inputs of state of technology (resulting in obtainable performance) and costs cannot be separated in making a technological appraisal.

The functions of the requirements analyst at the user location is not only to recognize the need for a systems improvement through analyzing the future environment threats, but also to aid in the systems planning process by formulating new operational concepts which can utilize the improved technological elements. Thus, during the first phase of the systems planning process, called "mission analysis," the new environment is compared against the current and programmed systems. System deficiencies are thus determined and a set of desired system characteristics generated. In addition, new operational concepts employing the new state of technology may be generated.

The set of system characteristics resulting from the mission analysis should really be called "desires" (or goals) rather than "requirements." As will be seen later in the work process, only when the cost of these system characteristics are related to obtainable performance can the system planner determine which of the set of "desires" can be included as part of the set of system specifications or "requirements."

Functional Analysis

This next phase of the systems planning process is often called "functional analysis"; it is during this phase that the functional aspects of the problem are examined, and various operational tasks that need to be performed are translated to system functions which must be implemented. Flow models are generated which make explicit the flow of activities, materiel, and information within the system as it performs its functions. Such flow models aid the analyst to better understand the type of system performance that is to be obtained.

System Synthesis and Evaluation

Various combinations of technological elements which will be available by the time the system is needed are combined in an attempt to implement the functions to be performed. The output of this phase consists of a set of feasible systems, each having specific system performance characteristics, including capacity, reaction time, and reliability. During the system cost-effectiveness evaluation, the total cost of each system alternative over the entire system life is determined, as well as the amount of benefits or system effectiveness each alternative provides.

A most important factor to be taken into account in evaluating system effectiveness is the set of higher level organizational considerations which the decision-makers who are involved in reviewing the proposals intuitively feel are important. These would include meaningful information regarding national (or corporate) policies and goals, existing contingency plans or future environments anticipated. Thus it is extremely important to properly interface in some fashion with the decision-makers, or their staffs, to obtain such information. The output of this evaluation function then consists of the effectiveness of each system and its resulting costs under the conditions or contingencies which might occur.

Coping with Uncertainties

The systems analyst must realize that the results of the analysis are only as good as the validity or accuracy of the data used. Thus sensitivity analyses are performed to determine the extent to which uncertainties in the operational environment or the performance or cost of the system elements will affect the effectiveness of each system. If it is determined that uncertainties critically affect the evaluation results, steps must be taken to obtain more accurate information about these particular characteristics.

Creating Additional Alternative Systems

Systems planning is an iterative process. During the process of evaluating the various system alternatives, the systems analyst is constantly alert for any other system possibility which may offer some improvements or may overcome some of the shortcomings noted in the systems being evaluated. As will be seen, analysis can aid in systematically creating more efficient system alternatives.

Steps Involved in the Systems Planning Process

It is not possible to indicate an exact step-by-step approach which can always be used in systems planning, because of the large amount of iteration involved (e.g., tentatively completing some steps and then going back to re-modify steps previously accomplished). This type of iteration is represented by the three feedback loops on the right side of Figure 2-1. Further, the various steps which are taken during the entire study effort have been summarized and listed in Table 2-1. These are the steps which will be further described and followed in the cases contained in subsequent chapters.

THE WORK PROCESS: REQUIRED SKILLS

As discussed previously, people in five different areas contribute to the systems planning process:

1. Operations or requirements analysts;
2. Technological specialists and synthesizers;
3. Cost analysts;
4. Systems analysts;
5. Decision-makers.

The skills which the contributors to the systems planning process offer are as follows. The operations or requirements analyst represents the system operation (e.g., Air Defense Command, Field Artillery Unit, Manufacturing Department), and, in general, understands better than anyone else in the

Table 2-1. *Steps in Systems Planning*

1. Understanding particular job to be performed (Mission analysis).
   a. What is the problem?
   b. What is the objective?
   c. How present system would be used?
   d. Inadequacies of present system.
   e. Operational constraints.
2. Understanding the system activities (Functional analysis).
   a. Operational concepts.
   b. Functions to be performed.
   c. Determining factors of interests.
   d. Developing relationships.
3. Creating system alternatives.
   a. Subsystem elements involved.
   b. Performance characteristics.
   c. Operational concept used.
4. Identifying other system competitors.
5. Building the systems evaluation model.
   a. Determining measures of the objectives.
   b. Quantifying the environment.
   c. Relating system performance and environment characteristics to the measures of the objectives.
6. Data estimation.
7. Exercising the systems evaluation model.
8. Estimating system cost.
9. Coping with uncertainties.
   a. Competitive.
   b. Technological.
   c. Cost.
   d. Chance.
10. Creating additional alternative systems.
11. Selection of preferred system.

planning process the missions of the using organization, the operations to be performed, and the deficiencies of the current system as it relates to the projected competitive threat. The operations analyst creates new operational concepts, and evaluates the feasibility of new concepts brought to him by the system technologists.

The system technologist and synthesizer, who is familiar with what systems performance characteristics the technological elements of the system will provide, configures different combinations of these elements in different system alternatives.

The cost analyst can predict what each part will contribute to the total system cost. It is he who can help make explicit the cost-sensitivity relationships of the various system functions or performance characteristics (i.e., the cost impact of achieving various system performance levels) and can thus aid the system synthesizer in eliminating the less efficient system designs.

The systems analysts, some of whom work at the design level, can translate the operational activities desired from the system into system functions which must be implemented by the technological hardware elements and procedures. Systems analysts also operate at the systems effectiveness level; hence they provide a "capping effort" to the systems planning process. In this role, these analysts must interface with the operations personnel, the decision-makers, and the systems designers to properly determine the system effectiveness (i.e., its ability to perform the intended mission).

Finally, there is the decision-maker. He should be available to provide an appropriate set of considerations which he feels should be built into the system effectiveness evaluation model used in choosing the preferred system. As mentioned previously, what makes this part of the task difficult is that there actually is no one decision-maker, but a hierarchy of decision-makers who are involved in passing on recommendations up through the higher levels for their decisions and action. However, means must be developed for taking all of these into account. (For further details, see Chapter 4.)

SUMMARY

To summarize the systems planning work process just described, first, five different types of skills are required for systems planning. Second, the analysis process is a step-by-step job (Table 2-1), but the work process itself is iterative (Figure 2-1). System concepts which are feasible are generated and evaluated first in terms of system performance versus cost, and then in terms of system effectiveness versus cost. The deficiencies are then noted and an improved redesign created. This iterative process continues until additional analysis will not yield much higher improvements in the design as compared with additional analytical time required.

Finally, the system design chosen as a preferred solution is actually only a compromise among the following factors: (a) the level at which the job will be performed, (b) the time when a feasible system will be available, and (c) the costs or resources required for implementation over the system life. Here is where the decision-maker's judgment in choosing among the various alternatives must be employed.

3

Systems Analysis as Part
of the Management Process

As indicated in Chapter 2, the objective of systems analysis is to provide meaningful information to a decision-maker faced with the problem of allocating resources among a number of system proposals which have been generated, and whose total costs exceed the amount of funds available. In this chapter we shall develop a set of information needs in the form of questions which should be answered if the decision-maker is to rationally select among proposed systems planning alternatives, and properly allocate the resources available to the organization. The problem and some of the management principles involved in this decision-making process are described by examining the management information system developed by one large organization (the Department of Defense) in coping with their problems of medium and long range planning and acquisition of large, complex systems. As the reader will see when following the problems which led to the development of this management information system, many of the same system planning problems faced by DOD are also faced to some degree by managers of all organizations, and many of the approaches to solutions which were developed by DOD are also applicable to decision-makers involved in capital budgeting problems within other types of organizations.

Two main aspects of decision-making are explored: (a) problems involved in planning and budgeting of resources among competing lower level units of the organization, and (b) the application of a rational, systematic approach to the planning and budgeting tasks. This systematic approach was developed as follows:

1. Determining the over-all objectives of the organization so that program overlap among competing components of the organization can be reduced.

2. Establishing a long range plan for the organization containing the approved systems and total resources required over time to develop, procure, operate, and maintain each of the systems.

3. Developing a structure of the total systems acquisition process and division of the process into primary phases so that key decision points can be established for initial system approval and subsequent confirming approvals as additional information becomes available.

4. Delineating appropriate information required by the decision-maker in approving proposals and confirming previous approvals of systems at the key decision points.

SYSTEMS PLANNING AND BUDGETING IN THE DEFENSE DEPARTMENT

The objective of the discussion in this section is to lay the groundwork for obtaining the proper information for high-level decision-makers, and to show how systems analysis can aid in providing such information. The description which follows is in the form of a case study; it describes the development of a management information system (originally developed by Dr. Charles Hitch, Assistant Secretary of Defense under Robert S. McNamara, during the Kennedy-Johnson Administrations). To better understand how the current management information system was developed and how it operates with respect to systems planning and budgeting, we shall first examine the DOD management information system which was in place prior to 1961. This analysis will serve to uncover the decision-making problems and the type of information available at that time.

Pre-1961 Management Problems

We shall first focus on some of the problems involved in planning and budgeting for new defense systems in the pre-1961 DOD administrations. These problems have been excellently described by Hitch, who also worked very closely with the system prior to the one which he developed under the Kennedy-Johnson Administrations.* The key elements of this procedure are now described.

Each of the three services (i.e., the Army, Navy, and Air Force) planned for its medium and long range system improvements, essentially independently of the others. Thus, each service accumulated a series of "requirements" which were really a set of their desired programs for the next four to five years. These requirements were measured in terms of weapon systems or military units needed to accomplish military missions, but were

---

* This section is based on Hitch's own analysis of the problems of the pre-1961 planning-budgeting system and the need for the Planning-Programming-Budgeting system subsequently created by Hitch. These are described at length in his book *Decision Making for Defense*, 1965.

completely separated from the financial implications which would result from satisfying them. However, the Office of the Secretary of Defense had control of the budgeting procedure through the Assistant Secretary of Defense (Comptroller). The Comptroller compiled the budget, but only for the next fiscal year. This budget was compiled in terms of required input resources, listed in categories of personnel, operations and maintenance, procurement, military construction, research development, test, and engineering, each as a separate item of the budget. The difficulties occurred when each year it was found that not enough funds would be allocated to DOD to satisfy all of its plans for the year. To resolve this situation, the following procedure was applied.

1. The President indicated DOD's allocation of the national budget based on such factors as the international situation and pressures from the nondefense federal agencies.

2. Each service was given its own separate share of the total budget which, in general, in the middle to late 1950's, turned out to be about 47% for the Air Force, 29% for the Navy, and 22% for the Army. Since these budget allocations to each of the services were never enough to meet the list of requirements requested, each service was asked to assign its own priorities to each of its program requests and thus submit two lists for priorities to the Comptroller. The primary budget list would then contain the highest priority programs whose total was not to exceed the service fund allocation. An addendum budget was also submitted containing the lower priority programs, which would be used if the DOD budget were raised.

3. In assigning its own priorities, each service gave the highest ratings to those programs which emphasized their capability in a single service mission area. Lowest ratings were assigned to programs designed to provide a capability useful to joint service missions.

As an example of the last point, the Air Force concentrated on the strategic mission with its bombers and missiles, but assigned lower priority and correspondingly provided less funds to the Tactical Air Command or for airlift capability, both of which are required to support Army ground operations. Moreover, of the total allocation received, the Tactical Air Command further reduced resources available by allotting some of its fighter-bombers to the strategic nuclear mission, rather than to close air support for limited war.

The Navy concentrated its efforts on Polaris submarines as well as on nuclear attack aircraft from Navy carriers. On the other hand, lower priority, and hence only small monetary allocations, were provided to antisubmarine warfare or Navy escorts.

The Army concentrated on increasing the number of divisions available, while keeping at an absolute minimum the amount of equipment and supplies needed to maintain these divisions in the field in the event of a crisis. Studies indicated that equipments and supplies for only a few weeks of combat operations were available in support of these Army divisions.

This emphasis on essentially decentralized service operation was contrary to General Maxwell Taylor's observations on the subject as presented to a Congressional Committee in 1960. He indicated that in recent history, major engagements have been fought not by each service as a separate group, but by *task forces* having joint, interservice participation. Thus, he concluded that one might view the Defense Department as consisting of three separate services whose primary role was to provide trained manpower and equipments to joint service task forces, sized to meet a given contingency or operation.

While this decentralized planning by each of the services may have led to an increase in certain capabilities of each service as improvements in technology occurred, it also resulted in a large overlap or duplication in certain parts of the over-all DOD capability. Up through World War II, it was perhaps satisfactory to indicate that the Army's role was in land warfare; the Air Force's role was combat operations in the air; and the Navy's mission was on the sea, in the air above the sea, and below the surface of the sea. However, with the vast improvements in technology, interservice conflicts arose concerning whether a missile was actually an unmanned aircraft, and as such, part of the Air Force operations, or whether it was an extension of field artillery, making it an Army responsibility. Such decisions were made, but only after some duplication of effort on such programs as the Air Force Thor versus Army Jupiter Intermediate Range Ballistic Missile programs. As indicated previously, other products such as bombers, missiles, Polaris submarine missiles, carrier-based nuclear attack aircraft, tactical nuclear fighter-bombers, competed and overlapped in requests for strategic funds.

Some apparent duplication of these overlapping programs was worthwhile since a mixed force structure containing units of each of these systems provides force flexibility and is one means of coping with enemy uncertainty. However, it soon became evident that someone at the highest level would have to determine how many of each competing element would be needed to make up a unified mixed force structure. Such decisions for choosing among interservice competing alternatives could only be made on the basis of information which resided within each of the three services, and such information was not available to the Office of the Secretary of Defense in sufficient detail or enough in advance of the final budget review for this office to have much effect on the decision-making process.

In addition to the problems of interservice overlap and decentralized service control of program priorities, a third problem was the lack of a strong effort by each of the services to relate military requirements with their cost implications. In his book, *Decision-Making for Defense,* Hitch relates that in the history of warfare, military decision-makers have always had to decide whether to allocate given resources to a large number of simple weapons or to a smaller number of higher quality weapons. Up through World War II, the emphasis had been on quantity (e.g., divisions of foot soldiers, large numbers of bomber aircraft, etc.). However, with the advent of the atomic bomb, the emphasis had shifted to relatively small numbers of highly sophisticated weapons. Thus, in allocating a fixed resource, such as any given DOD budget level, the relative effectiveness of a small number of supersonic bombers, for example, must be compared with a much larger number of Mach 1.0 bombers costing the same total amount.

Perhaps one of the main reasons for this lack of emphasis on such a fundamental economic principle was the lack of availability of credible and complete cost information for the military planners. Several reasons may account for this; first, the service organizations themselves were divided by function (one group for Research and Development, one group to manage production contracts, and another to manage operations, maintenance, and logistics, etc.). Thus, while each group attempted to procure development contracts on the basis of lowest cost for performing its own function, such suboptimization rarely guaranteed that the lowest *total cost* system was being procured. Thus, one reason that reliable total cost estimates were rarely obtained was that over the complete life of the system from research and development through system procurement and operations and maintenance, no one group was given responsibility for estimating total cost. A second problem was that the cost and time estimates that were obtained from industrial contractors tended to be optimistically low, resulting in large procurement overruns of many weapon system programs. This has been documented by a Harvard Business School Report * which indicated that in twelve key weapon system programs, the average total expenditures were 3.2 times the original cost estimate and 1.36 times the original time estimates. While high uncertainty is to be expected in predicting the cost and time required to build a technological system never built before, the planning and procurement procedure itself seemed to exert pressures which encouraged contractors to submit optimistically low estimates. Most contracts for development and procurement were "cost plus fixed fee" (CPFF), and since any overrun would be borne by the government, there was little risk to the contractor if the actual costs were higher. Also, after the contract was underway, if development problems arose leading to cost and time

---

* Scherer and Peck, *The Weapons System Acquisition Process,* Harvard Business School, 1962.

overruns, there was actually little the government agency could do, since it might cost more to cancel the contract in mid-phase and award it to some other contractor than to continue on with the original contractor. This low risk, accompanied by the pressure to remain competitive with other contractors who might also bid low, added to procedural problems.

One other aspect contributed to the problem. In the effort to rapidly procure hardware, planning may have been insufficient, resulting in incomplete, or unfeasible specifications which later had to be changed, thereby increasing contractor costs.

A third fiscal deficiency of the system was that all plans and budgets were focused on only one year's programs and expenditures, ignoring the expected cost of the program in subsequent years. This led to many "false starts." For example, a program would begin with small R & D efforts, with the hope that the following year's service budget might be larger, enabling the program to be continued into the next, more expensive phase of development. When finally this program grew in cost, and the higher service funding hoped for did not materialize, there was nothing that could be done except either to stretch out the programs or to cancel those programs felt to be of lesser value than others.

Post-1961 Management Changes

These then were the problems which faced McNamara following his appointment as Secretary of Defense in 1961. His background, experience, and management attitude enabled him to encourage and implement a new philosophy in the field of defense decision-making. This experience included the application of statistical and other quantitative techniques to Air Force cost control programs during World War II. This was followed by an application of these methods to the improvement of the cost control and logistics system of the Ford Motor Company. He subsequently became the general manager of the Ford Division and had just been appointed President of the Ford Motor Company when he was appointed Secretary of the Defense Department by President Kennedy. His background and temperament led him to the desire to take an active lead in the management of the Defense Department, but he soon realized that what he needed was a large improvement in timely information in order to properly weigh the various alternatives recommended to him by various services. One of McNamara's first appointments was his selection of Charles J. Hitch as Assistant Secretary of Defense (Comptroller).

Hitch was an economist by training and his career included teaching at Harvard, followed by his service as Chief of the Economics Division at the RAND Corporation. It was in this latter position that he was most concerned with methods for improving the management of the Defense Department and his book on the subject * became a pioneering milestone in the field of defense management. Hitch, the economist, found a responsive listener in McNamara, the business man, when he suggested an approach for coming to grips with the management problems previously described. Thus began the development of an evolutionary management information system designed to more closely couple the objectives of the defense establishment to the assigned DOD resources within the federal budget and to provide decision-makers with timely information needed to implement the planning and budgeting tasks. The key portions of this management information system were:

1. The DOD Planning-Programming-Budgeting System which provided a structure for connecting the Planning and Budgeting processes. This system included two main facets: (a) the development of improved cost data banks for storing the actual equipment costs and for better estimating the costs of proposed systems; and (b) the development of the Five Year Force Structure and Financial Plan which made explicit the currently approved force structure in terms of DOD systems, equipment, personnel, etc., and the total annual costs which they require during the five year period.

2. The establishment of a formalized system acquisition process as a series of steps from system planning, system development and system procurement, so that changes in the approved program could be evaluated and acted upon. Key decision review points were established with specific information requirements also established so that at these review points the worth of continuing the program could be confirmed.

Management Information Requirements

All of these procedures serve as methods for providing a decision-maker with timely information for selecting one or more system proposals on the basis of the system benefits and cost. The process of system evaluation, illustrated in Figure 3-1, basically compares each system alternative against an

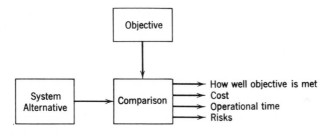

Figure 3-1. The system evaluation process.

* Hitch and McKeon (1960).

objective and determines the value of a system based on information found by asking six key questions:

1. What is the system objective?
2. When will the proposed system be operationally available?
3. How well does the proposed system meet the objective?
4. How much will it cost to implement the system?
5. Is the proposed system the best way of meeting the objective?
6. What are the risks and uncertainties involved in obtaining the stated performance on schedule and at the estimated cost?

The changes in the Defense Department management approach instituted by McNamara and Hitch were all aimed toward providing answers to these six fundamental managerial questions.

THE IMPROVED DEFENSE DEPARTMENT MANAGEMENT INFORMATION SYSTEM

In this section is presented a description of the key elements of the changes made in the systems planning management information system of the Defense Department in order to show how these are used to answer the six managerial questions indicated previously. While the DOD management system is still dynamic and evolving, we shall discuss the current status of the DOD management system to show how it is being used for purposes of management planning and control. As will be seen in this section, the management information system concepts described have wide application not only to DOD but to many other nondefense organizations. For example, the planning-programming system is now being applied to all nondefense government agencies. As Charles L. Schultze, Director, Bureau of the Budget commented:

"The new planning-programming-budgeting system will be capable of making a major contribution to greater efficiency in the allocation of resources, and thus will increase the benfits derived from the Government's many activities. It will provide the information and the analyses needed by Government managers as the basis for an improved ability to make rational choices among the alternatives offered." *

In describing the improvements made, emphasis is placed on the key principles of systems analysis which may be applied to any organization in at-

* A further treatment of the programming system, describing actual and potential applications in the nondefense activities of space, transportation, education, federal health expenditures, and natural resources activities, is found in David Novick (Ed.). the RAND Corp., *Program Budgeting: Program Analysis and the Federal Budget,* U.S. Government Printing Office, 1965.

tempting to set up a management information system which will aid in the task of systems planning and deciding upon system alternatives which involve conditions of uncertainty.

Structuring the Total Defense Program

Two major problems faced the new DOD administration when it first assumed its responsibilities. In dealing with the overlap of weapon system programs among the various services, an immediate question which had to be answered was: Is the DOD program mix a proper one? Specifically, how well did the current programs contribute to DOD objectives with respect to the DOD resources they required? All incoming managers new to an organization are faced with this immediate problem. To help answer these questions, a structure which related system programs to organizational objectives needed to be developed. This was done by first listing all of the various approved programs of the services which were currently being financed (i.e., all ways that the organization was spending funds). These programs were then structured in some way to include the relationships which existed among various programs. Two types of program relationships were recognized.

1. Programs could be complementary to one another; that is, each was needed to perform a given mission. An example of this would include a bomber aircraft, the bomber crew, the bomb, and any other element needed to get the bomb to its target.

2. Programs could also compete with one another in performing the same mission objective. Here, examples are bombers, ICBM missiles, and Polaris type missiles when directed at the same target. However, even these similar forces may be constructed to form a complementary set of mixed forces. For example, ICBM's may be targeted to destroy enemy bomber defenses so that the bombers have a greater probability of reaching their targets.

One possible structure which shows the relationship of DOD programs is shown in Figure 3-2. Several observations may be made from such a structure:

1. Certain programs may contribute to certain types of defense capability, but have questionable value in providing other types of capability. For example, long range missiles may provide a deterrence for general nuclear war, but may be of little benefit in a limited war. Recognition of the specialized value of various weapon systems leads to classification of DOD objectives according to whether they provide a capability in the two different areas of warfare, as shown in Figure 3-2.

2. Each of the two areas of warfare shown are subdivided into the com-

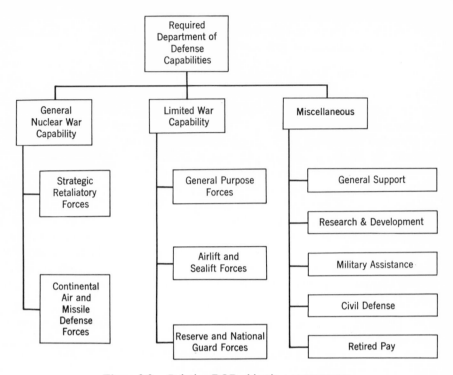

Figure 3-2.   Relating DOD objectives to programs.

plementary elements supporting the over-all objective. For example, airlift and sealift forces contribute to the limited war capability.

3. There exists a miscellaneous category containing those DOD programs which contribute to both of the other two categories. For example, resources are required to maintain the headquarters and staff of the Office of the Secretary of Defense and those of each of the services. This is a required "overhead" activity and no attempt need be made to divide the required resources among the other programs. The same is true for those research and development programs which have general defense application, however those R & D programs which support a specific weapons system program are included within that program.

Based on the structure in Figure 3-2, nine major classifications called "program packages" have been designated, each containing a set of major DOD system programs, each of which is called a program element. The current DOD program element structure is shown in Table 3-1. This structure has undergone several phases. For example, Program I formerly consisted of the strategic retaliatory forces containing all strategic offensive systems, and Program II was continental air and missile defense forces, containing all strategic defensive systems. However, since there was high interrelationship between these two programs, they were combined in 1966. On the other hand, some elements, such as airlift and sealift, are kept as separate programs, for administrative or other purposes even though interrelated to some other program (general purpose forces), since the Secretary may desire to view how much is being spent in this area.

Table 3-1. *DOD Program Element Structure*

PROGRAM I—STRATEGIC FORCES
 1 1 Offensive Forces
 1 2 Defensive Forces
 1 3 Civil Defense
PROGRAM II—GENERAL PURPOSE FORCES
 2 1 Unified Commands
 2 2 Forces (Army)
 2 3 Forces (Navy)
 2 4 Fleet Marine Forces
 2 5 Forces (Air Force)
 2 6 Other
PROGRAM III—SPECIALIZED ACTIVITIES
 3 1 Intelligence and Security
 3 2 National Military Command System
 3 3 Special and National Activities
 3 4 Activities (other)
 3 5 Military Assistance
PROGRAM IV—AIRLIFT AND SEALIFT
PROGRAM V—GUARD AND RESERVE FORCES
PROGRAM VI—RESEARCH AND DEVELOPMENT
PROGRAM VII—LOGISTICS
PROGRAM VIII—PERSONNEL SUPPORT
PROGRAM IX—ADMINISTRATION

Program I–Strategic forces consists of all offensive weapons such as bombers, missiles, and Polaris missile submarines in Program I-1; whereas Program I-2, strategic defensive forces, consists of manned and unmanned interceptor surface-to-air missiles, the Ballistic Missile Early Warning System, SAGE air defense system; Program I-3 consists of civil defense. Program I also includes the command organizations associated with these forces.

Program II–General Purpose Forces consists of force oriented program elements other than those in Program I, including the command organizations associated with these forces, the logistic organizations organic to these forces, and the related logistics and support units which are deployed or deployable as constituent parts of military or naval forces and field organizations. This program consists of six major subdivisions whose basic mission is to cope with limited war: (a) unified commands, (b) forces (Army), (c) forces (Navy), (d) fleet marine forces, (e) forces (Air Force), and (f) other.

Program III–Specialized Activities consists of missions and activities directly related to combat forces, but not a part of any of the forces listed in Program I or II, on which independent decisions can be made. This includes resources for primarily national or centrally directed DOD objectives for intelligence and security: specialized missions such as weather service, aerospace rescue/recovery, and oceanography. This program consists of five major subdivisions: (a) intelligence and security, (b) national military command system, (c) special and national activities, (d) activities (other), and (e) military assistance.

Program IV–Airlift and Sealift consists of airlift, sealift, and other transportation organization supporting the limited war mission. This includes command logistic and support units organic to these organizations.

Program V–Guard and Reserve Forces elements are arranged by program (strategic forces, general purpose forces, specialized forces, airlift and sealift, logistics, personnel support, and administration) in order to facilitate relating guard and reserve forces to the active forces.

Program VI–Research and Development program includes all research and development activities which are not related to items which have been approved for procurement and deployment. The cost of R & D related to such items will appear in appropriate elements in other programs.

Program VII–Logistics consists of supply and maintenance that is not organic to other program elements. It includes nondeployable supply depots and maintenance depots.

Program VIII–Personnel Support consists of training, medical, and other activities associated with personnel, excluding training specifically identified with another program element, housing, subsistence, medical, recreational, and similar costs that are organic to another program element (such as base operations).

Program IX–Administration consists of resources for the administrative support of departmental and major administrative headquarters, field commands and administrative activities (not elsewhere accounted for), construction support activities, and miscellaneous activities not accounted for elsewhere.

The next step in analyzing these programs is to total all costs by year pertaining to each of the program packages, as a function of all of the subsystems which make up each of the program elements in the program package. This includes all costs of research development, test and engineering, production of operational units, and operation and maintenance for each of the next five years. (These costs are discussed further in Chapter 9.) The resulting structure is called the Five Year Force Structure and Financial Plan, which is the approved five-year plan for DOD. The creation of the Programming System and the Five Year Force Structure and Financial Plan enabled the Secretary of Defense to identify all on-going or approved competing and complementary program elements which contribute to a particular mission area. Also, while the previous budgeting system dealt with only the next year's financial requirements, the current planning, programming, budgeting system still yields the same information, but also shows the cost implications of the current and future planned programs which have been agreed upon for the next five years. Thus, peaks or gaps in either the force or the cost of the force in any mission area can be readily identified.

Changing the Defense Program

It was recognized that periodic proposals for changes in the total defense program as made by planners from the different services would have to be evaluated to determine the worth of the proposal with respect to its costs and risks. Such changes might include the start of a new weapon system development. In addition, existing weapon system programs currently in the state of development would normally progress into the system production phase which would require the approval of additional expenditures. McNamara realized that the type of information which he would need to properly evaluate these changes would be a function of where in the system acquisition process the proposed change was. For example, since programs tend to become increasingly more costly as they progress from development to production to operation and maintenance, it is essential that the manager be more certain about expected system benefits and costs. For this reason, the system acquisition process was formalized and explicitly divided into four phases as shown in Figure 3-3. The four phases of the systems acquisition

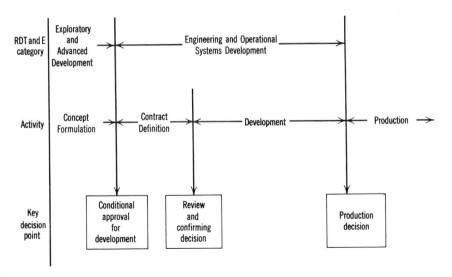

Figure 3-3. RDT&E process during systems acquisition.

process were called *concept formulation, contract definition, development,* and *production* (or *acquisition*). Such a division allowed the establishment of three key decision points at which time specific information would be required for the program to advance to the subsequent phase. The specific types of information requested by the Secretary of Defense were delineated in a DOD directive,* which states the objectives of the first two preliminary design phases, and further indicates the information required to be submitted to the Office of the Secretary of Defense (OSD) by the proposing component following completion of these phases.

Since this procedure has general applicability in the development and acquisition of nondefense programs as well, the following extracts of DOD Directive 3200.9 are now presented.

DEFINITIONS

"*Concept formulation* describes the activities preceding a decision to carry out engineering development. These activities include accomplishment of comprehensive system studies and experimental hardware efforts under exploratory and advanced development, and are prerequisite to a decision to carry out engineering development."

This phase consists of the initial analytical studies sometimes called the Exploratory Planning portion of systems engineering,† systems planning, or advanced systems planning.

Following the development and approval of a technically feasible system solution, the more detailed level of system design, called *contract definition*, is performed, normally by an industrial contractor.

"Contract definition is that phase during which preliminary design and engineering are verified or accomplished, and firm contract and management planning are performed."

OBJECTIVES

"The objective of concept formulation is to provide technical, economic and military bases (through experimental tests, engineering, and analytical studies) for a conditional decision to initiate engineering development. Conditional approval to proceed with an engineering development will depend on evidence that the concept formulation has accomplished the following prerequisites:

1. Primarily engineering rather than experimental effort is required and the technology needed is sufficiently in hand." (If not, an exploratory or ad-

---

* DOD Directive 3200.9, "Initiation of Engineering and Operational Systems Development," July 1965.
† A. D. Hall (1962), Chapter One.

vanced development program is to be undertaken to prove the feasibility of the technology).

2. The mission and performance envelopes are defined.

3. The best technical approaches have been selected.

4. A thorough tradeoff analysis has been made.

5. The cost-effectiveness of the proposed item has been determined to be favorable in relationship to the cost-effectiveness of competing items on a DOD-wide basis.

6. Cost and schedule estimates are credible and acceptable.

"The over-all objective of contract definition is to determine whether the conditional decision to proceed with engineering development should be ratified. The ultimate goal of contract definition, where engineering development is to be performed by a contractor, is achievable performance specifications, backed by a firm fixed price or fully structured incentive proposal for engineering development. Included in this over-all objective are subsidiary objectives to:

1. Provide a basis for a firm fixed price or fully structured incentive contract for engineering development.

2. Establish firm and realistic performance specifications.

3. Precisely define interfaces and responsibilities.

4. Identify high risk areas.

5. Verify technical approaches.

6. Establish firm and realistic schedules and cost estimates for engineering development (including production engineering, facilities, construction and production hardware that will be funded during engineering development because of concurrency considerations).

7. Establish schedules and cost estimates for planning purposes for the total project (including production, operation, and maintenance)."

This management system has been applied only to specific research and development projects which relate to a specific defense system included in one of the mission area programs (i.e., Programs I, II, III, IV). All general research and development programs are included in Program VIII, R & D, and are not subject to the same larger context value analysis. In addition, it was intended to apply the DOD directive to only the more costly programs, which would more logically require OSD's attention. Hence the following threshold levels were established:

APPLICATION

1. "All new (or major modifications of existing) engineering developments and operational systems developments as [previously] defined . . . estimated to require total cumulative Research, Development, Test and En-

gineering (RDT & E) financing in excess of 25 million dollars, or estimated to require a total production investment in excess of 100 million dollars, shall be in accordance with this directive unless specific waivers are granted by written approval of the Director of Defense Research and Engineering.

2. Other projects may be required to be conducted in accordance with this directive, in whole or in part, at the discretion of the DOD component or as directed by the Director of Defense Research and Engineering (DDR & E)."

DOD Management System in Practice

Various authors, particularly in the popular press, have attempted to evaluate the DOD management system and decision-making process, presenting its benefits and weaknesses.* While it is not intended to provide such an evaluation in this chapter, it is of interest to indicate the usefulness of the DOD management system by referring back to the list of the six key management questions originally posed and to note how well the DOD Management System provides answers to these questions.

1. *What is the system objective?* The DOD programming system has explicitly divided all DOD elements into nine key program packages, each having an explicit objective from which some quantitative measure(s) may be derived. Thus, any organization which is proposing some advanced system must indicate its system objective by explicitly indicating to which program package the system contributes.

2. *When will the proposed system be operationally available?* An estimate of the schedule for contract definition, development, and acquisition phases leading to initial and full operational capability is made during the concept formulation phase and this schedule must be credible and acceptable.

3. *How well does the proposed system meet the objective?* An evaluation of the proposed system must be performed as part of the cost-effectiveness analysis required in the concept formulation phase. This evaluation indicates in as quantitative a way as possible how much the system contributes to the over-all objectives of the program package. In addition, all of the assumptions and contingencies which were considered in this evaluation must be made explicit.

4. *How much will it cost to implement the system?* An estimate of the total costs expected to be incurred each year over the total life of the system is also made during the cost-effectiveness analysis of the concept formulation phase. Such costs include all elements of RDT & E, maintenance, and

* D. Novick (1964).

operational maintenance. Costs are not limited to monetary (dollar) units, but include all scarce resources required, such as scarce skills of people, and expected casualties.

5. *Is this the best way of meeting the objective?* The term "best" implies two things: (a) that a number of alternatives were examined, and (b) that the preferred solution was chosen on some defendable basis considering both system performance and cost. With respect to (a), the systems planner must show that he has considered a number of alternative approaches (including the "do nothing" option); otherwise he cannot leave the concept formulation phase, since the six prerequisites for doing so include the performance of a cost-effectiveness analysis of the alternatives considered. In addition to this, many times interorganizational competition can be used to initiate the creation of other desirable competing alternatives for performing the same objectives. One example of this has been described by a high-level OSD official: The Navy was proposing an increase in the size of their carrier-based fighter force for limited war operations in South Vietnam; they sent their proposal to OSD indicating that it would require lower costs for the same number of combat sorties as compared to the Air Force's tactical fighter operations. Their analysis indicated that these cost savings were achieved by using available aircraft carriers which could be moved to different locations for different contingencies, whereas the Air Force would have to construct new airfields and require defense elements, a more costly logistics supply system. To attempt to confirm the Navy study, OSD sent it to the Air Force Chief of Staff for comment (hoping to arouse their competitive spirit). The study was subsequently forwarded to the RAND Corporation whose analysts soon noted that the Navy proposal omitted certain cost elements, and that the Navy costing system itself differed from the Air Force costing system. This difference in costing procedures was then resolved, not only for this proposal; but from a longer range point of view, this served to unify cost procedures for subsequent Navy and Air Force competing proposals. In addition, the Air Force then set up a task force to re-examine Air Force limited war operations to see how they could be improved, using the Navy proposal as a basis for comparison. Undoubtedly, future limited war operations will consist of a mixed force using both Air Force and Navy fighter units. Thus, this strategy of attempting to promote interservice competition where two services both have overlapping capabilities cannot help but produce more creative plans and efficient operations.

The preferred solution is generally taken to be that system alternative which accomplishes the objective at lowest total cost. The task of evaluating the system alternatives many times is not a simple one. Consider the analysis of two aircraft, the first having a speed of Mach 2.0 and the second Mach

2.5, but which are identical in all other performance characteristics. The total cost of each is 2.0 million and 3.0 million dollars respectively over its entire operating life. Which is the better buy?

This problem, identical to many other problems in which one system clearly has superior performance characteristics to another system while costing more, illustrates why cost-effectiveness analysis is required. Further discussion of this problem and the selection criteria which can be used is contained in Chapter 4.

6. *What are the risks and uncertainties involved in obtaining the stated system performance on schedule and at the estimated cost?* There are four important sources of uncertainty which can cause the actual system results obtained (in system performance and required cost) to differ from that predicted. These are:

Technological uncertainty: Because of technological or development difficulties it may take much longer and cost much more to develop and produce a system containing the system performance characteristics originally predicted.

Cost uncertainty: The final system costs may differ from that originally estimated, not only because of technological uncertainty, but also because of inaccuracies in the cost estimating relationships used.

Competitive uncertainty: Because the competitive environment may be different from that originally assumed (e.g., quantitatively or qualitatively superior), our own system effectiveness may be correspondingly lower than estimated.

Chance or statistical uncertainty: A systems planner can never guarantee results any time a nondeterministic or random process (involving chance) is involved.

Each of these uncertainties can lead to changes in the actual total system cost required to obtain a given level of systems performance or effectiveness by a given operational time period.

The big change which was made by OSD in formalizing the total systems acquisition process was the insertion of the two stages of systems planning requiring systems analysis (i.e., the concept formulation phase and the contract definition phase) before engineering development of the system could proceed. This was done to minimize the risks always present in the development of a system having high technological content. By requiring the three prerequisites needed to leave the concept formulation phase (i.e., the establishment of technological feasibility, the cost-effectiveness value of the program is favorable compared with other DOD alternatives, and the schedule is feasible) certain of these risks were reduced. For example, the cost/time risk is minimized by insisting that technological feasibility has been established. If this feasibility has not been established, the system proposer may request an advanced development authorization for this purpose. When advanced development has been completed, the program may then advance to the next phase. By insisting upon programs whose cost-effectiveness is favorable compared with alternatives, the proposals of lesser value are filtered out.

The contract definition phase reconfirms the value of the proposal by awarding a more detailed study contract to two or more industrial contractors who are qualified to do the actual engineering development. More than one contractor is chosen to improve the competitiveness of the program. Each contractor now analyzes the system in a more detailed fashion than was done during concept formulation. Each contractor highlights the technological problems that are involved and arrives at a fixed-price or incentive-priced type of proposal showing the schedule that can be maintained, with a firm estimate of the cost. Thus, the time and cost risks are further reduced since the contractor indicates he would do the engineering development at this fixed price. This reduces the previous deficiencies of optimistic costing which industrial contractors may have done in the past. In addition, the contract definition phase permits "technical transfusion" (i.e., the contracting agency may meld together technologically superior aspects of several proposals into one proposal). A renegotiation of costs for the new proposal specifications would then take place with each of the contractors. It is only at this point that the much costlier phase of engineering development can begin.

While technological and cost uncertainties are minimized by the two planning phases of concept formulation and contract definition as previously mentioned, competitive uncertainty is coped with in the concept formulation phase by analyzing the various contingencies which may occur at the initiation of a competitive, potential threat, and performing suitable sensitivity analyses. Randomness or chance, which results in statistical uncertainty, is covered by probabilistic analyses in the concept formulation phase so that the decision-maker may see the extent to which chance may be significant in the operation of the system. Finally, the system planner can show by means of his analyses that he has chosen the design which will minimize these uncertainties through being least sensitive to the above factors.

An example of the influence of chance or statistical uncertainty which is encountered in nondeterministic system problems is contained in the following simplified illustrative problem. How many missiles, and hence what system costs, are required to destroy 100 separate but identical targets, if the probability of kill of each missile against the particular target is 0.7? It will be assumed that there does not exist any timely way of determining if a target has been destroyed by a missile or not, and hence the procedure to be followed will be to fire a salvo of one or more missiles at each of the 100 targets. The task to be solved is to determine the number of missiles required in each of the 100 identical salvos.

The problem may be solved as follows. If 100 missiles were purchased and fired with a salvo of one missile assigned to each of the 100 targets, the analyst can only say that the number of targets which will actually be destroyed can vary anywhere from zero to one hundred. Thus to absolutely guarantee that the entire 100 targets would be destroyed would require procuring an infinite number of missiles.

The systems planner must then determine a measure more meaningful to the decision-maker. One such measure would be the number of missiles required to achieve an expected number of targets destroyed, where the expected value would be the average number of targets destroyed if the offensive mission were performed a large number of times. A more sophisticated measure is used later in Chapter 7, but for now we shall use the average or expected value of targets destroyed as the measure. The planner could then rephrase the question to: How many missiles would be required to destroy an average of, say, 95 of the 100 targets?

Using this measure, the systems planner can indicate that if 100 missiles were fired with each missile aimed at one of the 100 targets, 70 targets could be expected to be destroyed with 30 targets expected to survive. This result is illustrated in Figure 3-4.

If 200 missiles had been purchased, a salvo of two missiles could have been fired at each of the 100 targets. In this case, on the average there is an "overkill" of 70 per cent of the 70 missiles already destroyed, and an additional kill of 21 targets (0.7 times 30 surviving targets), yielding a total of 91 expected targets destroyed, and 9 surviving. This is also shown in Figure 3-4. Similarly, the third hundred missiles provide an increment of only 6.3 expected targets destroyed, and the fourth hundred missiles provide less than two targets expected to be destroyed.

These results, as plotted in Figure 3-4, illustrate the concept of "marginal effectiveness" (i.e., each succeeding unit of missiles provides an incremental benefit or effectiveness which is less than the previous unit and there is some point at which it becomes less desirable to continue allocating resources in this fashion compared to either using the resources in other ways for accomplishing the same objective, or for accomplishing other objectives). Thus, decisions must be made on the benefits of the *next unit* of resource, as opposed to the ratio of *total benefits* obtained to *total cost*.

Of course total missile system costs required are *not* directly proportional to the number of missiles for several reasons. First, missile costs versus volume number procured is not a linear function because of the many one time fixed "set-up" costs of procurement. In addition, some elements of the

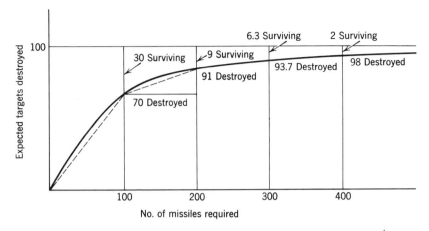

Figure 3-4. Missiles required for varying levels of target destruction.

weapon system, such as command and control, are either independent of the total number of missiles or must be repeated with each specified organizational quantity of missiles (e.g., one missile squadron headquarters is required for each squadron of missiles acquired). Hence, the analyst should use such information to convert number of missiles to total cost and prepare a plot of effectiveness versus cost for the decision-maker, as shown in Figure 3-5. This information relating effectiveness to cost can be used by the decision-maker to determine the incremental effectiveness obtained for incremental costs expended. Obviously, this nonlinear characteristic changes as different system alternatives are configured in different force combinations. However, the analyst and decision-maker must analyze these various combinations of force structure to determine when diminishing returns occur so that limited resources may be better allocated to another program. Decisions of allocating funds among the different programs (or mission objectives) are, in general, made on an intuitive basis since many political considerations are also involved and there are no current methods of analytically relating certain missions [e.g., our ability for coping with general war

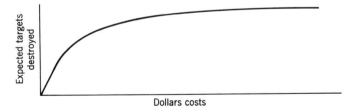

Figure 3-5. Missile cost for various levels of target destruction.

(Package I) and limited war (Package II)]. Other programs may relate more closely, such as our ability for coping with limited war (Package II) and air lift and sea lift (Package IV). Thus, the decision-maker must look at the marginal effectiveness of any proposal and intuitively determine if it is high enough compared with the needs of another program package.

II

FOUNDATIONS OF
SYSTEMS ANALYSIS

4

Developing the Systems Analysis Procedure

The various steps performed in conducting the concept formulation phase of systems planning were described in Chapter 2. The objective of this chapter is to develop an over-all procedure for evaluating a system alternative in terms of its effectiveness to accomplish an objective, and its cost. In particular we shall explore the role which models, particularly analytical models, play in systems planning. We shall discuss what a model is, and why models are used in problem solving or systems planning. The various types of models will be discussed, including those which describe a system, the job to be done, and the environment in which the system operates. The two models used for evaluation are described: (a) the effectiveness model which enables the analyst to predict how well the system will perform a job in a given environment, and (b) the cost model which predicts total system cost requirements. Examples of each model are given. Finally, the interrelationship of these models in performing a systems analysis, using the systems planning steps discussed in Chapter 2 are developed.

BASIC TERMINOLOGY

Systems planning involves the following key terms:

Analysis leading to the creation or *synthesis* of *system* alternatives, their subsequent evaluation, and the selection of the preferred alternative. Before considering the various types of models and their uses, let us define these key terms that are used in systems planning.

Analysis: The dictionary defines analysis as "the separation of a whole into its component parts; an examination of a complex, its elements, and their relations." Thus, the key concept involved in analysis is the separation, dissection, or partitioning of a complex problem into its smaller pieces

45

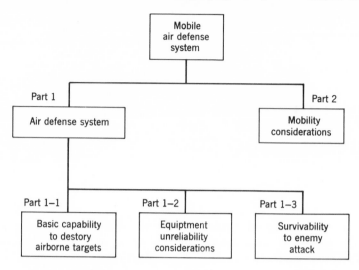

Figure 4-1.   Partitioning the mobile air defense system problem.

which are more readily solvable, where the final solution is the summation of these smaller solutions joined in the proper fashion to include the interrelationship of the parts.

For example, in an analysis whose objective is to evaluate alternative mobile air defense systems, one could partition the analysis into several fundamental parts initially treated as being independent as shown in Figure 4-1.

Part 1: System's ability to perform the air defense mission once the system is installed into position, omitting considerations of mobility.

Part 2: The mobility aspects which would modify the capability of the fixed system described in Part 1. These aspects would include the benefits of system mobility such as its ability to be emplaced in different locations as required. It would also include the degree of reduction of the fixed system's operational capability during the time the system is being moved and installed in its new location.

Similarly, the system's ability in Part 1 can be subdivided into several smaller parts to allow for separate, smaller, and less complex analyses of:

Part 1–1: Basic capability of the system to perform the air defense mission omitting any other considerations.

Part 1–2: Malfunctions caused by equipment unreliability.

Part 1–3: System survivability following enemy attack.

Thus a complex problem can be subdivided into a set of smaller problems, each of which can be more easily analyzed separately.

However, even though separate, smaller analyses can now be made, the analyst cannot overlook two problems:

1. A way of combining the separate results will have to be formulated.
2. Factors involved in one part of the analysis may still impinge on another part.

One example of this interrelationship of parts is that the mobility considerations of Part 2 may now subject the system to high shock and vibration, which may reduce the reliability of the equipment (Part 1–1) as compared with a fixed system.

Synthesis: Synthesis comes from the root meaning "to put together," and consists of the composition of parts or elements so as to form a whole. Thus synthesis may be thought of as the opposite of analysis. For example, in electric network analysis, one is given a combination of circuit elements (the system) connected in a specific fashion and is asked to determine circuit performance (the output) when connected to an electric source (the input). In the converse problem of network synthesis, one is given the circuit performance desired and asked to find that combination of circuit elements whose performance will closely approximate the desired performance when the input is applied.

The objective of analyzing a problem is to determine a preferred solution to that problem. The specification of the elements which make up the system solution, including the element interconnections and how the system is to operate, is called the system design. As indicated in Chapter 2, the process of analyzing the problem and creating a preferred system solution to that problem is called the systems planning or design process. Obviously this process involves both systems analysis and synthesis.

Essential Elements of a Systems Analysis

While the term "systems analysis" has already been defined in Chapter 1, and consists basically of system evaluation and creation of additional alternatives if needed, we shall now examine its component parts. Hitch lists the following elements as essential in systems analysis.*

"An *objective* or a number of objectives.

"Alternative means (or *"systems"*) by which the objective may be accomplished. (These may consist of either different sets of system elements or components or different strategies of using a given system.)

"The *"costs"* of resources required by each system.

"A mathematical or logical *model* or models: that is, a set of relationships among the objectives, the alternative means of achieving them, the environment, and the resources required.

* E. S. Quade (Ed.), 1964, pp. 13–14.

"A *criterion* for choosing the preferred alternative. The criterion usually relates the objectives and the costs in some manner; for example, by maximizing the achievement of objectives for some assumed or given budget."

These elements of a systems analysis and their interrelationships are illustrated in Figure 4-2 and form the basis of the analytical procedure described and applied in the remaining portions of this book. Note that the following additional elements have been delineated to accompany Hitch's elements:

There is a *job to be done* which is the basis of the objective. Each system alternative is expressed as the *system design model.* The model which determines how well each system accomplishes the objective while operating in the *external environment* is called the *system effectiveness model.* The model which determines the total cost of each system in meeting the objective is called the *cost model.*

Each of these elements is a model and will be discussed in greater detail in this chapter, but first we shall examine the term "model."

Definition of a Model

A model can be defined as an explicit representation of some phenomenon or problem area of interest, including the various factors of interest and

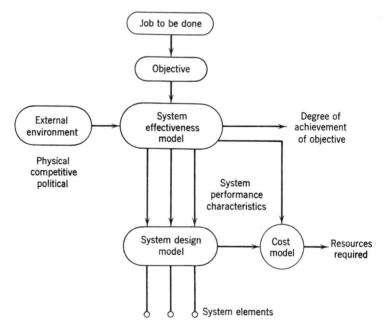

Figure 4-2. Models required in systems evaluation.

their relationship, and is used to predict the outcome of actions. Thus, a model is some analog or imitation of the real world. Note that this definition is a rather broad one, and so includes both qualitative and quantitative models.

Types of Models

R. D. Specht indicates: *

"We can classify models according to:
1. Purpose—training, study, etc.
2. Field of application—strategic, tactical, logistic etc.
3. Level—from national policy to base operations.
4. Static or dynamic.
5. Two-sided or one, conflict or not.
6. Degree to which mathematics is used.
7. Use of computers—how much and how.
8. Complexity—detailed or aggregated.
9. Formalization—the degree to which the interactions have been planned for and their results predetermined."

He further develops his own model classification scheme which he states "is as unsatisfactory as any other." His scheme consists of five man categories of models:

   I. Verbal;
  II. People—as an integral part of the model;
 III. People and computers interacting as a part of the model;
 IV. Computer;
  V. Analytical.

Each category has two subcategories "according to whether or not an active opponent is involved and conflict is an essential part of the model."

Thus a verbal model is a word description of some occurrence, activity, or system. A war game exercise simulating an actual battle would exemplify a people model. A command post exercise in which decisions are made by people but a computer is used to represent part of the problem, such as an enemy nuclear attack and our own force response, is an example of a people and computer model. He further classifies computer and analytical models as mathematical models. Some mathematical models, such as Ohm's law, are deterministic; however, other models which describe random processes are probabilistic in form, which introduces three other model classes. If the results of the analytical model are expressed as a probability distribution it

* E. S. Quade and W. I. Boucher (Eds.), 1965, pp. 288, 289.

is sometimes called a "probabilistic model." If only the expected value of the result is provided, the model is sometimes called an "expected-value" model. If the probability distribution is obtained by drawing random samples from each of the probability distributions associated with the operations involved, the model is sometimes called a "Monte Carlo simulation." Details of these models are provided in Chapters 6 and 7. While many of the models which we shall use are mathematical models, our definition of a model as "an analog of the real world" would also include the following examples of models:

1. A mechanical drawing;
2. A scale model of an aircraft for wind tunnel measurement;
3. A map;
4. An electrical schematic;
5. A flight simulator for training pilots;
6. A scale model of a skeleton;
7. Pictures of man's nervous system, bone structure, and blood system.

Notice that each model on the list is either a two-dimensional drawing or three-dimensional representation of some physical object; the specific purpose of each is to describe some real object(s) occurring in nature or created by man. It is important to note that the specific model(s) used relates to the specific problem area under investigation. For example, when trying to determine the preferred route to travel by automobile from one place to another, one model of the situation would be a road map that indicates not only all the existing roads but also the characteristics of each. Since the criterion for selecting the preferred route may be shortest time, shortest distance, lowest cost, or most beautiful scenery, or some combination thereof, a road map is a good model for gaining insight into the solution of this problem.

On the other hand, if one were operating an airline company and wished to increase business volume by conveying travel information to prospective passengers, two models would be required: (a) a map that shows what cities are interconnected by the airline company, and (b) a time schedule showing departure and arrival times of all flights in all cities served by the airline.

Finally, in planning an interurban highway, the planner would require additional maps that contain both topographical terrain features and property locations so that property values could be inserted. Thus, the cost of condemnation actions could be calculated when considering different highway alternatives. Note that each of these three problems required a map but each map was of a different type, depending on the particular characteristics important to the problem.

Many of the models just described were descriptive in nature and showed elements (such as type of highway), and their relationships (such as which highways interconnect given cities). Another type of model also used in systems analysis involves some form of transformation between a system input and output. An example of this would be the transformation of a modulated radio wave which enters a radio receiver and is converted to sound as an output. Such models are used to describe and understand a system, such as a manufacturing process.

In this book, first the various models needed to perform the function of systems evaluation, as illustrated in Figure 4-2, are described. The process of obtaining numerical results from these models, using probability theory, combinatorial techniques or Monte Carlo simulation, is defined as "model exercise" and is described in Chapter 7. Obviously the form of the model is related to the method of model exercise, but this is discussed later in Chapter 7.

Describing the System

The first model to be considered is the system design model which provides a description of the system alternative under consideration. In doing so it is appropriate to first consider the question of what a system is.

The term "system" has several different meanings, because in today's system-oriented environment, everyone seems to be working on systems. There are radar systems, data processing systems, communication systems, production control systems, and market research systems. Can all these truly be called systems? Also, how do systems and subsystems differ?

One can attempt to answer these questions by starting with the real world which is composed of things (such as equipment) and people, related to one another through some organizational structure. A table of an organization type model which illustrates this is shown in Figure 4-3. This model shows the collection of elements which make up the United States national establishment, going down through the services, the Air Force commands, and finally down to air defense equipments.

The air defense system shown is composed of an interceptor weapon system which is provided initial guidance information by a command and control system. This in turn contains, as shown, a radar system which can be thought of as a collection of transmitters, receivers, and antennas. It also contains a data processing system which includes input/output equipments, and data storage devices, and central processing equipments, as well as communication systems and display systems. Can they all be systems?

The dictionary defines the term "system" as follows:

"An aggregation or assemblage of objects united by some form of local interaction or interdependence; a group of diverse units so combined by nature or art as to form an integral whole, and to function, operate, or move in unison and, often, in obedience to some form of control; an organic or or-

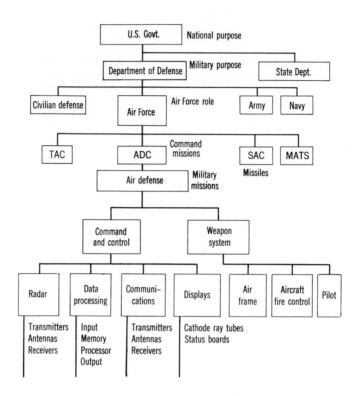

Figure 4-3. Hierarchy of systems.

ganized whole; as, to view the universe as a system; the solar system, any telegraph system."

In a report of the Systems Science and Cybernetics Group of the Institute of Electrical and Electronic Engineers (IEEE), the term "system" is defined as follows: *

"A system is a collection of interacting diverse human and machine elements integrated to achieve a common desired objective by manipulation and control of materials, information, energy, and humans.

"With this definition, it is possible to think of systems as characterized by

* System Sciences Symposium of the IEEE, 1964.

their objectives or by their elements, or by their major technology. Thus, one has command and control systems, naval systems, or electromechanical systems."

The universality of this definition has an advantage in that it indicates that *anything* can be viewed as a system. For example, the IEEE report indicated that even a table can be classified as a system since it consists of four legs and a table top (the elements), all joined together to achieve a common, desired objective (holding up objects). In fact they indicated that perhaps the only thing they could not classify as a system was a table leg, but even here a molecular physicist might take issue with this statement. Thus, the universality of the definition may have the disadvantage of not being too meaningful.

The author's approach is that, since everything can be classified as a system, the definition has some drawbacks. A better approach might be to recognize that the only objective of considering the concept of a system is to solve a problem or fill a need. Hence, one might conceive of a "systematic (or rational) approach for handling complex problems and making decisions."

One can accept and work with the dictionary and IEEE definitions of the term "system" if one supplements these definitions with the concept that there really exists a "hierarchy of systems" as illustrated previously in Figure 4-3. By constructing such a structure for any given problem, the systems planner can explicitly identify the three levels whose elements are of greatest importance to him. The first level is that immediately below him, and contains those elements of the problem which the systems planner has under his immediate control. Thus, in the case of the design of a command and control system, the elements include radar sensors, data processors, communications, and displays. As will be seen later, it is the primary function of the system designer to achieve a balance between the system performance of these elements (i.e., the degree to which these elements meet the command and control objectives and the resources required to obtain this performance). Obviously, the performance and cost characteristics of any of these elements are composed of, and hence implicitly contain, the performance and cost characteristics of all of their subelements. Thus, the systems planner may delegate the design of the subelement echelons such as the radar components to other technical specialists thereby easing his span of control problems.*

---

* The systems planner must, of course, assure himself that he will, in fact, obtain the performance and cost characteristics predicted by the first echelon levels. Prediction of such risks can be asked for from the first echelon levels. This point is considered in Chapter 9.

Constructing this hierarchy also permits the system designer to more clearly identify the objective of the next higher level (i.e., the air defense mission). This is important since the lower level objectives must be consistent with those of all higher levels. Lastly, the hierarchy also identifies those elements with which the command and control system designer must interface; that is, the aircraft weapon system.

Thus the system design model contains:

1. A description of all system elements required for its operation and support, including hardware, software, facilities, and people. Such a description would follow the structure illustrated in Figure 4-4.

2. The policy, strategy or operational concept that interconnects the elements and is followed in the actual system operation.

3. Key system factors, characteristics, or attributes which describe the system and its elements.

While parts one and two describe the characteristics of "things" (system elements) and how they are to be used, the third part of the model includes the set of system performance characteristics. For example, the performance characteristics of a bomber would include its speed, penetration

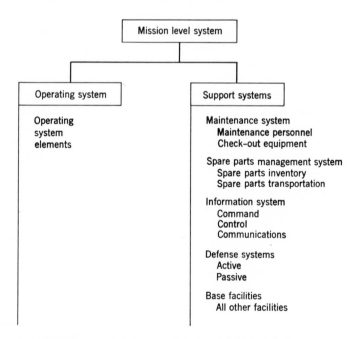

Figure 4-4.   Mission level system and its subsystems.

range, altitude limits, bomb-carrying capacity, vulnerability to air defense, etc. Obviously, the system design model may go down to any level of detail which might be required.

FACTORS INVOLVED IN SYSTEMS PLANNING

The various factors which must be considered in a system planning problem may be divided into three classes: (a) those factors which are directly controllable by the system planning organization, (b) those only indirectly controllable by the system planning organization, and (c) those not controllable by the organization. The latter class is defined as the external environment.

Assume you are given the assignment of improving the Air Force Semi-Automatic Ground Environment (SAGE) Air Defense Control System whose elements are shown under "Command and Control" in Figure 4-3. We shall now examine these classes of factors for this example.

Directly Controllable Factors: These include all elements of the SAGE Air Defense Control System directly under the control of the systems planner. These elements consist of radar, data processing, communications, and displays, all elements of an information system.

Indirectly Controllable Factors: These are the parts of the higher-level system which must interface with the directly controllable parts of the system but yet are not under the direct control of the systems planner. These elements include manned and unmanned interceptors, and Army surface-to-air missiles, the Joint Chiefs of Staff, and the President.

External Environment: These factors are all other uncontrollable elements to be considered in the study. These constraints include those pertaining to the physical environment, such as weather, atmospheric pressure, temperature, etc., and also include the potential enemy or competitor.

Recognition of these classes of factors aids the systems planner in coping with them, as shown in the following.

Dealing with These Factors

Directly controllable elements are those under the direct control of the systems planner and are specified in the system design model. As emphasized in subsequent chapters, the systems planner's task is to divide the available resources among these elements which will yield maximum effectiveness for this given level of resources. With respect to the indirectly controllable elements, the systems planner can obtain and accept as fixed constraints the performance characteristics of these elements as given to him by those organizations having direct control over them. This is done for most of these elements.

However, in certain cases the analyst may discover that if certain indirectly controllable elements of the problem can be modified, the same higher level objective can be met at lower cost than by solving the problem as originally presented to the analyst; many examples of this are cited in subsequent chapters. Thus the systems planner has a professional obligation to attempt to modify these other elements, either by influencing the interfacing organization or by indicating the situation to the higher-level systems planner who has direct control over both elements.

The external environmental factors can be dealt with first by making explicit the way those factors are being considered in the study. Some of these factors, such as physical environment, are not controllable by anyone, but their values have a high degree of certainty (if only in probabilistic form) associated with them. Other factors involving the performance characteristics of the competitor's or potential enemy's system not only contain a high degree of uncertainty, but also an assurance that the competitor shall attempt to operate his system deliberately to minimize the effectiveness of the system under study.

Thus, the various factors to be considered in this category include the different operational situations which may be encountered. These are called "contingencies." A qualitative description of each contingency is called a "scenario," which S. Brown * defines as a "statement of assumptions about the operating environment of the particular system we are analyzing." Generally the scenario consists of a verbal description indicating how the mission starts, the type of weapon (systems) the competitor may employ, and the number and mix of his weapons, and the performance characteristics of the competitor's system, and how he may operate his system. In general, an analysis must be performed for each of the different scenarios which may occur.

DESCRIBING THE OBJECTIVES

There are two types of objectives with which the systems planner must be concerned. The first can be called the planning objective, project objective, or problem objective. This objective is the reason for initiating the planning effort in the first place, and in general, involves efforts for improving one or

* E. S. Quade and W. I. Boucher (Eds.), 1965.

more system performance characteristics (such as increased aircraft speed, range, payload, reliability) or reducing system cost.

The second type of objective can be called the system objective and consists of the function or mission which the system performs.

System Effectiveness

One primary purpose of a systems analysis is to determine quantitatively how well a system alternative meets an objective. The answer to this question, as indicated previously, basically involves creating an interrelationship among performance characteristics (obtainable from the system design model), the environment in which the system will operate, and the objective(s) of the mission. Thus, the effectiveness of the system can be defined as the degree to which the system can perform its mission or function while operating in the external environment.

To illustrate the importance of the environmental factors on how well the system objective is met (i.e., system effectiveness), consider the problem of determining the effectiveness of obsolete World War II B-25 light bombers in today's environment. The measure of the bomber's effectiveness could be defined as its ability to destroy targets. To analyze this situation, one must look at the type of warfare involved in the mission: is this general nuclear war, counterinsurgency, or limited war against an enemy posssessing an air defense system? Thus, one part of the problem is investigating the mission to be performed, the objective of the system within this mission, and the environment within which the mission is to be performed. The other side of the problem is to consider the characteristics of the bomber. One could measure the characteristics of the system (the bomber), independent of the job to be performed. In the case of the B-25, these include low speed, low penetration range, low altitude, low vulnerability to enemy air defense, and low bomb load carrying capacity, as constructed in the system design model. These performance characteristics would be found to be poor for one type of mission such as general nuclear war, yet may be excellent for another, such as counterinsurgency.

In this latter application, low speed may not be a problem since the air defense may consist of only small arms ground fire. In fact, since the targets may be camouflaged targets of opportunity, low speed may be not only satisfactory, but desirable. If the targets are close to the air base, limited bomber range may not be a problem.

The Operational Flow Model

One of the more difficult parts of a systems analysis is to properly determine the proper objectives. One reason is that the system performance characteristics are many times treated as multiple objectives resulting in a difficulty of obtaining one "objective function" which can be used for system "optimization." Further, just as any particular system is only part of a hierarchy of systems (Figure 4-3) with each system level having its own specific system objective, there must also exist a hierarchy of system objectives. Hence what the systems planner needs is a framework which will make explicit this hierarchy of objectives and will relate these to the various system performance and environmental characteristics. Such a model is called the "operational flow model" and consists of the series of activities or events which the system must perform in accomplishing the *entire* (higher level) system objective. Each of these activities can be represented in quantitative form by a series of measures describing the input and output of each activity as well as the activity's input to output transformation (called the transfer function of the activity). The primary purpose of the operational flow model is a means of evaluating how well each system alternative performs the objective. Hence the operational flow model can be used as the system effectiveness model. In addition, by examining the specific system objective and then constructing a hierarchy of higher level objectives and measures, additional possibilities for solving the problem can often be perceived. An example of how the operational flow model is constructed and its usefulness in solving a problem is contained in the discussion of the following problem.

Elevator Delay Complaint Problem

A new forty story office building was open only a few months when the building superintendent received a number of complaints from the building tenants. They felt they had to wait too long for the first floor elevators when they arrived at the building in the morning. After the superintendent confirmed the problem, he tried several possible solutions.

Various adjustments were made to the automatic elevator programmer but the improvements were insufficient to cope with the peak morning traffic. A man was next assigned to manually regulate the elevators in accordance with the incoming morning traffic. This, too, proved unsatisfactory as measured by the number of complaints continuing to reach the superintendent. The superintendent then made contact with the elevator manufacturer who sent an analyst to investigate the situation.

The analyst began by collecting environmental data to quantify the amount of traffic demand as a function of time during the peak morning hours, and to measure user waiting time. Finally, several possible solutions were suggested by the analyst. The first was to replace the existing elevator programming unit with a more complex unit having more flexible features. One such feature would permit two of the elevators to concentrate on the first twenty floors, whereas the other two would be express to the twentieth floor and beyond. This unit would cost $7,000. A second possibility was to install a fifth elevator on the outside of the building. This would be all glass-enclosed and, in addition to its functional utility, could be used as a source of advertising. However, this alternative would cost a total of $50,000.

At this time a third suggestion was made which solved the problem at a cost of $300: a television receiver was installed near the elevators. This permitted the tenants to watch a TV program while they were waiting for the elevator to return to the first floor. While some students of creativity may suggest that this is a good example of the application of intuition or inspiration in solving a problem, the point I wish to make from this example is not the sheer brilliance of the idea which solved the problem, but that there exists a systematic approach in which such alternatives can be uncovered and created. Perhaps the first step in this systematic approach would be the construction of the operational flow model which contains the series of activities or events to be performed in accomplishing the *entire* job and meeting the planning objectives. This model, shown in Figure 4-5a, illustrates that the activity under study ("waiting for elevator") is only one activity of a higher level objective to get an individual from the building entrance to his office. The other major activities include "riding in the elevator" and "walking from elevator to office." Associated with each of these three activities is the key measure of time taken to complete each activity (as shown in Figure 4-5a). Since our highest level objective being considered is to reduce the number of complaints to the superintendent, and since we intuitively feel that there is a relationship between waiting time and number of complaints, we must indicate this relationship as the final transfer function of the operational flow model. This model is completed when we determine the quantitative expression of each of the transfer functions as a function of the key system performance and environmental characteristics shown in Figure 4-5a (on the right and lefthand side respectively).

However the purpose of this example is to show how this framework could be used for both systems evaluation and the creation of other alternative approaches for accomplishing each of the objectives in the hierarchy. Let us first concentrate on the submodel which relates the total time taken for the elevator service (which includes the waiting time, the primary measure considered thus far) to user dissatisfaction (as measured by the number of complaints, the higher level measure). Thus, as shown in Figure 4-5a, a certain amount of time is required for the total elevator service. Each user then implicitly evaluates this service received in terms of various performance factors such as the quality characteristics of the ride (smoothness, crowdedness, etc.), time required (total and waiting time), for the ride, as well as how the user's time is occupied. If the user's evaluation of these characteristics is satisfactory, he may register his satisfaction directly by

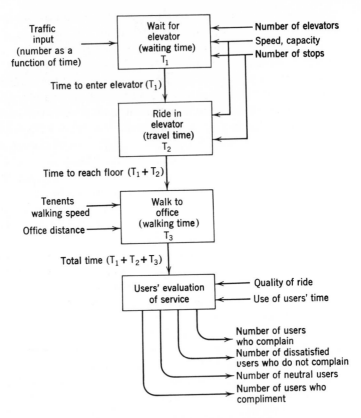

Figure 4-5a.   Operational flow model.

praise (but generally indirectly by not complaining) or, on the other hand, if he evaluates the service as low, he may become annoyed enough to complain to the building superintendent or to others. Thus, the higher level objective of this project is "waiting passenger satisfaction," and its measure in this case is the "number of complaints to the building superintendent." * The question of meeting this objective now translates to reducing the number of user complaints, which means reducing the number of users who are annoyed because of wasted time. The user evaluation function for this problem

---

* In this case, the data relating to the measure is directly available. In other cases, this may not be so. For example, customers in a supermarket may be unhappy with the service but instead of complaining to the service manager, they discontinue their patronage of the store. This might be the case in a supermarket whose checkout lines are very long at peak hours and the store very probably would lose customers. Sometimes a manager may have to conduct customer surveys to uncover the conditions and to determine what percentage of customers are approaching the dissatisfaction point.

may be thought of as a function (called a utility function) which relates the degree of user dissatisfaction as a function of waiting (which appears to be the key variable of interest in the problem as registered by the stated complaints). Such a utility function is shown in Figure 4-5b. As indicated, complaints (or a given percentage of complaints) start to come in when the waiting time exceeds t_1. Customer dissatisfaction with the service may begin at some value lower than t_2 but this dissatisfaction may not be great enough to evidence itself by a tangible indication, such as a complaint.

Thus the primary function of the operational flow model is to indicate the relationships which exist between each of the factors under the system planner's control and the system objective. In this way the planner can determine the worth and the resulting cost of any proposed system improvement. Hence the model permits the systems planner to balance the total system, as will be described in greater detail in Part V.

Another important use of the operational flow model is that many times it proves beneficial in uncovering alternative, complementary ways of accomplishing the higher level objective, even though the systems planner may be constrained by the inability to satisfy a particular lower level objective. For example, a person's dissatisfaction with waiting time is, in general, dependent on how that waiting time is utilized. A long ride on a train may not cause dissatisfaction if the time is spent in enjoying the scenery, the club car, a leisurely meal, reading a good book, etc. This leads to the concept that user dissatisfaction may be reduced if the user time spent in waiting is used pleasantly or profitably. This now opens a completely new view of the problem and provides another variable which the systems planner may be able to use to solve his problem. For example, other possibilities for using the time pleasantly or profitably might include the use of a TV or radio receiver to

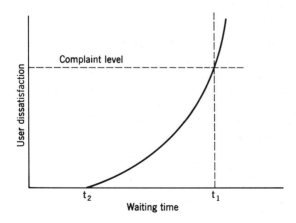

Figure 4-5b.   User utility function.

provide the morning news or other programs, movies, recorded music, etc., each capable of reducing the number of complaints to a certain degree, even if the waiting time is unchanged. Note, however, the sensitivity of the analysis to the measure chosen. If the office workers were all employees of the same firm, "manhours wasted" instead of "customer satisfaction" might have been a more appropriate measure, and the analytical approach would have been different.

WSEIAC Approach to Measuring Effectiveness

In 1964 the United States Air Force Systems Command created a Weapon System Effectiveness Industry Advisory Committee (WSEIAC),* consisting of representatives from industry and the Systems Command, whose mission was to examine the problem of evaluating systems. Some of the conclusions which they reached are as follows: systems effectiveness can be defined as a measure of the extent to which a system may be expected to achieve a set of specific mission requirements and is a function of three primary components: availability, dependability, and capability.

"*Availability* is a measure of the system condition at the start of a mission and is a function of the relationships among hardware, personnel, and procedures.

"*Dependability* is a measure of the system condition at one or more points during the mission, given the system condition(s) at the start of the mission, and may be stated as the probability (or probabilities or other suitable mission oriented measure) that the system (a) will enter and/or occupy any one of its significant states during a specified mission and (b) will perform the functions associated with those states.

"*Capability* is a measure of the ability of a system to achieve the mission objectives; given the system condition(s) during the mission, and specifically accounts for the performance spectrum of a system."

Note that with the WSEIAC definition of system effectiveness, one can determine the effectiveness of any type system. For example, the effectiveness of a system which performs a mission could be measured in terms of the probability of performing the mission to a given degree; (e.g., the effectiveness of an air defense system could be measured in terms of the probability of destroying at least N aircraft in a given engagement). Similarly, the effectiveness of a system which performs a function could be measured in analogous fashion. Thus, a radar could be measured in terms of the probability of detecting a given size aircraft by a distance of 100 miles, for example.

---

* Weapon System Effectiveness Industry Advisory Committee (WSEIAC), Final Report of Task Group II, "Prediction–Measurement (Concepts, Task Analysis, Principles of Model Construction), January 1965, AFSC-TR-65-2-Vol. II, p. 3.

Thus the WSEIAC definition of system effectiveness handles the measurement of any system in the hierarchy of systems. In all cases, each system must be placed in its operational environment, and operated in accordance with the specific environmental conditions which have been established in the analysis. Note that by this definition the system effectiveness is always measured in a probabilistic fashion.

One main advantage of the WSEIAC approach is that it forces the analyst to focus upon the three main components of effectiveness: availability, dependability, and capability, which are, in turn, functions of the various system performance characteristics as shown in Figure 4-6. The reader will note that with the WSEIAC definition as given, the factors of system reliability and survivability must be considered under both availability and dependability since the system may malfunction or have been destroyed by enemy action, either before system operation begins or during the time when the system is operating and performing its mission.* Such a chart can serve as a useful checklist for the analyst when constructing the operational flow model, and, hence, helps to prevent the omission of any of the factors.

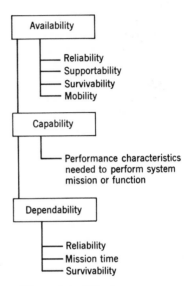

Figure 4-6. Effectiveness components.

THE INTERRELATIONSHIP OF MODELS IN A SYSTEMS ANALYSIS

Having described the various models which can be constructed as part of systems analysis, we shall now summarize the procedure for using these models as part of a formalized systems planning work process, such as indicated previously in Table 2-1. This is the work process which is used to structure the case situations examined in the remainder of the book. While the main steps in the analysis are listed, this work process should be consid-

* This is for the general case of a system not in continuous operation, such as a missile or an aircraft. A continuously operating system such as an electric utility generator has an infinite mission time; hence, the component of availability contains dependability.

ered an iterative one. The analyst proceeds with a few steps, and may then go back as new data become available.

The procedure for implementing the steps in systems planning is modeled in Figure 4-7a showing the interrelationship of the models just described, and provides a better understanding to three types of problems. These problems are the "problem as given," an expansion of this problem, called the "problem as understood," and a final contraction of this problem called the "problem to be solved." We shall now describe why the device of identifying three related problems is a helpful one in systems planning.

The systems planning process begins with the Problem As Given (PAG) which consists of a statement of the problem as originally presented to the systems analyst, and which initiates the analytical effort. The PAG is sometimes described as a "felt need," since in general, it is a qualitative descrip-

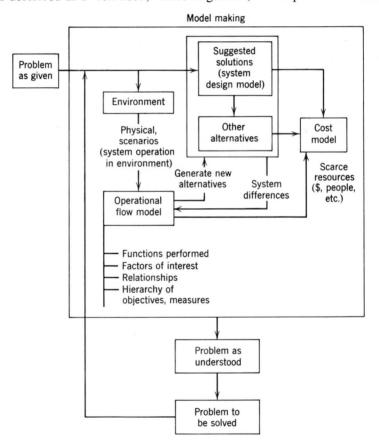

Figure 4-7a. Model making in systems analysis.

tion of some intuitive feelings on the part of a decision-maker that some problem exists in his area which he feels requires some systems analysis support. For example, in the SAGE Air Defense Control System, the problem could have been given as follows: "Explore various ways of improving the effectiveness of the SAGE System." The PAG by the decision-maker could also have been accompanied by one or more suggested solutions; e.g., "Explore ways of increasing the communications capacity of the SAGE System by using digital data links rather than voice communications."

At this point the systems analysis can go in either of two directions as the analyst attempts to develop the Problem As Understood (PAU) (i.e., to gain further understanding of the PAG). The PAU is, in general, an enlargement of the PAG and consists of those factors which, while not identified in the original statement of the problem, are identified by the analyst during his preliminary analysis of the problem. As shown in Figure 4-7a, he may move to the right and focus on the current system in use and any suggested solution(s) which may have been proposed for the stated problem. This is done through the vehicle of the system design model by structuring the elements of the suggested solution (if one has been given) and the performance characteristics which are achievable, or which can be achieved, within the suggested solution.

The scope of the analysis can then be broadened to place the solution to the stated problem in the higher level context by constructing a higher level system design model which explicitly includes all elements (such as the lower-level suggested solutions to the stated problem) which are required or can be made available to meet the mission objective (in this case the air defense mission). The higher level system design model is useful not only in identifying competitors (i.e., other alternatives), but in making explicit the mission(s) which the system performs.

Continuing with this first path of systems planning, the analyst can examine the suggested solutions and focus on the functions to be performed by the system, the flow of operational events or activities (including mission objectives and quantitative measures) and the relationship of the system performance characteristics to these operational activities. This analysis serves as an input to the second of the two parallel approaches in developing the PAU, as shown in Figure 4-7a. This approach is to focus on the mission and its objectives by constructing the environment, including scenarios, and finally, the operational flow model. The environment in which the system is to operate includes the physical aspects of the environment for each of the various operational contingencies under which the system may have to operate. Scenarios which verbally describe each contingency (i.e., possible type of operation) are constructed. Each scenario includes a description of the various conditions which may exist during the years of system operation and leads to the construction of the operational flow model, relating system activities or events to be performed with the system performance characteristics and the environmental characteristics. This model is used as a basis for evaluating the effectiveness of the system and also provides a means of generating new system alternatives which have not previously been suggested.

Note that the work process could have gone from the PAG directly to the environment and then to the operational flow model as indicated in Figure 4-7a. This approach focuses first on the mission and the job to be performed before examining the system alternatives which may have been suggested. Since both steps have to be performed during the work process, and there is iteration between the steps, the order in which they are performed may not be too important. We tend to follow the first path, and do so, in general, for the case situations described in subsequent chapters.

The Problem To Be Solved (PTBS) as finally derived is, in general, a simplification of the PAU, arrived at either by making simplifying assumptions, by treating certain factors as constants rather than variables, or by completely eliminating some factors or less likely scenarios from consideration. This simplification procedure is adopted to keep the problem within manageable bounds with respect to the analytical resources, such as manpower, computer time, and total time to perform the analysis. Obviously, the more simplifications or omissions that are made in reducing the scope of the problem to be solved, the less credible the analysis may be to the decision-maker. In general the systems analyst must exercise care in balancing the accuracy or credibility of the analysis against the resources allocated to the problem. This is part of the art rather than the science of systems analysis and is discussed in greater depth in Chapter 16, and is illustrated in the case situations described later.

Let us emphasize that the entire process described is an iterative one. The style I find successful is to construct crude models initially and then refine and enlarge these to include greater detail during each iteration. Iteration with the decision-maker is a necessary part of a systems analysis in order to ensure validity and credibility regarding the logic of the analysis and the appropriateness of the factors included, with respect to the resources available to do the job. It is particularly important to obtain the approval of the decision-maker concerning how the scope of the problem can be reduced (i.e., the parts of the broader analysis that will be eliminated in going from PAG to PTBS) to keep within the analytical resources available. Only by explicitly indicating to the decision-maker(s) the type of assumptions and omissions that will be made and the resulting analytical resources required can credible models be constructed and a valid analysis performed. Obviously the analyst must exercise care in making the proper simplifications to the model, so that key variables are not eliminated. Doing this properly is part of the art of systems analysis.

The WSEIAC Committee, previously cited, took a somewhat related approach to systems evaluation.* Their process (modified slightly to use the same nomenclature previously developed in this chapter) can be viewed as the following series of tasks, illustrated in Figure 4-7b, the first five of which are quite similar to those represented by Figure 4-7a.

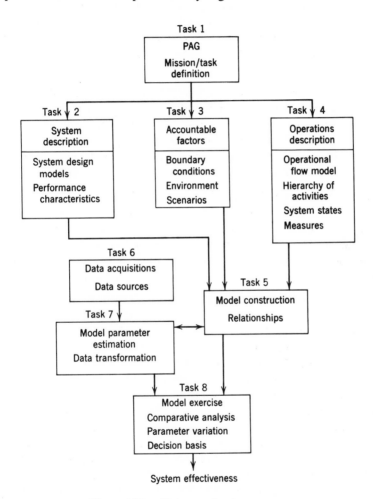

Figure 4-7b. System evaluation process.

* WSEIAC report AFSC-TR-65-2, Vol. 2, "Prediction–Measurement (Concepts, Task Analysis, Principles of Model Construction)," pp. 5–16, and AFSC-TR-65-6, Final Report, pp. 10–19.

Task 1: Mission Definition. This process begins with the mission definition which is a precise statement of the PAG, including the intended purposes of the system. (The environmental conditions, including scenarios, are described under Task 3.)

Task 2: System Description. This description should contain not only a description of the particular system under study, including the performance characteristics, but also a block diagram showing its relationship to its complementary systems and any other competing system which can perform the same mission. Thus, the system description will also include the hierarchy of systems which will provide the performance of a mission (or missions for a multipurpose system).

Task 3: Identification of Accountable Factors. This should include not only all of the assumptions and boundary conditions made, but also the assumptions of the external environment, particularly the scenario or description of the various environmental conditions under which the system is expected to operate. If there is more than one scenario to be considered, these should be listed accordingly.

Task 4: Operations Descriptions. Given the hierarchy of systems, the operational flow model which will be used as the basis of the effectiveness evaluation is constructed, including the various events which must take place, the hierarchy of measures, and indicating the various system performance and environmental characteristics which relate to each of the submodels.

Task 5: Model Construction. This is a twofold task. The first step includes the work of constructing the system design model and the operational flow model. The second step is to determine the quantitative relations which are a set of mathematical and/or logical equations which connect the various factors involved in each of the operational flow submodels.

To accurately determine the parametric relationships among the factors and the data needed to obtain quantitative results, the analysts should assume some scenario of interest involving one of the systems under consideration and permit the system to interact with the environment, including an assumed enemy force through the scenario. The analyst then considers each of the detailed events involved in the total operations displayed in the operational flow models previously constructed to determine step by step the specific factors involved and how they are related. This is best done by the analyst constructing a "test hypothesis," which is a relationship expressed in parametric form, for each event involved. This relationship is based on the analyst's knowledge of the type of operation which, of course, comes from his observations of or experience in this type of operation. The analyst then determines the data which he feels can be obtained within his time allocation and then makes any necessary assumptions needed to simplify the problem so that he can satisfy his time and resource constraints. These data are obtained from Tasks 6 and 7.

Task 6: Data Acquisition. This task indicates the data to be acquired and used as the basis of numerical inputs to the operational flow model, the location of these data, and the difficulty in obtaining them.

Task 7: Model Parameter Estimation. Using whatever data or information which is available or can be made available within the time and resource constraints of the project, numerical estimates of the pertinent parameters are made. The estimation process might include the use of known equations which relate similar appropriate factors, statistical techniques, or subjective estimates based on judgment. The estimation process used, of course, determines the amount of error involved in the input data used. The output of this task should include an indication of the method by which the data will be extrapolated to estimate model parameters.

Task 8: Model Exercise. Model exercise is the task of combining each of the submodels involved in the operational flow model and obtaining a quantitative output for a given input. Methods for model exercise include various mathematical techniques and simulation. The exact method chosen is a function of the amount and accuracy of the available data, the accuracy of the desired results, and the amount of resources available for performing the analysis.

COST MODEL

The previous discussion has centered on the models used for measuring the effectiveness of alternative system designs. It is also necessary to determine all of the resource expenditures or costs involved in implementing each of the system alternatives under consideration. Thus cost models must be developed which can provide this output. Every time a consumer makes a purchase where he has a choice of optional features such as in buying a new home or a new car, he comes in contact with one type of cost: investment, in which the investment cost model shown in Table 4-1 is affixed by law on the left rear window of every new car. This model lists the price of each element of the automobile, including options. The total sum is the cost of the delivered automobile, and obviously does not state the cost of operation or maintenance of the car. Other cost models are needed for this. Further discussion of cost models is in Chapter 9.

Table 4-1. *Auto Investment Cost Model*

Basic 4-door sedan	$2486
Automatic Transmission	193
Power Steering	72
Power Brakes	47
Radio	94
Heater	86
Bucket Seats	102
Transportation from Factory	68
TOTAL	$3148

CRITERIA FOR SYSTEM SELECTION

Given that the effectiveness and cost of each particular system alternative can each be measured separately, the systems analyst is still faced with the problem of how to combine these two factors to reach a final selection. If an economic measure of effectiveness can also be chosen, system cost can be included in the effectiveness measure and a true optimum can be selected. Consider the example of increasing the degree of automation in a manufacturing facility which produces television sets. Here, if company profits over the next three years, as an example, were chosen as the effectiveness measure, and many different manufacturing system improvements were considered, it would be possible to structure the effectiveness versus system cost, as shown in Figure 4-8. Company profits would be based on such factors as assumed market demand and total time required for paying off the capital appropriation for the new facility. Note from Figure 4-8, profits increase as

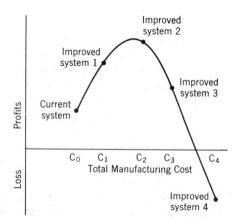

Figure 4-8. Company profits for various manufacturing systems.

system improvements (and costs) increase, but then decrease (and even go negative) with further cost increases. Here the planner can indicate an optimum operating point (i.e., near C_2) for the particular set of conditions assumed.

However, there are many other system planning problems where the measure of effectiveness cannot be measured in economic terms, but in benefits (or effectiveness). For example, as illustrated in Figure 4-9a, we might

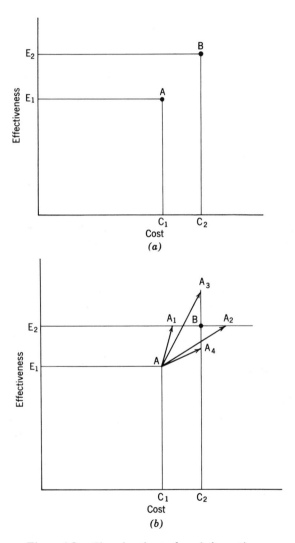

Figure 4-9. Choosing the preferred alternative.

have a situation where System B provides a higher level of effectiveness than System A but costs more.* The system selection problem is to find which is the better system (i.e., whether the additional amount of effectiveness is worth the added amount of cost). It is impossible to answer this question, except on a purely intuitive basis (which may be wrong or difficult to defend), without resorting to either of the two source selection criteria used in a cost-effectiveness analysis:

1. Specify a level of effectiveness which all systems must meet, and select that system which meets this level at lowest total cost. This criterion is called "pivoting on constant effectiveness." Thus, if E_2 is chosen as the comparison level of effectiveness, and the effectiveness of System A increased accordingly, its new "operating point" on Figure 4-9b might be either at A_1 (lower cost than B and hence selected), or A_2 (higher cost than B and hence rejected).

2. Specify a level of cost which all systems must not exceed, and select that system which provides the highest level of effectiveness. This is called "pivoting on constant cost." Thus, if C_2 is chosen as the comparison level of cost, and the cost of System A is increased accordingly, its new "operating point" in Figure 4-9b might be either at A_3 (higher effectiveness than B and hence selected), or A_4 (lower effectiveness than B and hence rejected).

Many times there is no one level of performing the objective which can be stated as an absolute requirement. For example, consider the analysis of alternative defense systems (both passive and active) for protecting a country against general nuclear attack. Here the measure used could be "population surviving," and it follows that the more funds spent for defense, the more population would survive. Hence the decision-maker needs to be presented with the function which relates the level of effectiveness and system cost for the set of "efficient solutions" as shown in Figures 4-10a (pivoting on constant effectiveness) and 4-10b (pivoting on constant cost). These figures also indicate not only the preferred system which provides a given level of effectiveness (or cost) for the scenario considered, but also indicates what other systems were also considered. The decision-maker can thus choose the appropriate level of effectiveness based on this type of information and the pressure for funds for other nonrelated objectives. This demonstrates a limitation of systems analysis. Unless multiple objectives can be interrelated in some fashion, a decision of how much money to spend to satisfy any one objective, such as in the defense example previously described, must be based on judgment. However, systems analysis can provide both the most

---

* Note that the decision is straightforward if we have a dominant case where one system provides more effectiveness at a lower cost.

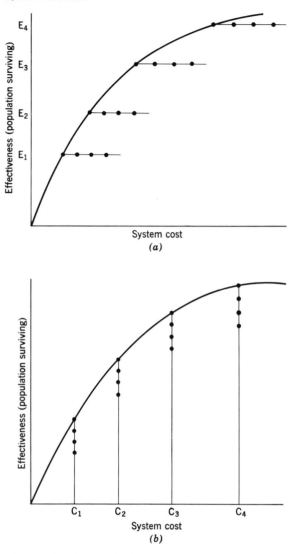

Figure 4-10. Structuring the set of efficient solutions: (*a*) pivoting on constant effectiveness: (*b*) pivoting on constant cost.

efficient solution for any level of effectiveness (or cost), and the function which shows the additional benefits obtained for additional cost expended, as shown in Figure 4-10a (again under the set of assumptions stated in the models constructed. Varying these assumptions is a very important part of the analysis as is discussed in Chapter 8.)

HOW MODELING AIDS THE PLANNER

The previous remarks were intended to indicate that there is a systematic approach to problem solving (systems planning) which can be used in the evaluation of systems and in the generation of new system designs. It is a process in which systems analysis plays a large role. Let us now discuss how models help us to do these tasks.

What Does a Model Do?

Models help the systems analyst gain greater understanding of the problem and the phenomena involved in the problem. This is done by converting a real world situation into a series of pertinent factors and their relationships. Thus, by working with the model instead of the real system or situation, it is possible to achieve certain advantages.

Modeling offers a method of evaluating a system (i.e., predicting the expected performance of the proposed system), without actually building and operating the system. There are several reasons why it may not be possible to actually completely operate a particular system:

1. It may be impossible to perform a complete system test. A complete evaluation of an ICBM missile system, such as the one described in Chapter 5, involves an exchange of nuclear weapons which is to be avoided.

2. On a smaller scale, an actual full-scale test might be too disruptive, as in the example of actually building and testing one or more proposed redesigns of a production line.

3. System performance data for each of several possible situations which might arise may be very limited. If the system involves some random processes, the use of limited data may result in inaccurate predictions of system performance. Analytical models permit the utilization of many related quantitative sciences such as mathematics, probability and statistics, and other scientific methods, to supplement limited test data.

4. In a similar fashion there may be performance data available regarding only some parts of the system. For example, reliability tests of subsystems may be available, or warhead lethality tests of some particular warhead may be available. Thus, if some way could be found to combine and use these data, the results would be some indication of expected system effectiveness.

5. The system may not be physically available (i.e., only a conceptual or preliminary design may be available, hence there is no hardware to test). Some way is needed to use whatever related data which might be available

to make an early prediction as to whether the system shows sufficient promise to proceed on with hardware development.

6. In each of the above cases, it may be too costly in dollars as well as time to physically build more system alternatives and test these in sufficient detail than to obtain sufficiently accurate predictions through modeling.

Modeling offers a method of generating many alternative system improvements, which is a systems planning function. By being able to analyze the functions to be performed, particularly in determining the dominant system performance characteristics which are critical in determining system effectiveness, the analyst can generate a series of physically realizable methods for implementing these functions and thus create a new set of alternative ways of implementing the system and solving the problem. Then cost-performance tradeoffs can be used to decide which performance characteristics provided are worth their cost.

Modeling provides various personnel involved in the problem with an excellent means of communicating with one another. This provides the various decision-makers, technologists, and analysts with a better common understanding of the problem, the job to be done, the key factors being considered, and the assumptions being made. This is particularly important when the project is large in scope and involves many people. Modeling offers a better means of integrating the efforts of many specialized talents, each of whom has different interests in the problem.

How Does Modeling Do This?

Modeling accomplishes these things by attempting to apply the scientific method including inductive and deductive reasoning to the problem, as outlined in Figure 4-11. The first step in the inductive method, for example, is focusing on some particular problem and observing some phenomenon or operation. A scientist, for example, looks into a microscope or through a telescope or he observes some phenomenon in nature, such as when Galileo dropped the different weights from the Tower of Pisa or when Newton supposedly observed an apple falling.

1. "Observe" the system in use.
2. Construct key system operational activities.
3. Develop measures describing "how well" activities are accomplished.
   ("Hierarchy of measures")
4. Determine key system and environmental characteristics representing system effectiveness.
   ("Performance / accountable factors")
5. Test, revise, validate models.

Figure 4-11.   Constructing system evaluation model.

The scientist then generates a hypothesis (i.e., a model in systems analysis terms), including all those factors and their interrelationships which he believes are relevant and are needed to describe the phenomenon to the accuracy he desires. For example, Ohm's law considers only the factors of current, electromotive force (or voltage), and resistance; the frequency of the voltage was not considered. Similarly, Newton considered relationships among the mass of a body, distance traveled, and time, but omitted air resistance. In systems analysis, it is up to the analyst to determine, through his own knowledge as well as the intuitive judgments of other experienced people involved on the project, which factors are deemed to be of highest importance.

The third step is to test the relationships for validity. This is done by gathering data which will support (or disprove) the hypothesis. In the physical sciences, experiments are performed in the laboratory to obtain such data after which the hypothesis is validated, rejected, or modified.

In analogous fashion the systems analyst constructs the operational flow model which will be used to evaluate system effectiveness, by observing the system in operation either entirely, if available, or by subsystem if these are as yet unconnected, or mentally if the subsystems are still unavailable. In the latter case, the analyst must construct a verbal description or model of the future environment (the scenario) and the system operational procedures which he obtains from the system synthesizers and users who would operate the proposed system in the future. The operational flow model then quantifies the verbal models.

Unfortunately, when analyzing the expected performance of large, complex systems, it is not always possible to obtain all of the experimental or test data needed to validate all of the models and submodels (i.e., hypotheses or relationships) involved. For example, some of the data may be in the possession of a competitor, as in the case of the number and characteristics of equipment belonging to a business competitor or a potential enemy. In addition, some data are obtainable only by conducting experiments which are too costly, such as building prototypes of large expensive systems, or by conducting tests, such as atmospheric nuclear explosions, which are prohibited by international agreement. For the cases where complete test data are not available, deductive reasoning must be used and the only test of validity available is the test of logic, reasonableness, or credibility. Hence, systems analysis uses inductive reasoning to determine the proper validated transfer functions of the models, where test data are available, and deductive reasoning represents the transfer functions of the models where few data are available. The question always arises concerning how we know how much of the defense systems planning analyses performed is really valid. The only answer available is to wait until the recommended system is built and tested, which may take five or more years. However, even this is not a true test for several reasons. First, the environment may have changed during the intervening time, thus changing the actual system effectiveness from that which was predicted. Second, one of the rejected systems, which was never built, may actually have turned out to be better than predicted, hence making it the best system; but since it was never built, the decision-maker may never know this. The key to the questions of whether these models (and hence, the systems analysis) is valid or not is the following: when system proposals are presented to a decision-maker, a decision must be made. (The options of doing nothing or delaying the decision is, itself, a decision.) Hence, the use of a logical framework for inserting all available information including data as well as the seasoned experience, judgment, and intuition of decision-makers, technologists, and operational personnel is the best that is available as an aid in making that decision. This is what systems analysis provides. How this is provided is now discussed through examples.

III

AN ILLUSTRATIVE EXAMPLE:
MISSION ORIENTED
SYSTEMS ANALYSIS

The material in the following chapters consists of a series of cases designed to illustrate the principles of model building in systems analysis as it applies to the evaluation and planning of large, complex systems. The specific cases presented are fictional and, to comply with security restrictions, contain no real numerical data. However, they are based on problems which could occur and the analytical approach presented illustrates the type of systems analysis which might be performed under circumstances stated. Several points about this case material should be mentioned.

First, there are several ways that case illustrations might be presented. Sometimes all the data are given and the problem solver attempts to assimilate the data in some fashion which will yield some solution to the problem as given. Unfortunately, the systems analyst's world is not that convenient. Most times in dealing with complex problems the real problem (the PAU) is rarely the original one he was presented (the PAG). In addition, much time is spent in deciding what data are important and accumulating such data. For this reason all case illustrations in this book will start with some "felt need," and very little data will be presented initially. The solution which follows will show "an approach" which was (or could be) followed in performing the study. It should be emphasized that these cases have not been written as complete case histories showing the oscillations which invariably occur in study, including mistakes made, dead-end approaches, etc. On the other hand, an attempt is made, wherever possible, to indicate alternate approaches which could be followed in performing the analysis, taking into account some of the constraints which the systems analyst is under, including analytical resource scarcity (time, manpower, computer time, scarcity of data, etc.), boundary conditions, and value scales of the hierarchy of decision-makers. Emphasis is given to the iterative approach to analysis, illustrating that the only way to satisfactorily perform an analysis is to present initially alternative analytical approaches which are possible, and through a series of iterations with the decision-makers and others involved in the problem, "zero in" on the approach which can satisfy all of the above boundary conditions. Such an approach will result in an analysis which has been "custom designed" to the particular set of conditions, the set of values of the people involved, and the analytical resources available.

Second, some readers who may be familiar with some of the case illustrations presented may be tempted to find one of either of the following criticisms with the solutions presented:

1. Too many variables were considered.
2. Some variables were omitted that should have been included.

For these readers, I offer the following response. First, these case illustrations have been constructed primarily to demonstrate application of systems analysis principles to planning, and not to report actual technical results or case histories, which is outside the scope of this book. However, in response to the first criticism, it was my intent to include as many variables in the models as possible to demonstrate how such variables can be included, if desired. The reader, unlike the course student, does not have an opportunity to ask how a particular factor would be included in the analysis and, hence, I feel it is better to err on the side of inclusion rather than omission. Remember that just because a factor is initially included in a model, and it later turns out that the factor will not affect the result to a large extent, the factor does not have to be included in the final model exercise or numerical calculation. In addition, inclusion of the factor in the model also offers an additional benefit. Since each of the factors in a model can be used as an aid in creating new system alternatives not previously constructed, early inclusion of many factors aids in the creative process and, hence, increases the probability of obtaining highly efficient solutions.

In response to a criticism of omission of any factors, let me reiterate that the primary objective of each problem is to illustrate the *principles* of quantitative analysis in systems planning rather than showing *concrete solutions*. Of course, I could fall back on the comment attributed to a systems analyst from the RAND Corporation: "There is no such thing as a good model, there are only better models." Thus, if the reader feels that he would like to use the models developed but feels some factor should be included, add in the factor and you are on your way. Remember that the practice of systems analysis is still an art and each analyst constructs the various models based on the judgment of both himself and others involved in the project.

There will invariably be some factors which may be very important for the decision-maker's consideration but which cannot be included in a quantitative analysis, based on our knowledge of the art at this time. While such factors will not be considered in this book, they would normally be considered in a qualitative fashion and be included in the systems analysis report under a section entitled "Qualitative Considerations".

Lastly, let us note that the case solutions will intersperse the specific solution with the key principles of analysis and the reader will have to cope with two tasks simultaneously: to understand the case and to understand a new technique as it is introduced. I believe this is the best way of teaching this material, namely to develop the need for a technique, describe it, and then illustrate its application.

5

The Strategic Force Planning Case: Qualitative Aspects

The first case situation has a number of objectives. The primary objective is to illustrate how to perform a systems analysis at the mission level, following the procedure described in Chapter 4. In this regard the reader will notice what appears to be redundancy, but this is intended to illustrate the iteration that can and should occur in a systems analysis. For example, this chapter consists of a verbal description of the problem, culminating in the development of an over-all qualitative structure to the problem, with Chapter 6 focusing on the development of models used to quantify the system performance characteristics, the environment, and to determine system effectiveness, thus expanding on the previous work. Chapter 6 gives heavy emphasis to operations analysis methods of gathering data to generate probabilistic models of the many random processes involved in this problem, and the use of applied probability theory to represent these. Chapter 7 describes the two primary methods of model exercise (i.e., combining the numerical transfer functions of the operational flow model to determine system effectiveness), the Monte Carlo simulation method, and various analytical methods which could be used. Since the output of an analysis is only as good as the logic of the models used and the accuracy of the data employed, Chapter 8 explores the various uncertainties associated with the analysis: technological, statistical, and competitive, and shows how the models developed in the analysis can be used to generate new system concepts which have not been previously identified. Lastly, methods for presenting the results of the analysis to the decision-maker are also discussed. All cost aspects of the problem are included in Chapter 9.

The final objective for those interested in defense analyses is to demonstrate how the various elements of a defense system can be interrelated quantitatively and thus set the stage for proper subsystem planning. The specific problem to be discussed is concerned with the planning of strategic forces which are used to prevent general nuclear war, and was chosen for

83

several reasons: first because it contains many factors and is sufficiently complex to show the need and importance of building models "custom-designed" to the intuitive views of high-level decision-makers. Second it is a particularly good vehicle for illustrating how to deal with random processes and competitive environments which are involved in many complex problems in both the defense and non-defense areas. Because of the many articles which have been written about this significant problem in the popular press, even the nondefense oriented reader will understand the problem background.*

Problem As Given

This case begins with the following problem statement:

"It is now 1972 and you are a systems analyst with a firm which provides technical support to the Missile Systems Division of the Air Force systems development command. You have been assigned to a strategic planning study and learn that recent intelligence estimates indicate that the Soviet Union appears to be increasing the size of its ICBM capability. Since they have already developed high energy propulsion systems for their space program, it is believed that they have the capability of combining these high energy rocket engines with a high megaton warhead and a more accurate guidance and control system so as to produce a more lethal missile. In light of this intelligence information, we are concerned that our hardened Minuteman silos and missiles may soon become vulnerable to these new offensive weapons.

"To counter this possible shift in the balance of power, the Missile Systems Division is proposing two possible methods for improving the strategic missile force. The first consists of replacing each of the present Minuteman silos with a superhardened silo located nearby which can withstand much greater blast overpressure. The second proposal consists of converting the present Minuteman missiles to a mobile system by mounting them on motorized truck carriers and continually moving them over highways in the less populated sections of the country. This is similar to the Minuteman railroad system which was tested a number of years ago. The technical feasibility of each of these proposals has been demonstrated by the missile system designers.

"Conduct a cost-effectiveness analysis which complies with DOD Directive 3200.9 † and compare each of the two system alternatives with other DOD competitors for resources."

---

* The case described in this chapter is a hypothetical one. However, much of the philosophy behind the case is drawn from public testimony of Secretary McNamara before the House Armed Services Committee in 1965 and thereafter.
† As described in Chapter 3.

Developing The Problem As Understood

Initial investigation of the problem indicates the following aspects: At any particular point in time, there exists a set of American targets felt to be of value to the Soviet Union. These targets are of two types. The first includes all military bases and weapon locations; the second includes all industrial centers. The locations of all targets (see Figure 5-1) are known to the Soviet Intelligence agency to some accuracy. This set changes with time as weapons and bases are added to or eliminated from the American force structure. The relative ranking or value of each of these targets within its subset can also be established. The basis of this value measure could be industrial capacity or population density in the case of industrial centers. It could be weapon lethality capacity in the case of weapon centers. Conversely, there also exists a similar set of Soviet targets whose composition, location, and value are known to the United States to some degree of accuracy.

The intelligence estimates of the Soviet weapon threat level which exists presently or at some future period of time is illustrated in Figure 5-2a. However, such estimates are always a *subjective* estimate based on individual interpretations of a set of available intelligence data. Hence there is some range of uncertainty regarding the exact number of enemy missiles which will be available as a function of time. The effect of such uncertainties will be dealt with in Chapter 8. In addition, some estimate of the capability of these missiles in terms of lethality (size of warhead expressed in megatons) and warhead guidance accuracy can be provided by intelligence. As indicated in Figure 5-2a, the new threat is predicted to rise at some future time, t_1, and to reach full operating capability at t_2.

Intuitively, it is believed that the projected increase in Soviet missile capability increases the threat to American cities or American weapon systems,

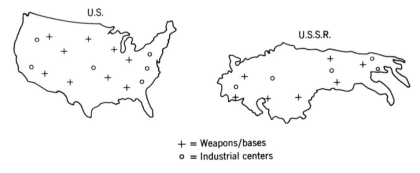

+ = Weapons/bases
o = Industrial centers

Figure 5-1.   Target sets.

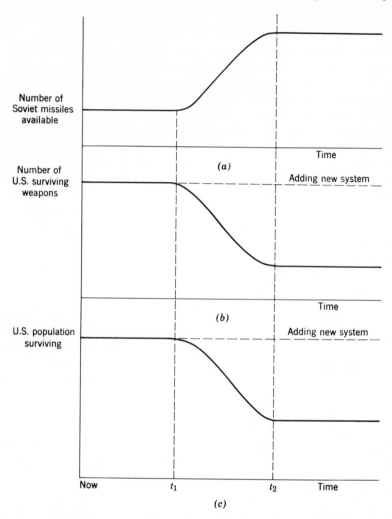

Figure 5-2. Decreased effectiveness resulting from increased threat.

or both, depending on how these are targeted. The predicted damage to the U.S. targets caused by the assumed increase in Soviet ICBM capability is also illustrated in Figures 5-2b and 2c as a function of which year such a full surprise attack would come. If the weapons were launched against the American fixed ICBM sites, the resulting number of surviving ICBM's would decrease as the enemy force increases with time, as shown in Figure 5-2b. Similarly, if the Soviet weapons were launched against American cities, the expected destruction to our population or industrial capability could be illustrated as shown in Figure 5-2c. This intuitive feeling that the new intelligence estimates may result in a problem for the United States is what initiated this study. *How much* of a problem is the question that the study should answer.

Adding units of one of the American system alternatives being analyzed would have the effect of increasing the American effectiveness level, perhaps even up to the original level, as shown in Figure 5-2. Of course, the action must be started sufficiently before t_1 so that the improved systems could be phased in proper time to meet the new threat.

The objective of the systems analyst is to quantify the functions shown in Figure 5-2 which were obtained intuitively by the military officers associated with this problem. To do this, the analyst must at a minimum evaluate the two new proposals presented by showing what the American effectiveness would be if no action were taken (the lower curves), how much the effectiveness would increase if either (or both) of the proposals were implemented, and the costs and risks associated with each option.

Understanding Proposed Solutions

The two system alternatives proposed are: (a) superhardened missile silos, and (b) mobile missile systems.

Note that the purpose of each of the proposed systems is to increase system survivability by reducing the vulnerability of each system to enemy attack. Hence, before describing the characteristics of each of these systems, the specific characteristics of the enemy attack to be examined should first be made more explicit.

Understanding the problem is a two-pronged effort, as illustrated in Figure 4-7a, and discussed in Chapter 4. One part is systems oriented and involves an understanding of the elements of which the system is composed and the performance characteristics it offers. A second part is operations oriented and focuses on the job to be done and the environment in which the job is to be performed. In conducting an analysis, the analyst invariably oscillates between both parts as he gathers information and gains increased understanding of the problem. In the analysis which follows, we shall initially focus on system description, and then go on to the operational job which the system is to perform.

TYPE OF ATTACK ON SYSTEM

Various types of enemy attack mechanisms could be considered and those chosen should be made explicit. It is assumed in this analysis that we are primarily concerned with an enemy nuclear warhead attack from either an enemy missile or bomber, as opposed to saboteurs, for example. Enemy target destruction or disability could occur from various effects such as blast overpressure, gamma radiation, thermal shock, or EMP (electromagnetic pulse). For illustrative purposes, let us consider how one would analyze the effect of blast overpressure. A similar type of analysis could also be conducted for the other kill mechanisms.

Target destruction is a function of three other primary factors:

1. Warhead lethality;
2. Warhead miss distance;
3. Target vulnerability.

The relationship among these three factors is now described qualitatively. How the quantitative relations are obtained is described in Chapter 6.

Figure 5-3 indicates that the blast overpressure which would exist on a point target * varies in some decreasing fashion as some function of detonation distance from the target and the size of the warhead detonated. Figure 5-

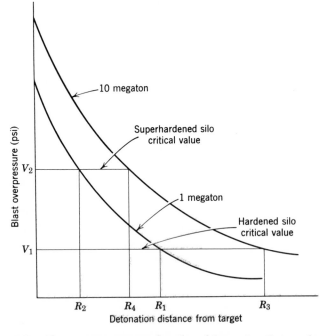

Figure 5-3. Blast overpressure as a function of detonation distance from target.

* This analysis treats the target as a point. However, an area target could be resolved into a series of points and each analyzed in the same fashion.

3 illustrates qualitatively the overpressure provided by two different war-head sizes (e.g., one megaton and ten megatons). It can be assumed that any given target has associated with it some critical design overpressure (e.g., 5 psi, 50 psi, or 150 psi) such that target destruction may be defined as occurring when the actual overpressure equals or exceeds the critical design overpressure. The distance that this occurs for a particular warhead is called the "radius of vulnerability." Thus, by analyzing the various points of an area target, one could construct the area of vulnerability of the target with respect to a given size warhead. Obviously, the area of vulnerability of a point target is the circle using the radius of vulnerability. Thus, as shown in Figure 5-3, the hardened silo missile would be destroyed if a one megaton warhead exploded within a range of R_1 from target. On the other hand, the superhardened silo missile would be destroyed only if the one megaton warhead exploded within a range of R_2. The other two critical ranges R_3 and R_4 for a ten megaton warhead are also shown. Obviously, the accuracy of this type of data is dependent upon the amount of data available from nuclear tests.

Thus, it is possible to determine the vulnerable area (a circle of vulnerability for a point target) around each of the silos based on a given warhead size. These data can be combined with the enemy warhead accuracy to determine the likelihood of the warhead landing inside the vulnerable area, thus destroying the target; this is done quantitatively in Chapter 6, taking into account all sources of warhead inaccuracy.

SUPERHARDENED MISSILE SILO SYSTEM

Having obtained some qualitative understanding of the mechanism by which reduced target vulnerability could be obtained for the superhardened missile system, one can now investigate further details regarding this system. For example, how much additional hardening is to be achieved and by what means? The system designer indicates that he has considered various possibilities of hardening such as increasing the depth of the current silos, or reinforcing the sides further. He also considers the possibility of constructing a new silo which is of much greater depth in the vicinity of the old silo so that the same communications lines may be used.

Operational Concept

The systems planner indicates this system would operate in the same manner as the current ICBM system, the only system change being the construction of new silos. In addition to obtaining a detailed description of the concept of operation, the systems analyst constructs part of the system de-

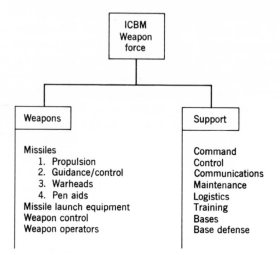

Figure 5-4. System design model.

sign model of this ICBM weapon force represented by the structure of Figure 5-4.

The system design model which describes "things" (system elements) and their relationships is of value in making explicit all of the system elements that are required for operation of the total weapon system. This model may be checked against the description of the concept of operation to make certain that no system elements have been overlooked and to check for feasibility of operation. In addition it is useful for determining what system performance characteristics should be considered in the evaluation.

The system design structure illustrated is a convenient way of viewing all types of weapon systems. It divides the weapon force into two major components: (a) prime mission equipments or weapons, and (b) all other support of the weapons. In this case, the hardened missile silos would be included as a subelement under *bases*. A more detailed system design model containing system elements of a lower level (and greater level) of detail could also be constructed. This more detailed model would be used for costing purposes by helping the analyst determine those system elements which are not currently at hand and will have to be procured and maintained in operation.

The final part of the system design model construction would include identifying the various performance characteristics of the systems which are to be evaluated. These characteristics would include reliability, missile accuracy, and particularly, the improved system survivability to enemy attack as discussed previously. Quantification of these characteristics is discussed in Chapter 6.

MOBILE MISSILE SYSTEM

The same system design model previously developed in Figure 5-4 can also be used to describe the mobile system. Again, the total system force can be structured into two main parts: (a) the weapons themselves (the operating system) and (b) the support elements. Before determining which elements of the existing ICBM system may be used as part of this new proposal and which must be procured because of the greater problems in operating the system, let us discuss the operational concept of this system. While there are several versions of this system that are possible, one in particular is described.

Operational Concept

Increased survivability of the missiles and their launching mechanisms will be provided by mounting them on trucks that will continuously move in a random fashion over a large terrain, making them difficult to target. Sufficient distances between system elements will be maintained so that no one enemy warhead can destroy more than one ICBM (bonus kills). However, by van-mounting these missiles, their vulnerability is increased (i.e., critical design overpressure decreased) because they are now above ground. Therefore, the same type of vulnerability analysis discussed previously would have to be performed to determine the proper separation between targets to avoid bonus kills.

Upon tactical alert and receipt of a higher level command to prepare for fire, the missile launcher will be erected, missile earth location will be determined, specific target coordinates inserted into the missile, and the missile will then be fired upon command.

Further System Discussion

Of course, many technologically feasible versions of this system exist, including the one of mounting each missile and launcher on a continuously moving railroad car; this was tested some years ago. Each version, however, depends on random movement as the means of gaining increased survivability. Such random movement presents great difficulty to an enemy since he would have to keep track of many moving targets before he could fire as it is too costly to fire over a broad area randomly. Reconnaissance satellites could provide some of this information but the accuracy of the target information would be a function of satellite sensor accuracy, its data rate (time between looks at target), target movement velocity (speed and direction), and missile flight time to target. If this target positional accuracy were not high enough, an enemy would probably not attack these targets at all but might change the nature of his attack, such as firing on fixed installations (e.g., command centers). This degree of positional accuracy could be examined quantitatively.

Compared with the fixed missile system, a random movement type of operation results in many more operational problems which are discussed below:

1. Mobile missiles are subject to more shock and vibration than a fixed missile. There is a need to examine whether the existing Minuteman missiles could still be used (thus saving costs). If so, one could expect the reliability, and probably other performance characteristics such as accuracy, to be reduced as compared with a fixed missile unless appropriate modifications are made.

2. Because of the difficulty of finding large areas of terrain which will not block mobile communications at least occasionally, the command and control system problems are increased, making positive control of all weapons at all times more difficult. Geographical terrain analysis would have to be undertaken to determine the extent of the control problems.

3. There is a greater problem of locating one's own position accurately and rapidly prior to inserting proper commands to the missile. One way of overcoming this problem is to use previously surveyed sites. Other ways could also be developed.

4. A greater logistics and support problem is associated with the mobile system. This includes not only first-echelon maintainability of equipment (i.e., performed right at the operational site), but also food handling, etc.

5. Personnel problems should be greater with this system since people are continually moving. Consequently, the length or percentage of duty hours may have to be decreased.

All of these aspects can and must be reflected into the various system performance characteristics by the system designer.

TRIGGERING THE COST-EFFECTIVENESS ANALYSIS

Decision-Maker's Options

When the Secretary of Defense receives these two proposals (superhardened or mobile missile system), he has the following decision options available to him:

1. He may buy a number of superhardened missile silos;
2. He may buy a number of mobile missile systems;

3. He may buy some of each;

4. He may take no action at the present time because he has concluded that his present posture will be good enough for even the increased threat. This may be because other systems (e.g., bombers, Polaris missiles, etc.) provide sufficient capability.

5. He may wish to consider some other alternative system for accomplishing the mission.

To aid in uncovering these other alternative systems which should be considered within the scope of this problem, a higher level system design model as shown in Figure 5-5 should be constructed. This model is basically an organizational chart which indicates the structure of the services as well as those weapon systems in operation or under development, which contribute to the strategic war capability.

The main purpose of building this higher level system design model is to determine which alternative system not presently under consideration should be regarded along with the two missile system proposals being made. For example, this model may show that other systems already in the inventory can provide some functions which do not have to be duplicated by the two systems under study. These functions may include information sources, bases, defenses, etc. One other reason for building the higher-level systems design model is to note other elements which require interfacing with the systems under examination.

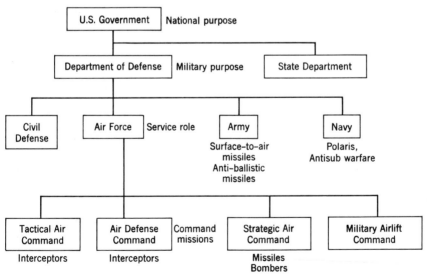

Figure 5-5. Hierarchy of systems.

In addition to these other alternatives, the decision-maker has another option. He may buy a mix of any of the above systems, with various quantities of each type within the mix.

The Need for a Cost-Effectiveness Analysis

We shall summarize this problem from the decision-maker's viewpoint to emphasize why a cost-effectiveness analysis is required in this case, what information the final products of the analysis should contain, and how the analysis might be structured.

This strategic planning study was initiated because a new threat was uncovered by the intelligence group. This new threat may lead to a corresponding reduction in American force effectiveness, which can be corrected only by additional systems. The question now to be solved is which of the system proposals submitted by the different government agencies (or the other options open to the decision-maker) can best overcome the possible mission deficiencies, as compared with the resource allocation required?

One of the reasons for the decision-maker's problem is specialization and the many options it provides. Specialized organizations are created within the federal government, as they are within any large organization, to provide greater efficiencies through the employment of specialized systems for accomplishing specific missions assigned to them. Those specialized organizations and the products they provide, as applied to this problem, are illustrated in Figure 5-5. This figure can be constructed either by working down from the top (the President) to the system level under consideration (Air Force Missile Systems) or by working up from this level. The model indicates the existence of various government organizations for coping with the problem of general nuclear war. The Defense Department is one such organization with various defense components; the State Department, containing nondefense means such as diplomacy and foreign aid, is another which can improve this nation's posture for avoiding nuclear war. Other nondefense departments not included in this model also exist. However, these were not included since we intend to concentrate on the defense problem.

The Air Force, Army, and Navy, which exist within the over-all Department of Defense, have their own military commands and weapon systems (as indicated in Figure 5-5) to cope with the problem of general nuclear war. Since these various weapon system competitors contribute to the same mission, the higher level decision-maker's systems planning task of finding the preferred mix or combination of systems which together provide maximum mission effectiveness for a given level of resources expended (or lowest cost for a given level of effectiveness) is a difficult one. Several factors contribute to this difficulty:

1. In general, there is always more than one way to accomplish a mission or an objective. In this example, various weapon systems are available (i.e., SAC missiles, bombers, carrier-based aircraft, and Polaris submarines and missiles).

2. The total costs required to fund all of the new proposals which each of the services generates each year far exceed the total amount of resources available to the Defense Department. Even if the Defense Department budget were doubled or tripled, this problem would not be eliminated, since the decision-maker would still have to choose between funding more new proposals, or buying more units of the existing systems.

3. Military problems such as the one described are now so complex that even experienced military commanders concerned with these problems can no longer agree on what the proper mix of weapon systems should be when resources are limited. This problem of obtaining resolution is difficult, since, generally, only limited information is available, and the experiences of these officers have been different. The problem is particularly difficult when considering problems for which their experience is limited, such as the problem of general nuclear war.

The basic problem of the higher-level decision-maker is that he needs some better way of determining the value of a system proposal before large, costly system engineering efforts are committed. While a program may intuitively appear to be of value, intuitive opinions are no longer adequate as the sole justification. Thus, complex problems of this type require a framework on which the intuition and judgments of experienced decision-makers in the various specialized decentralized organizations can be brought together. This is what the cost-effectiveness analysis helps to provide. In the analysis which follows we shall obtain and structure information which will provide answers to the six management questions previously mentioned.

Consider the first question: What is the system objective? In his testimony to Congress on various occasions,* Secretary of Defense McNamara indicated that strategic offensive and defensive weapon systems are designed to accomplish two mission objectives: (a) to prevent general nuclear war, and (b) to limit the destruction to the United States and allies if general nuclear war occurs.

Many times it is possible to gain greater insight into system objectives by analyzing all of the elements which contribute to a mission. This was done by constructing the mission objectives model shown in Figure 5-6, which relates those systems dealing with general nuclear war shown in Figure 5-5 to the stated mission objectives as follows:

* Testimony to House Armed Services Committee, 1965.

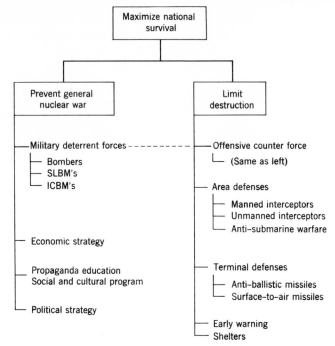

Figure 5-6. Mission objectives.

1. The strategic offensive forces primarily contribute to the objective of preventing general war—the deterrence mission.

2. The strategic defensive forces primarily contribute to the objective of limiting destruction to the United States and its allies if general war occurs.

3. Nonmilitary approaches such as the economic strategy of foreign aid can be utilized to help prevent war. These were explicitly listed so that we could specifically inform the decision-maker that while we are aware of them, we plan to exclude them from further consideration in this analysis, unless he desires us to include them.

4. Some of the offensive forces also can contribute to the damage-limiting mission by being programmed in a counterforce fashion (i.e., to destroy enemy offensive forces which could otherwise destroy allied targets, such as industrial centers).

The next step is to find some quantitative means of measuring how well each of these two missions is accomplished.

The definitions used by DOD to quantify these two objectives are summarized in Table 5-1. The easier definition to understand is that of damage limitation. Its measure is the amount of damage (or damage limitation) to the

United States and allied population and industrial capacity. Deterrence is the more difficult concept; since it cannot be simply defined, it is explained below.

Table 5-1. *Quantifying the objectives*

• PREVENT GENERAL WAR (DETERRENCE)

    CONVINCE ENEMY THEY WILL SUFFER
    ASSURED DESTRUCTION AS A VIABLE
    SOCIETY IF THEY ATTACK U.S.
    OR OUR ALLIES.

    MEASURE: ASSURED DESTRUCTION
    TO: ¼ TO ⅓ OF ENEMY POPULATION
    ⅔ INDUSTRIAL CAPACITY.

• IF WAR COMES

    LIMIT DAMAGE TO U.S. / ALLIED
    POPULATION AND INDUSTRIAL
    CAPACITY WHILE OBTAINING A FAVORABLE
    MILITARY AND POLITICAL OUTCOME.

    MEASURE: AMOUNT OF DAMAGE
    LIMITATION.

Measuring Deterrence

By definition one might assume that the United States has deterrence today since we are not currently engaged in a general nuclear war; but will such deterrence continue to exist with the new threat? And how can deterrence be measured quantitatively?

McNamara has attempted to define and quantify these objectives as follows: *

"The strategic objectives of our general nuclear war forces are:

1. To deter a deliberate nuclear attack upon the United States and its allies by maintaining a clear and convincing capability to inflict unacceptable damage on an attacker, even were that attacker to strike first.

2. In the event such a war should nevertheless occur, to limit damage to our populations and industrial capacities.

"The first of these capabilities (required to deter potential aggressors) we call "assured destruction" (i.e., the capability to destroy the aggressor as a viable society, even after a well planned and executed surprise attack on our

* Secretary McNamara's testimony to House Armed Services Committee, 1965.

forces). The second capability we call "damage limitation" (i.e., the capability to reduce the weight of the enemy attack by both offensive and defensive measures and to provide a degree of protection for the population against the effects of nuclear detonations."

Several comments regarding this definition are in order to illustrate the quantification logic used and to make certain assumptions explicit:

1. It is not possible to absolutely prevent or deter one country from attacking another since the initiative for such an attack is solely in the hands of the attacker. However, a *rational* enemy will be deterred from a first attack if he believes the consequences of such an attack will be completely unsatisfactory to him, as, for example, his destruction as a viable society. This logic serves as a *proximate* definition of deterrence for the United States and can apply as well to any other country having our set of survival values. However, if this logic cannot be assumed for another country (such as China), another definition of deterrence must be generated, based on their set of values.

2. Destruction as a viable society is defined as destruction of one-quarter to one-third of the country's population and two-thirds of their industrial capacity; the country thereby ceases to be a world power for some long period of time. One could argue about the degree of simplification. For example, all population and industrial capacity are treated uniformly (e.g., agricultural destruction is not explicitly considered). In addition, there is some correlation between population density and industrial capacity density.

3. The term "assured destruction" should also be commented upon. The term "assured" encompasses the many uncertainties which would arise in a nuclear exchange, including the potential enemy capability such as the number and capability of his strategic forces (warhead accuracy and lethality), as well as other of the enemy's system performance characteristics. It also includes the uncertainties in our own systems, including the uncertainties in human performance in time of general war. These uncertainties can be factored in by applying a low or pessimistic estimate of our own capabilities (see Chapter 8), and a high or optimistic estimate to the enemy's. Thus it is possible to measure our deterrent capability indirectly by estimating analytically our ability of rendering a given level of destruction to a potential enemy under the worst conditions to us and the best conditions to him. These conditions include the enemy attacking first.

4. How do two competitors "convince" each other that each has an assured destruction capability? One means is to make public certain information regarding the force, such as the large number of weapons available, so that the other side may become somewhat uncertain that it can successfully attack without its own society also being destroyed.

Since previous studies have been performed by the higher-level decision-maker and his staff of analysts, in the analysis which follows, the objectives and measures indicated above will be assumed as part of the "problem to be solved."

Consider the second question: When will the system be operationally available? A system development plan showing the various milestone dates involving RDT & E, production, and initial and full operational capability in the field would provide the answer to this question. Presumably only those systems which would meet the date of the anticipated threat would be considered for further evaluation in this study.

Consider the third question: How well does the system meet the objective? This question will be answered by evaluating each of the systems against each of the two objectives in terms of the measures described in the first question (i.e., Soviet damage and American damage).

Consider the fourth question: How much will it cost? This question may be answered by providing the *expected value* of the cost of the preferred system for a given level of effectiveness or, conversely, as is actually suggested in this problem, a fixed level of cost can be assumed and the effectiveness of each system calculated. Uncertainties in cost due to any number of factors are treated in the sixth question, below.

The answer to this question will also depend on how many units of each system are needed. The question of need is a relative one. A decision-maker can always obtain additional effectiveness by allocating additional resources. His primary interest focuses on those characteristics that relate effectiveness and cost. The systems analyst establishes this relationship. This relationship is shown graphically in Figure 5-7a where American population surviving is plotted as a function of additional increments of resource (dollar) allocation. C_0 is the cost of the present capability when operated over a fixed period of time (e.g., five years operating life). C_1 represents additional expenditures above C_0 (e.g., \$2 billion more as an example). Since we wish the analysis to pivot on equal cost, all systems (F_1, F_2, . . . F_5) will be designed to cost C_1 and the points plotted indicate the effectiveness which can be achieved by each of the system alternatives which were evaluated. Obviously, F_3 is the preferred system (if only the assumptions made in the analysis were considered) since it provides the highest effectiveness for the costs expended. Similarly, C_2 represents an additional expenditure over C_0 and the resulting effectiveness values are also indicated. Now the decision-maker can relate the increase in lives saved with the corresponding increased expenditures required. Considering only these factors, the decision-maker would want to choose that system which yields the highest effectiveness for a fixed amount of cost.

The data could also have been structured on a constant effectiveness basis

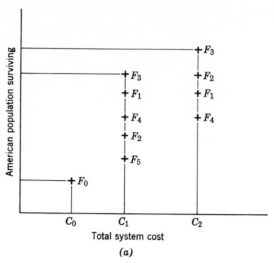

(a)

C_0 = Cost of operating present capability

C_1, C_2, C_3, C_4 = Additional costs associated with the
additional effectiveness obtained

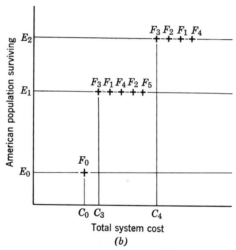

(b)

F_i = The level of effectiveness and cost for
a configuration of the ith force (system)

Figure 5-7.   Providing structured information for decision maker.

as shown in Figure 5-7b. In this case each system was configured for a specific level of effectiveness, E_1 or E_2, and its total system life cost calculated. Now the decision-maker would choose that system which provides the given level of effectiveness at lowest cost.

The previous discussion also shows how to answer the fifth question: Is this the best way of meeting the objectives? In this analysis, each of the decision options of interest (including those of mixed forces) will be configured for some constant level of total system cost. Hence, the best system will be that which provides the highest level of effectiveness for that cost.

Consider the sixth question: What are the uncertainties involved? Four main types of uncertainties must be considered.

1. *Technological.* This involves the capability of the industrial contractor to manufacture system elements which, when operational, will provide the performance characteristics predicted. Failure to provide any of the technological elements which should be included in the system will result in either: (a) a reduction in system performance and hence, effectiveness, (b) delay in the time the system becomes operational, and/or (c) a requirement for additional funding to attain the originally predicted performance characteristics and reduce the time delay.

2. *Cost.* Additional uncertainties arise because of inaccuracies in the cost estimating relationships used to predict total system life costs of a system which may have never been built or developed before.

3. *Enemy.* Systems analysis involves many types of enemy (or competitive) uncertainties. These include the magnitude of the enemy's forces, the performance quality of his forces, or the strategies applied to utilizing these forces in response to various threats. Each of these factors can result in reduced effectiveness of our system.

4. *Chance.* The exact effectiveness of a system cannot be predicted with even the best analytical skills. Random fluctuations (i.e., "good luck" or "bad luck") are involved during the operation of any complex system. What an analyst can do is to predict statistically how effective the system will be when it becomes operational. The results can be expressed in probabilistic form as shown in Chapter 7. The other uncertainties will be considered in Chapter 8.

DEVELOPING THE SCENARIO FOR GENERAL NUCLEAR WAR

Once each system alternative has been described and the system objectives and measures determined, the analyst must next find a way of relating the key systems performance characteristics to the mission objectives. These relationships are made explicit by the analyst through his construction of the operational flow model which he will use as the vehicle for quantifying the events involved in the operation of the system. To construct a quantitative model of the system operation, however, the analyst must first obtain a qualitative description of the system as it would be expected to operate in its environment(s). The environmental aspects of this qualitative operational description is called the "scenario."

Creation of a scenario for general nuclear war is one of the most difficult aspects of this systems analysis, and calls for great imagination on the part of those concerned with predicting how a system may be used at some future time. The analyst obtains information regarding possible scenarios by discussing such possibilities with different decision-makers and operational personnel, each of whom may have one or more environmental situations in mind. Thus, there will probably be no uniform agreement on any one scenario, and the analyst may be forced to deal with a wide range of possibilities. Incidentally, this is a common problem which the analyst also must face in gathering data which involve the intuitive judgment of others. As indicated in Chapter 8, the analyst copes with these uncertainties, particularly in the early phases of the analysis, by making explicit all data and information he uncovers, including contradictory opinions.

The difficulty in dealing with scenarios stems from the fact that since there never has been a general nuclear war, there is no historical precedent on which to base such predictions. However, there have been events in recent history which could be extended into one or more possible scenarios for general nuclear war. Consider, for example, the Cuban Crisis in 1962 when both sides were presumably close to a global conflict. A theoretically realistic scenario, representing an extension of actual events, could be developed as follows:

The Soviet missile-carrying freighters are steaming towards Cuba. The United States intercepts one of the ships and warns its captain not to proceed past a certain geographical location. The Kremlin in Moscow contacts the United States President and issues an ultimatum that if this, or any other Soviet ship is molested or damaged, dire consequences will occur. The Soviet ship proceeds into the forbidden area, and the Navy cruiser responds by firing a shot across its bow as a warning to turn back. However, the ship continues to penetrate deeper into the area, is fired upon by the cruiser and is consequently disabled or sunk. In retaliation, a Soviet missile is fired on a small American air base.

As the crisis was developing, United States forces were placed on strategic alert. When the United States is informed of the Soviet missile launch, the United States quickly retaliates by launching a missile at a similar Soviet target. Single moves such as this are continued, perhaps with gradual escalation, while both sides attempt, through diplomatic channels, to negotiate a stop to the conflict.

Another possible scenario, always mentioned, is the situation of the sudden, massive, sneak attack on American weapon systems and/or American cities, with corresponding American retaliation. A variation of this is the decapitation attack, a sudden attack on all American command centers, hoping to leave American forces without a unified leadership.

Another contingency that could be described is a United States preemptory attack based on a consideration of preventive war. For example, assume that the United States has authentic information from very reliable sources that an enemy has begun implementation of a plan for a nuclear attack to take place on the next day, and the United States decides to strike first. A decision-maker may ask for this variation to the basic scenario in order to predict, theoretically, the results on each side.

Another possible scenario might include consideration of an attack by a third nation against either the United States, the Soviet Union, or both countries. Under this scenario, the United States, if attacked, would strive to rapidly assess the damage on this country, identify the source of this attack, and prepare a retaliatory action.

Table 5-2. *Strategic Scenario Uncertainties.*

- ENEMY WEAPONS
   NUMBER, TYPE, PERFORMANCE

- DOES ENEMY ATTACK FIRST?

- STRATEGIC WARNING TIME AVAILABLE

- TACTICAL WARNING TIME AVAILABLE

- INITIAL TARGETS ATTACKED

- INITIAL WEAPONS LAUNCHED

- U. S. DAMAGE ASSESSMENT / RETALIATION

- ENEMY DAMAGE ASSESSMENT / RETALIATION

Structuring the Scenarios

Consideration of the many possible scenarios indicates that the logic of all of these scenarios can be structured to include the uncertainty factors involved, as shown in Table 5-2, and described below. One advantage in creating such a structure is that it can be used as a checklist in considering each scenario.

1. *What forces and targets exist on each side prior to attack?* It can be assumed that most of the Ur' ed States and enemy fixed target sets (i.e., cities, bases, fixed weapons) are known by both sides. Areas of uncertainty

do exist in regard to the number, type, and performance characteristics of the enemy weapons.

2. *Who fires first, and with what forces?* The magnitude of an initial surprise attack results in a reduction of the number of the other side's weapons surviving the first strike and available for retaliation.

3. *How much strategic warning will be available to each side?* Strategic intelligence information, such as unusually large military movements, can be used by each side as an indication of enemy intentions. This determines how much of the entire force can be alerted, dispersed, and made available to each side, and the extent to which passive defense measures, such as fallout shelters, can be utilized.

4. *How much tactical alert is available to each side?* Whereas strategic intelligence information is an indication that the enemy may be planning a military action, tactical alert indicators such as Ballistic Missile Early Warning System (BMEWS) radars and Distance Early Warning (DEW) line radars provide information that a military action is actually underway (disregarding false alarm electronic signals). While this information may provide shorter early warning time, it may be more credible than just strategic early warning information alone, and hence a more positive response may be taken such as increasing alerting actions compared to what may have been done using strategic alert information alone. Thus, the use of both types of early warning information will determine how much of the entire force structure under attack will be available for a counterstrike, and to what extent passive defenses can be used.

5. *What targets are initially destroyed?* This is determined by the attacker's choice of initial targets and the kill probabilities of the attacking weapons. These then determine the specific surviving forces available for a counterstrike.

6. *What is the assessment of the damage and the retaliation?* Information concerning the initial damage to the attacked country is useful in inferring the nature of the attack, thus leading to an appropriate military response. For example, do we conclude that the initial attack is on military targets only, industrial targets only, or both? "Flexible response" to avoid unnecessary escalation of a conflict is an American objective. Hence, the proper retaliation, in a timely fashion, must be determined, and may be preplanned explicitly, as a function of the type of attack to obtain the appropriate timeliness.

7. *What is the attacker's retaliation?* This begins a series of similar steps by the other side, followed by a set of similar, subsequent steps. Thus the total set of moves may be viewed as a series of actions and reactions (based

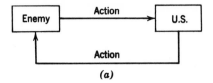

(a)

Figure 5-8. (*a*) Forces operational events.

on past actions) by both sides, as viewed in Figure 5-8a, in which the war operations follow the seven steps just described.

We shall now illustrate how the analysis of one possible scenario might take place. The scenario chosen is that of an enemy first-strike against American forces, including missile forces, with a subsequent American retaliation.* One key question of interest to the decision-maker is, will the present American force structure have an assured destruction capability, as previously defined, against the new future threat? If so, the United States possesses a deterrent capability; if not, presumably we need to add additional forces to obtain such capability. Hence the only test of deterrence is to perform an analysis which may be in the form of a war game and show the results of the game for each of the scenarios of interest.

To cope with the dynamic nature of the entire feedback system, the analyst needs first to break the feedback loops and thus obtain the two "open loop destruction models" as shown in Figure 5-8b. To obtain analytical results which have a suitable degree of accuracy, the analyst must then partition the total operations (performed within each destruction model) into a set of more detailed interrelated events, each of which can be analyzed in greater detail. Following this, the two models of Figure 5-8b will then be reconnected to bring in their interaction.

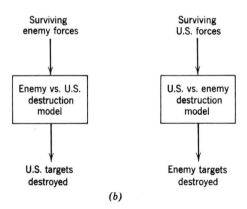

(b)

Figure 5-8. (*b*) The two operational transfer functions.

* The decapitation attack scenario is also considered in Chapter 12.

The total scenario involving the American destruction model of Figure 5-8 would consist of the following operational events:

1. Enemy launches a first strike against initially assigned targets and certain American forces survive, depending on factors such as target vulnerability and warhead delivery accuracy. These forces are the inputs to the "U.S. vs. enemy destruction model."

2. Of the surviving American forces, certain ones are available for use, depending on factors such as weapon reliability and maintainability.

3. Certain available American weapons are launched against enemy targets according to the particular American war plan selected.

4. Certain of the American launched weapons reliably arrive at their targets, while others fail to arrive because of equipment malfunction during their time of flight.

5. Certain of the reliable American warheads penetrate the coverage provided by the enemy antiballistic missile defense system. Others are destroyed by the enemy defense.

6. American surviving warheads destroy certain enemy targets. This implies sufficient guidance accuracy in locating the target, successful detonation, and that the target is still there. (For example, if the target were a missile, it may already have been launched.) This is the output of the model, which results in a new set of surviving enemy forces available for assignment to American targets as the input to the "enemy vs. U.S. destruction model," Figure 5-8a, where the same related activity as described above occurs. Some enemy forces are then assigned for the next enemy strike. Thus, the same cycle is repeated by the enemy and, in a "push-pull" fashion, subsequent plays are made by each side in a time sequence determined by the response times of each side.

DEVELOPMENT OF THE OPERATIONAL FLOW MODEL

The operational activities involved in a general nuclear war, previously discussed in verbal, qualitative form, must be converted into quantitative form if a calculation of expected system effectiveness is to be made. As indicated previously, the two objectives of the strategic forces (deterrence and damage limitation) were measured by the amount of assured enemy destruction and damage to the United States. These two interrelated objectives require the construction of two models which are interconnected in the feedback relationship already shown: a Blue versus Red operational flow model indicating how much Red (enemy) damage Blue (American strategic offensive) forces would inflict, and a Red versus Blue operational flow model indicating how much damage Blue (U.S.) would incur in the process. The framework of these two models, which will be operated in tandem are shown in Figures 5-9 and 5-10.

General Considerations

Several comments of general interest can be made regarding these operational flow models:

1. The Blue versus Red operational flow model consists of a series of submodels, each representing an event which is part of the total operational activity previously described. Each of these events is performed to some degree of success and this degree influences how successful the end result (i.e., damage to Red) will be. These events are given the following submodel titles in Figure 5-9.

(a) First-Strike Destruction and survival of certain *blue* weapons;
(b) Launch Weapons;
(c) Warheads Entering Red Defense Zone;
(d) Early Warning Detection;
(e) Destruction by Red Defense;
(f) Warhead Impact and Explosion.

2. This operational flow model is similar to one which would describe the manufacturing process of a product involving several steps. For the total product to be manufactured satisfactorily, each step must be performed satisfactorily (including the possibility of rework), and only subassemblies which pass the previous stage of inspection are permitted to advance to the next stage of manufacture.

3. The operational flow model in general shows not only the flow of successful units, but also the path which unsuccessful units follow. This would be particularly important in the analysis of a manufacturing facility where salvage and rework paths are utilized and must be analyzed. In the military case of Figure 5-9, the unsuccessful events such as airborne aborts or destruction by enemy defense, result in no salvageable units and thus the analyst need keep track of only the successful events. The only exception to this is "ground aborts" (i.e., unreliable weapons which cannot be fired, but which can be repaired by the maintenance crew).

4. Each of the five submodels can be considered as a function which transforms an input that can be quantitatively measured into a quantitatively measurable output. Note that the output of one function serves as the input of the succeeding function. Hence each of these submodels is called a transfer function. As shown later, the process represented by the transfer function may not be a deterministic one (i.e., one in which there is an exact out-

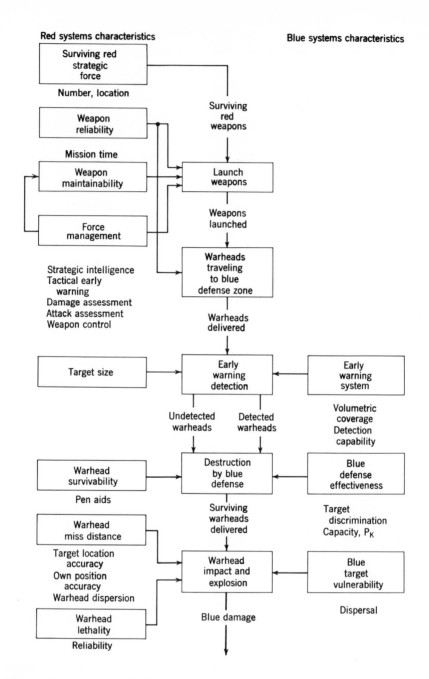

Figure 5-9. Framework of Blue vs. Red operational flow model.

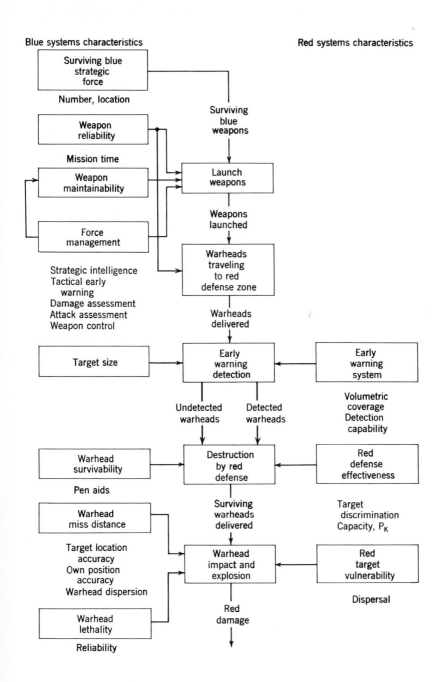

Figure 5-10. Framework of Red vs. Blue operational flow model.

Blue systems characteristics

Red systems characteristics

Surviving blue strategic force

Number, location

Weapon reliability

Mission time

Weapon maintainability

Force management

Strategic intelligence
Tactical early warning
Damage assessment
Attack assessment
Weapon control

Surviving blue weapons

Launch weapons

Weapons launched

Warheads traveling to red defense zone

Warheads delivered

Target size

Early warning detection

Early warning system

Volumetric coverage
Detection capability

Undetected warheads

Detected warheads

Warhead survivability

Pen aids

Destruction by red defense

Red defense effectiveness

Target discrimination
Capacity, P_K

Warhead miss distance

Surviving warheads delivered

Target location accuracy
Own position accuracy
Warhead dispersion

Warhead impact and explosion

Red target vulnerability

Dispersal

Warhead lethality

Reliability

Red damage

put for each input). Rather, a random process may be involved in which, for any deterministic input, the output may be described in probabilistic form. This is described further in Chapter 6.

5. As can be seen, the set of outputs make up a hierarchy of measures, any one of which serves as a measure of how well the system performs the previous set of events. Thus, the total operational flow model consists of a series of events, each of which must be performed satisfactorily if total success of the operation is to be achieved. These series of events correspond to the hierarchy of objectives mentioned in Chapter 4. Notice that in the hierarchy of measures the measure with the highest level of abstraction is the one applied to performance of the total offensive mission (e.g., Red damage).

6. The systems analyst must determine quantitatively each of the transfer functions (i.e., the explicit expression which translates each input into the subsequent output). The technique used is the "scientific method" described previously. The analyst uncovers those key characteristics or accountable factors involving the system and the environment which appear to pertain to this transfer function. Thus, in Figure 5-9 the performance characteristics of the Blue systems are shown on the left side of the appropriate transfer function and the environmental characteristics (Red) are shown on the right side of the appropriate transfer function. Having qualitatively indicated the factors which make up each of the transfer functions, the analyst must next validate the logic of these relationships and attempt to quantify them by accumulating appropriate operational and performance data and making the necessary extrapolations. Since data concerning the actual operations in a general nuclear war are not available, the analyst validates his model by interacting with the decision-makers (and/or their staffs who are assigned to this problem), operational system users, and the system designers.

7. As indicated in Figures 5-9 and 5-10 and in the discussion of the scenario, one cannot evaluate the American offensive mission apart from the enemy's offensive mission, since one move will depend on the results of the previous moves. Thus the final measure of Figures 5-9 and 5-10, "Red Damage," will be of two types: countervalue damage (to Red industrial centers) or counterforce damage (to the Red forces). The amount of the latter damage determines how much of the enemy's force will be available for his next move.

8. It should be noted that the operational flow models shown in Figures 5-9 and 5-10 have been deliberately overdesigned by taking into account many of the accountable factors which may not be used in the actual quantitative evaluation of the systems. For example, the use of early warning systems such as BMEWS, radars, and the DEW line and dispersal, were not

mentioned in the actual problem. However, it is realized that such early warning plays an important role in reducing the vulnerability of certain American targets such as SAC bombers. During the initial stages of model construction, it is wise to include as many of the factors believed to be important as possible. There are two reasons for this: first, while the decision-maker may not initially have asked that these factors be included, subsequent iterations with the decision-maker may reveal their importance and he may want to have them included eventually. Second, as shown later in the discussion of systems planning, even if these factors are not included in the actual systems evaluation, they may be of value to the system planner when he is trying to create alternative systems (see Chapter 9).

9. Note that the operational flow model could have been created by using the WSEIAC approach of Figure 4-6 as a checklist. In this case, the total model would have consisted of three separate submodels which would have been combined in series or cascade. As now described, the results would have been exactly the same, but the submodels might have been in a different sequence.

Referring back to the Blue versus Red operational flow model of Figure 5-10, the availability model would have been obtained from the residual, surviving force (covering initial survivability), and the Launch Weapons submodel (covering ground reliability and supportability). Mobility aspects would be included as a time lag in the Launch Weapons submodel.

The dependability model would have contained the "warheads entering defense zone" submodel (covering airborne reliability, as well as the aiming and fuzing reliability considerations contained in the warhead destruction submodel) and the early warning detection and destruction by enemy defense submodels (covering post-launch survivability).

The capability model would have contained the warhead destruction submodel (less the aiming and fuzing reliability considerations previously included under dependability).

If this checklist had been used, the systems analyst would have focused on survivability of all components of this system including the command centers needed to release weapons before they were fired. Such considerations have not been included in this study since we are assuming that command center survivability would be the same for all systems. In the scenario of a decapitation attack against command centers, this assumption may not be true. Such a scenario is analyzed in a related case in Chapter 12.

10. Note that the Red versus Blue model is the mirror image of the Blue versus Red model.

11. Note that the Blue versus Red model (or Red versus Blue) is really a model which portrays surviving forces acting on enemy targets after pene-

trating enemy defenses. Hence, this model is not restricted to the general nuclear war problem but can be used for any coı   at situation such as a limited war.

In conclusion, it should be stated that model building is an art, and there is no one standardized nomenclature for making explicit all of the factors of importance to the problem. The nomenclature used in the operational flow model of Figures 5-9 and 5-10 showed events as boxes and number of system elements as inputs or outputs of boxes. Another nomenclature might portray this in reverse. The modeling approach pioneered by Jay Wright Forrester,* for example, is one of the most thorough methods of modeling flows of material and information, emphasizing the various feedback loops which occur within a system and its environment. It is important to include all of the significant activities (to the degree of detail which will enable the analyst to quantify such things as rates of flow, probability of success of each activity, time lags, and geographical locations of system elements as a function of time). Of course, the dynamic problem of feedback between elements is a very important problem and while this is covered to a certain extent in this book, the reader is referred to Forrester for a more complete treatment of the subject.

How to determine the quantitative transfer functions of each of these models is now discussed in Chapter 6.

* J. W. Forrester (1961).

6

Developing Quantitative Relationships

After constructing in qualitative form the structure of the operational flow model which will be used to predict the effectiveness of each of the strategic systems (or force structures) under consideration, and validating the qualitative understanding of the problem with the higher-level decision-maker or his staff, the analyst must now attempt to develop the quantitative relationships which are involved in the model.

We shall now show how this is done by "walking through" an operational scenario of the events illustrated in the operational flow models of Figures 5-9 and 5-10, so that the detailed factors pertinent to each of the submodels may be structured to determine the transfer function of each submodel. In the discussion which follows we shall describe how to obtain the parametric transfer function for each of the identical submodels of Figures 5-9 and 5-10 and then show how to obtain a quantitative estimate of the parameters, based on different amounts of operational and other data pertaining to the performance of systems and the environment which might be available to the analyst. An examination of the uncertainties inserted into the models because of imperfect or incomplete information will be undertaken in Chapter 8.

LAUNCH WEAPONS SUBMODEL

The purpose of this model is two-fold:

1. To describe under what conditions certain weapons will be launched at an enemy, and

2. Given that the decision to launch weapons is made, to describe the time sequence of successful weapon launches.

Prewar/Prelaunch Operations

Before the start of the operations of interest, each side has separately deployed various amounts of its force structure of weapons, sensors, command centers, etc., to various geographical locations. As indicated in Figure 6-1, all weapons are initially available for launch (since there has been no attack as yet). However, no weapon can be launched without a direct command from the force management system, which can be modeled in greater detail, as shown in Figure 6-2. This system consists of a force commander (for each side) who has contact with the outside world through communications with the following sources:

1. Lower level commanders who may supply information regarding the environment and the status of their forces.

2. Sensors (both human observers or mechanized information sources, such as early warning radars) which provide information regarding both sides' targets and weapons. Such information might include movement or position of specific numbers and types of weapons or forces, and could be used for both strategic intelligence and tactical early warning of impending attack. The important point is that there is a specific accuracy and time lag associated with each type of information provided (which is generally more accurate and faster when observing our own forces rather than the enemy's).

Information is then inferred about the state of the external environment as gained from an analysis of the data presented by the sensor data and communications inputs. Such information provides strategic intelligence, tactical

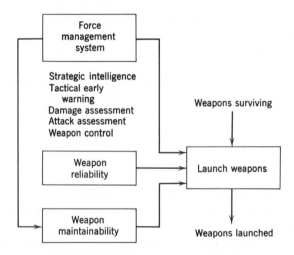

Figure 6-1.   Launch weapons submodel.

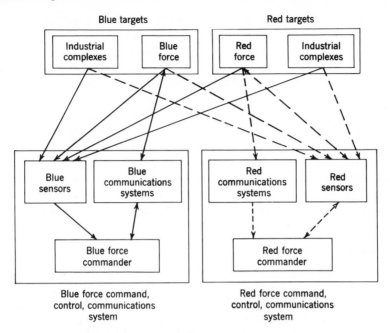

Figure 6-2. Strategic force management systems.

early warning information, damage assessment and attack assessment information, and is an important part of weapon control (i.e., the decision to utilize weapons). While it may be straightforward to quantitatively describe the information content and timeliness of the data supplied by the sensors and communications links, unfortunately, certain human tasks may be difficult to completely describe or measure explicitly. One such important task is human decision-making, encompassed by the term "force management" in the models of Figures 5-9 and 5-10. If one were to ask a force commander or his staff to explicitly describe how he would utilize his ICBM's in combination with the rest of the strategic force of SAC bombers, Polaris missiles, etc., the probable answer would be that it depends on the specific conditions with which he is faced at the time, and there are many conditions which could conceivably arise. In fact, each side generally configures various alternate war plan options which it would use, based on alternate contingencies (or scenarios) which each side feels may arise. This is generally helpful in reducing the time lags required for decisions, an important factor in real-life operations.

To attempt to model this situation, the analyst must gather whatever information he can find relevant to the set of contingency or operational plans which have been made for the employment of the force under different operational or environmental conditions (scenarios). This information would include the different decision rules employed, such as force allocation policies, weapon withholding policies, expected time lags in evaluating information and making decisions, etc. One way of obtaining some of this information is by attendance at operational exercises, or command post exercises, a simulation activity in which the force commander and his staff practice their jobs by receiving information as would be obtained from their systems. Each team must then evaluate their information and make a decision for allocating its forces, based on that information. Such a simulation technique can also be used as a research tool to make explicit the process by which commanders operate, or as a test bed for testing new procedures. A more extensive variation is competitive "wargaming" simulation in which different rules of command based on different possible situations may be tried and the effectiveness of each measured.

Launching Weapons

Given that a commander has decided to launch certain weapons against the opposing targets, he must now issue the alert order, so that weapon checkout can be initiated, and a weapon will be prepared for launch to a specific target (in this example we shall assume a missile launch). Thus, we are concerned with determining the following relationships for the launch weapons submodel:

1. What time is taken to decide to launch missiles (or other weapons)?
2. What time is taken to launch the weapons once the decision to launch is made?
3. What is the likelihood that each weapon launch will be successful?

To answer these questions, the analyst can model the various operational activities required to launch a missile, in parametric form, as shown in Figure 6-3a. Hence, the first performance characteristic to be quantified is the time required ($t_1 + t_2$) to perform all of these events shown in transmitting a readiness command (as a function of the number of missiles being commanded). The various times shown in Figure 6-3a may not be too important for the first wave of the initiating forces, but are very important for subsequent waves (when the opposing forces have established that the conflict has begun), since high time lags result in more forces being subject to incoming attack and possible destruction.

The second performance characteristic to be quantified is the number of a particular set of surviving missiles, located at a given base, which can be launched as a function of time, given that the readiness command has been received. This number is a function of the weapon maintenance support, missile reliability, the time taken to check out a satisfactory weapon and repair any found to contain a malfunction, and the launch rate.

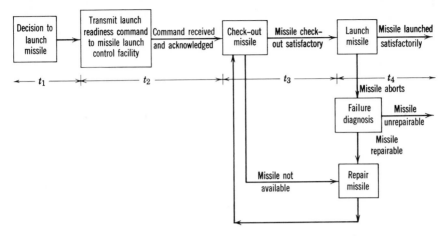

Figure 6-3a. Micro-model of launch weapon submodel.

Observation of the operations of a large weapon system such as the ICBM indicates that individual weapons become available as a function of time, as indicated in Figure 6-3b. Initially a certain number of weapons (N_1) are reported as being ready for operation. The rest ($N_2 - N_1$) are not ready for operation, either because they are in scheduled maintenance or in the process of being repaired due to a detected failure (unscheduled maintenance). This number is, of course, a function of the state of alert. As shown in Figure 6-3a, after the "get ready for launch" command is transmitted, increased numbers of these ready weapons are verified as being ready for launch; this requires some time following the initial decision to launch due to the time lags of communications and weapon checkout. Meanwhile, as shown in Figure 6-3a, those weapons which were not ready are being repaired, checked

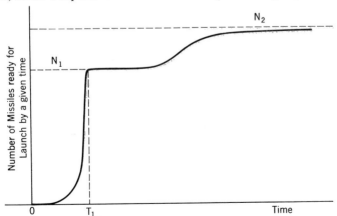

Figure 6-3b. Readiness for missile launch.

out, and are reported as being ready at a later time. Generally, weapons are launched in a salvo for a purpose such as a saturation attack. Assume that the initial salvo of weapons is launched at T_1. However, only a certain amount of these can be launched successfully. As shown in Figure 6-3a, some of the launch failures can be repaired and are available for subsequent launch at a later time, while some are unrepairable and lost. Lastly, the total number of weapons available can be launched at a rate which is a function of launch facilities (i.e., men and equipment), (indicated as maintainability in Figure 6-1).

Note that the same type of transfer function illustrated in Figure 6-3b is generic to the particular process described, and will apply to either the Red versus Blue model or Blue versus Red model. The only difference is in the specific numerical values of T_1, N_1, and N_2.

Gathering Numerical Data

We shall now consider the type of data which the analyst could gather for an existing operational system such as the Blue hardened ICBM system. The data could then be transformed appropriately for the superhardened and mobile systems, and finally for an enemy system.

In the case of an existing Blue weapon system such as the ICBM force, P_r, the readiness rate may be estimated by the analyst in either of the following ways:

1. Obtaining data at operational exercises. Here all pertinent data could be collected by weapon class. This type of data, however, is limited by the number of tests, and hence may not offer an accurate estimate of the pertinent parameters.

2. A larger set of data may be obtained from a recording of the long time history that each weapon of a particular weapon class was available or not available (under repair), as shown in Figure 6-4. If the analyst had such a time history of each weapon, he could then determine, as a function of time, the number of available weapons. One source of data collection error should be noted. Generally, periodic checks are made to detect a nonoperative system. In fact, since the checking procedure may itself induce additional failures, there is a force tending to reduce the rate of checking. When an inoperative system is found, there is a high chance that the failure occurred at an unknown time in the past; hence, the time between failures is less than the time between when the system became operational and when the failure was detected. Similarly, the time to repair would actually be longer than measured.

The difficulty with such an approach is that there are so many possible (but related) situations to analyze. For example, consider the task of launching 5 (or 10 or 100) missiles at their targets. If each of these situations was

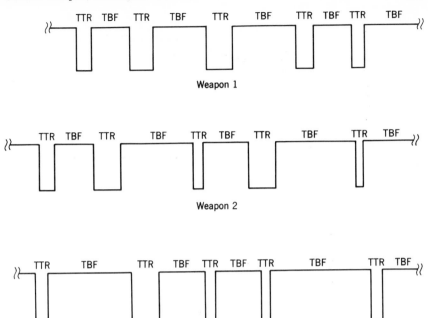

Figure 6-4. Weapon readiness for action.

tested individually, a sufficient number of tests would be needed for each situation. But certainly the results of firing five missiles (with from zero through five successes) has some relationship to the results expected from launching 100 missiles (with anywhere from zero through 100 being successful). An elaborate theory of probability and statistics has been assembled to deal with such analyses of random processes. We shall now apply parts of this theory to our problem of determining the pertinent quantitative relationships involved. For the reader who is not familiar with probability and statistics, Appendix I provides a review of some of the basic concepts which are used here. Application of these techniques in systems planning should be done by a professional statistician or a systems analyst trained in this field.

Constructing the Submodel

Drawing from the discussion in Appendix I on probability and statistics, we shall now construct the Launch Weapons transfer function in probabilistic form. (Recall that this model translates the number of surviving weapons into number of weapons launched as a function of time.) To do this, we con-

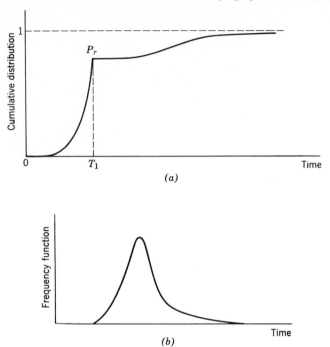

Figure 6-5. (*a*) Probability that weapon will be available by a given time; (*b*) time to report weapon is available.

struct a test hypothesis of the operations involved, as observed in some operational exercises of the existing Blue system. Such a parametric model is depicted in Figure 6-5. This shows that once the command to prepare for weapon launch at a designated target is transmitted (at time equal to zero), there is a certain probability, P_r, that a given weapon will report back on or before time T_1 that it is ready and available for launch. Here we are concentrating on weapon malfunctions rather than communications design. That is, we are assuming that time T_1 was chosen large enough to accommodate weapon checkout and status reporting. Now the analyst's problem is to determine the value of P_r for a given class of weapons (such as the ICBM's).

The most likely estimate of P_r (commonly called the readiness rate) may be obtained from an analysis of the amount of system "up time" and "down time," as illustrated in Figure 6-4. Since weapons may be needed at any given time, the readiness rate, P_r, is really that average fraction of the time that a weapon is operationally ready, or

$$P_r = \frac{\text{MTBF}}{\text{MTBF} + \text{MTTR}}$$

where MTBF is the "mean time between failure" and MTTR is the "mean time to repair," considering both scheduled and unscheduled maintenance. Obviously, the readiness rate will be a function of the ground checkout equipment, the size of the ground crew with respect to the number of weapons to be launched, and the degree of force alert, all included under the term "weapon maintainability" in Figure 6-1.

The time when a weapon has reported back as being available is also a random variable. Using the data available from exercises, a probability distribution (such as a normal distribution) may be fitted to the available data to approximate the actual distribution, as shown in Figure 6-5b.*

Weapons which report back as being not ready for launch due to an equipment malfunction will be ready at a later time, as shown in Figure 6-5a, when the weapon repair has been satisfactorily completed. This time may be quantified by statistically analyzing the "time to repair" using the data in Figure 6-4 to determine the frequency distribution of that fraction of weapons which can be repaired within certain time intervals (e.g., five-minute interval). Again, a probability distribution, may be fitted to the available data, as shown in Figure 6-5c. This probabilistic distribution of repair time may then be used to determine the time that an unavailable weapon would be available, making explicit the complete probability distribution of Figure 6-5a (i.e., the probability of having an available weapon as a function of time).

Next, data must be collected to indicate what fraction of weapons can be launched satisfactorily, given that they check out satisfactorily (P_1), and that the fraction of launches that fail can be repaired ($P_{r'}$).

Lastly, the time required for launch should be determined, as well as the number of simultaneous launches which can be accommodated, so that the total time requirements of the operations may be used in the analysis. This over-all factor is sometimes called the launch rate, and is a function of the number of launch facilities and launch crew available.

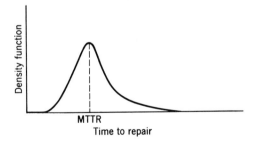

Figure 6-5c. Time required to repair weapon.

* See discussion on curve fitting in Chapter 7.

Inserting Numerical Data

All of the previous discussions have dealt with a procedure of generating parametric models which describe a set of operational activities involved in the total launch weapons submodel. The specific values of the parameters (such as P_r and P_l) are determined from operational data, which in the case of the current hardened system could be collected. If this data is not available, as in the case of the mobile ICBM system, best estimates can be made of the numerical values by using any type of related data for extrapolation purposes. If nothing else is available, considered judgment must be applied by comparing the new system against the unknown characteristics of an existing system. In the case of the enemy system submodel (i.e., Figure 5-10), available intelligence estimates and considered judgment are similarly utilized. The same approach also applies in estimating the numerical values used in the other submodels to be described.

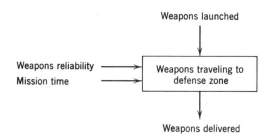

Figure 6-6.   Weapons arriving at defense zone submodel.

WEAPONS TRAVELING TO DEFENSE ZONE SUBMODEL

Given that a certain number of weapons have been successfully launched, as shown in Figure 6-6, we must now determine how many of these successfully reach the enemy defense zone (and hence are subject to subsequent enemy action). This can be obtained by finding the operational reliability of the weapon, where reliability is defined as the probability that a system does not fail during the period of observation of the activity (in this case the flight time to the beginning of the defense zone).

Since system malfunctions may occur in different ways, we shall now explore these different ways to determine how to satisfactorily model these to serve as a way of predicting when failure will occur.

We shall first examine a relatively simple device—an electrical switch which operates a household light. Suppose that an analyst wished to conduct an accelerated life test of a large sample of switches to gather data to be used to predict failures of this type of switch. The analyst would record data indicating the total number of switch operations completed before each switch failed. Analyzing this data would show that there was a low, rather constant failure rate over most of the life of this particular switch, with two exceptions: (a) the failure rate was higher during the early life of the switch, probably due to manufacturing defects; and (b) the switch "wore out" after a long time, as indicated by the higher failure rate near the end. From these data a probability distribution could be constructed, as shown in Figure 6-7, which could be used to predict switch life. Of course, other tests might be designed to include further key variables which the analyst thought to be important, such as the length of time switch is on and the amount of current passing through it.

Other types of equipment might have a similar type of probability density function using system life measured in time. In all of these cases the density

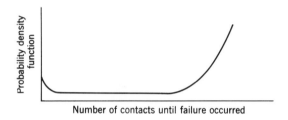

Figure 6-7. Equipment failure probability distribution.

function might be analyzed as consisting of the following three phases over the system life cycle:

1. The first phase consists of a relatively high failure rate which is probably due to manufacturing defects.

2. The second phase consists of a lower failure rate, probably due to some type of shock or strain placed on the equipment as it operates in its environment.

3. The third phase consists of an increasing failure rate which seems to occur as a function of equipment age or usage. In fact, many times a product is deliberately designed to have a certain amount of life. For example, the life of a tungsten light bulb is largely determined by the thickness of the tungsten filament used. Any variance in this life might be largely attributed to the on and off "shocks." Storage batteries are often sold with a warranty of 24, 30, 36, or 48 months, depending on the quality of the design. Again, the variance in this life might be attributed to a number of factors such as

electrical load, amount of stop and go driving, or temperature of the place of operation, all of which influence the load on the battery. The same analysis could be made for a rubber tire or a mechanical automobile clutch.

Greater insight into the problem of determining the type of probability distribution to use in predicting system failures may be found by examining the human mortality tables (Figure 6-8),* which are based on a large sample of data. Figure 6-8a is the cumulative probability distribution from which the probability density function of Figure 6-8b may be obtained by a process of differentiation. Figure 6-8c is the "conditional probability density

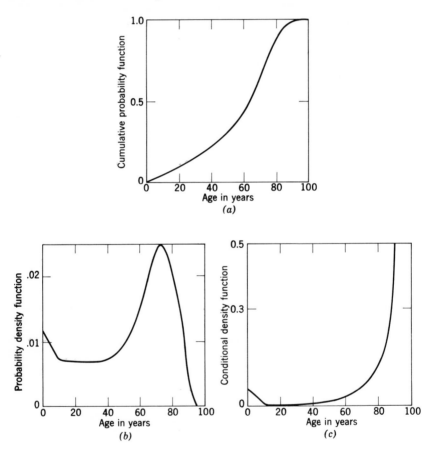

Figure 6-8.   Mortality probability functions.

* Tables 6-16, 6-18 and 6-19 are taken from D. J. Davis (1952), and modified to include infant mortality.

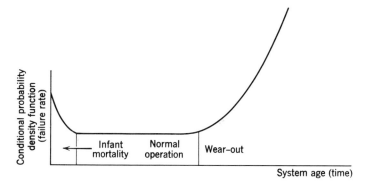

Figure 6-9. Equipment failure conditional probability distribution.

function," which may be defined as the probability of death during the next interval of time, Δt, given that the subject has survived up to time, t. The conditional probability density function shows in particular that there are three phases in the human mortality function:

1. Infant mortality, where the mortality rate is initially higher.
2. Natural life, where the mortality rate is much lower and fairly constant.
3. Old age, where the mortality rate increases rapidly.

These same concepts may be used to predict system equipment failure as shown in the conditional density function of Figure 6-9. Since equipment failure rates change with the age of the equipment, three stages are discussed, indicating methods for analyzing each stage:

1. *Normal operating life:* As indicated in Figure 6-9, the conditional density function may be assumed to be constant and independent of past activities. This phenomenon results in a probability density function (the reliability function) which is an exponential distribution, as shown in Figure 6-10:

$$R = e^{-t/m}$$

where R = Reliability, the probability of no failure during mission time, t,

m = Mean Time Between Failures.

The probability density function, f(t), is

$$f(t) = \frac{e^{-t/m}}{m}$$

2. *Infant Mortality Phase:* Conditional probability distribution is higher during this phase. While a probability density function could be generated, it is customary in the operation of large systems (e.g., defense systems where high failure rates are to be avoided) to "break in" the equipment. This involves operating it for an initial period of time to assure that the manufacturing defects which could cause early failures have been detected and eliminated.

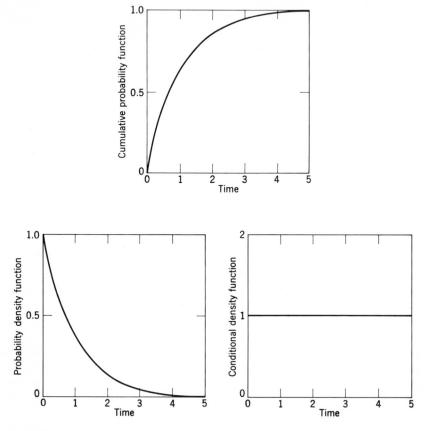

Figure 6-10. Cumulative probability, probability density, and conditional density functions for exponential theory of failure.

3. *Wear-out Phase:* This phase may be modeled by assuming a normal distribution for the probability density function, as shown in Figure 6-11. It is the practice however, when high failure rates are undesirable, to eliminate this phase through the employment of scheduled maintenance procedures. These procedures involve periodically testing and inspecting certain compo-

nents and replacing equipments whose tests show them to be in a marginal condition, or where the equipment failure rate has been noticeably increasing.

Application to Weapon Systems

In applying the above discussion of failure analysis to our missile analysis, we shall assume that we are operating the missile during its normal pe-

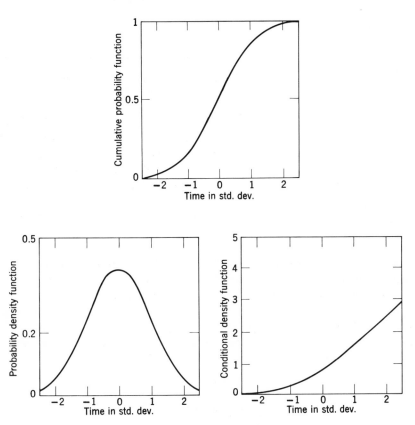

Figure 6-11. Cumulative probability, probability density, and conditional density functions for normal theory of failure.

riod in which the conditional failure rate is constant (i.e., there is both system break-in and scheduled maintenance to eliminate wear-out). Hence, an exponential distribution will be assumed to represent the system reliability, which may be found as a function of operating (flight) time from the cumulative distribution of survivability (Figure 6-12). The flight time is a function

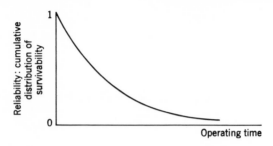

Figure 6-12.   Reliability of equipment as a function of operating time.

of the speed of the weapon, the flight path and its distance to the enemy early warning defense zone, shown in Figure 6-13, for both a bomber and an ICBM. This distance will be determined in the next submodel, early warning detection. In the case of the bomber, failure may occur at any time during t_1, hence t_1 is used in the exponential distribution to determine R_1 during this period. In the case of an ICBM, the flight path has both a powered flight phase of t_1 (during which failures can occur which would prevent continuation of the flight path to the defense zone), and a ballistic flight path of t_2 where the warhead is essentially out of control. Thus only t_1 is used to compute the reliability R of the missile to the defense zone. It should be noted, however, that other system failures which may have a later effect on target destruction may occur during time t_2, and these failures will be included in the "warhead impact and explosion" submodel.

Having then determined t_1 and the value of the weapon reliability to the enemy defense zone, the probability distribution for this model (warhead entering the defense zone) will be a binary distribution whose frequency function is shown in Figure 6-14. As indicated in the previous submodel, values of the reliability of flight to the enemy defense zone are calculated for each of the existing Blue weapons (i.e., bombers, ICBM's) based on the

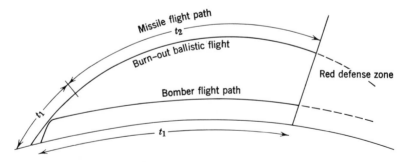

Figure 6-13.   Weapon flight path into defense zone.

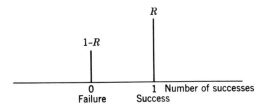

Figure 6-14. System reliability for a given operating time interval.

available data pertaining to system MTBF and the assumed exponential distribution. Values of reliability for the other Blue (and Red) systems must be determined through a subsystem reliability analysis or a comparative extrapolation, using whatever data is available to the analyst. Incidentally, if a weapon malfunction occurs and this information can be transmitted to the force management system, a replacement weapon may be launched.

EARLY WARNING DETECTION SUBMODEL

This submodel and its factors, depicted in Figure 6-15, are intended to determine the time that the opposing (say Blue) side first receives tactical early warning information about an attack. This warning information would be used to alert the Blue force and hence increase its state of readiness. In addition, the alert information might also be used for dispersal of the force and/or population.

The detection process is another random process which must be analyzed in probabilistic terms. One of the properties of any detector or sensor is its volumetric coverage (shown in Figure 6-16) for one particular instant of time since this coverage may change if the field of view of the detector

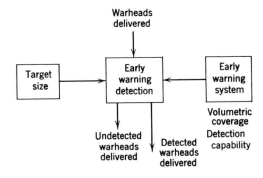

Figure 6-15. Early warning detection submodel.

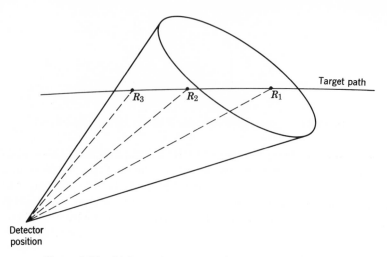

Target path

R_3 R_2 R_1

Detector
position

Figure 6-16. Volumetric coverage of early warning detector.

changes, such as in the case of a search radar. The probability of detection of the target (given that it is operating reliably) is a function of a number of variables, including (a) the performance characteristics of the detection subsystem elements, (b) the size of the target, and (c) the position of the target with respect to its volumetric coverage pattern, particularly its range (distance) from the target to the detector.

Hence, a given detection system and a given size target can either be measured or calculated by an analytical expression such as the "radar range equation." Such an expression provides the probability of target detection at a given range, as shown in Figure 6-17, which results in the binomial probability distribution of Figure 6-18.

As discussed previously, the availability and dependability of the detection system must also be considered when its effectiveness is being pre-

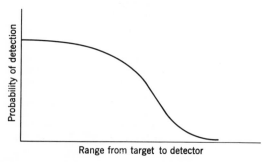

Range from target to detector

Figure 6-17. Probability of detection of early warning detector.

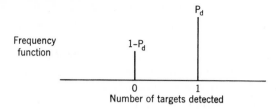

Figure 6-18. Probability distribution of detector for each target.

dicted. For a detection system, two main phenomena must be considered:
(a) equipment malfunctions leading to system unreliability, and (b) survivability to enemy attack (or other countermeasures). Early warning detection
systems must operate continuously, and hence are designed to have a high
total system reliability approaching unity through redundant equipments.
The estimated system availability and dependability should be determined
and inserted as part of the transfer function.

System survivability is a function of the enemy's decision and capability
to attack a target using lethal means, and of the system's ability to withstand
such an attack. More is said about analyzing these phenomena in Chapter 7.
Electronic countermeasures can also be used to confuse the detection system. This is discussed in the defense system submodel.

One other characteristic of a sensor system which the analyst (as well as
the systems planner) should consider in analyzing its performance characteristics might be encompassed by the term "information content," which
includes such terms as system "resolution" and "accuracy." All sensor systems as a class are designed to provide signals which can be used to infer or
predict information relevant to the following questions:

1. *What is it?* Is the target "friend" or "foe"? Is it a missile or penetration
aid? What are the physical characteristics of the object? Here we can extend our
concept of a sensor to include a physician's sight, and touch, and an electro-
cardiogram printout.

2. *Where is it?* Here the accuracy in measuring distance is important.

3. *How many objects are there?* Here the resolving power of the sensor is
important.

4. *Where did it come from?* Here a series of observations may be used in
some form of extrapolation to determine some past occurrence such as the
probable launch point of a missile.

5. *Where is it going?* Similarly, the series of observations may be used to
determine the probable target path.

The information process described in the preceding falls into a general
topic of "statistical inference," in which an individual is attempting to

make some determination or prediction based on a limited sample of data. Whether the "signals" are those available to a physician who is attempting to combine data observations into a medical diagnosis, or whether they consist of past sales reports and other data used to predict the next year's sales of a product, the approach is still the same as we have previously described in the section of constructing quantitative models. Obviously the larger number of different "signals" (i.e., variables) involved in the model, the more accurate the model (and hence, the prediction) may be.

Another characteristic of the detection system to be considered is the "false alarm rate," which is the probability that the detector will signal that there is a target within its volumetric coverage, when, in fact, there is none. This false alarm signal may be due to internal "noise" or some external environmental characteristic (such as a radar return from the moon appearing to be a missile return). For this reason a number of different detectors having some common detection volume are used to obtain cross-correlation with time of any detection signal received.

Lastly we are interested in the time lag required to transmit the detection information from the detection system to the force management command center for subsequent action.

Thus the output of this submodel consists of a set of detected enemy weapons (e.g., warheads) and a set of undetected enemy weapons which then proceed into the enemy defense zone where they are subject to attrition. We shall now discuss this operation.

DESTRUCTION BY ENEMY DEFENSE SUBMODEL

Given that a number of warheads had been successfully delivered to the vicinity of the enemy defenses, the defense forces then operate on these warheads, destroy some, and the remainder of the warheads survive to be delivered to their targets. This defense destruction submodel, and its factors, depicted in Figure 6-19, is intended to determine the number of warheads which will survive the enemy defenses, hence can be delivered to their designated targets.

The key performance measure of any defense system is its probability of kill against a given type target. This probability of kill, however, is generally a function of a number of factors. One of these is the lethal volumetric coverage, as shown in Figure 6-20, which is analogous to the early warning detection volumetric coverage discussed in the previous section. This volumetric coverage is a function of the following factors:

1. Flight profile of the defense missile employed. Each missile type has a maximum range beyond which its lethal defense mechanism cannot be ef-

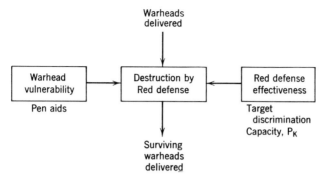

Figure 6-19. Defense destruction submodel.

fective, or certain regions such as low elevation angles, which cannot be covered.

2. Surveillance coverage, which again has a volumetric coverage limitation.

3. "Dead time," which is required to perform certain functions prior to missile launch, such as target acquisition, data processing, and target designation, once a target has been initially detected.

4. Target characteristics, such as speed, maneuverability, and vulnerability of the defense warhead.

5. Warhead lethality.

While the probability of kill of a single round of fire may be a function of the angle of fire, many times this probability of kill can be treated as a constant for all volumes inside the lethal zone.

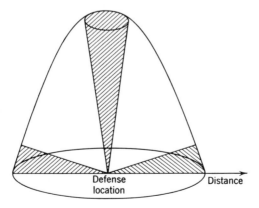

Figure 6-20. Defense zone lethal coverage.

Another performance characteristic to be considered is the maximum capacity of the system since all defense systems can be saturated at some point. One element involved is the maximum number of "rounds" (e.g., antiballistic missiles) available for a fire at a given location. A second performance characteristic is the firing rate (e.g., the number of rounds per minute) which can be fired at a target. A third characteristic involves the maximum number of targets which the system can handle at any one time. Each element of the system has a finite capacity and time required to do its task and, hence, is a potential "bottleneck" to the system. For example, a computer or a human operator can handle only a fixed number of targets at any one time; communication links have a fixed channel capacity. Figure 6-21a illustrates this system saturation phenomenon by relating the expected number of targets destroyed by the defense installation as a function of the number of targets simultaneously in the defense fire zone. Note that this figure implicitly assumes that there will be no "bonus kills," which means that the targets are not close enough to one another to allow one defense warhead to destroy more than one target. Several points should be noted in regard to this figure:

1. The transfer function of a perfect defense system, one in which all of the incoming targets would be destroyed, is a straight line at a 45° slope.

2. Any nonperfect defense system provides reasonable protection against small numbers of targets simultaneously being fired upon. As the number of simultaneous incoming targets increases, however, the slope of the curve (i.e., probability of kill) decreases until the expected number of targets killed might actually decrease as shown. This inefficient use of the system would occur if one or more of the system elements (e.g., operators) would attempt to operate in an overloaded condition. This condition can be alleviated and the system constrained to operate at its most effective level by changing the operating policy as follows: First find C_{max}, the maximum effectiveness level achievable by the system. This is the maximum number of targets (i.e., warheads) that can be killed as the number of simultaneous targets increases. Designate C_{max} as the maximum capacity of the system. Let us assume that this is 30, as shown in Figure 6-21a. The operational policy would then state that no more than the first 30 targets which enter the defense zone simultaneously will be entered into the system. Then, as each target is disposed of, a new target will be permitted to enter the system. In this way the maximum effectiveness would be achieved as shown in Figure 6-21a.

These data may be converted to a probability distribution function; Figure 6-21b shows how the defense system kill probability varies as a function of its load. The constraining operational policy previously described would permit this function to exist for targets whose number is less than or equal to C_{max}.

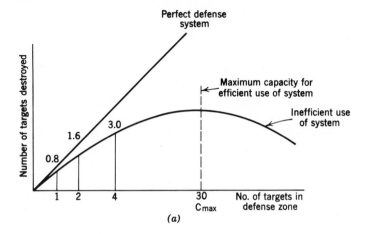

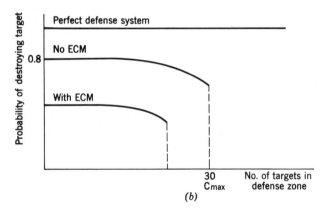

Figure 6-21. (*a*) Defense system transfer function; (*b*) probability distribution of defense system.

If enemy countermeasures (ECM) are employed (e.g., jamming or chaff), the operational effectiveness of the defense system may be reduced drastically, both in P_k and in C_{max}. The new defense system transfer function is also illustrated in Figure 6-21b.

The use of decoys can also reduce the effectiveness of the defense system. These penetration aids serve to effectively increase the number of targets which are in the defense zone at one time. If the defense zone has no means of discriminating decoys from true targets, the defense zone might be more readily saturated (i.e., approach the maximum level of 30 in the preceding example). In this case, not only would all of the targets and penetration aids above the first 30 be permitted to go through with complete survivability, but some of the defense missiles would be expended on the decoys in the group, increasing the total cost to the defense. There is, however, a cost to the attacker in using penetration aids, since the weight (or volume) used cannot be used for the warhead, resulting in a smaller and less lethal warhead. This is an application of systems analysis (i.e., the proper "mix" between warhead and penetration aids for a fixed weight or volume constraint on the mixed load in a missile). Alternatively, additional missiles containing only penetration aids and traveling in the missile wave could be used. Then the analysis would involve determining the total cost (warhead missiles and penetration aid missiles) of achieving a given level of target destruction, choosing that system mix which achieves the effectiveness at lowest total cost.

Lastly, the response time taken to perform certain events can be translated into operational rates (e.g., rate of fire, or launch rate of defense missiles) compared with the time that the targets are in the defense zone. These factors determine the total number of targets which can be operated upon.

Defense system reliability also determines the number of targets which can be destroyed. As indicated previously, defense system availability and dependability must also be inserted.

TARGET DESTRUCTION SUBMODEL

This submodel, depicted in Figure 6-22a, is intended to determine how many targets are destroyed by the given number of warheads which have survived the enemy defenses and have continued on to their targets. This transfer function may be obtained by combining the answers to three primary questions:

1. Will the warhead land inside the vulnerable area of the target?
2. Given that it does, will the warhead then detonate satisfactorily?
3. Given that it does, will the target still be at the targeted location?

A probabilistic expression can be generated for each of these random processes.

Warhead Accuracy

In Chapter 5 we described a method for determining the target destruction as a function of miss distance for a given warhead size and target vulnerability (Figure 5-3). The analyst must now gather data which quantifies this relationship for the particular size of the warheads and vulnerability of the targets under consideration in this study. The next problem is to deter-

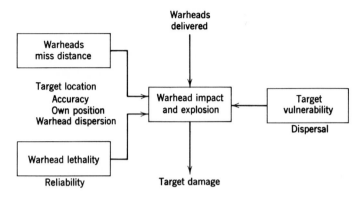

Figure 6-22a.   Target destruction submodel.

mine the probability of a warhead's landing within the target vulnerable area. To do this, consider a missile launched from point A on earth and directed to a target at point B on earth, as shown in Figure 6-22b. Since information regarding the geographical location of points A and B is needed to direct the missile, there are three sources of error which contribute to the missile not landing exactly at point B: (a) accuracy with which point A is known, (b) accuracy with which point B is known, and (c) warhead dispersion which is independent of the above accuracies.

We shall first consider warhead dispersion, assuming that the locations of A and B are known perfectly (zero error). The effects of actual errors in A and B are discussed later.

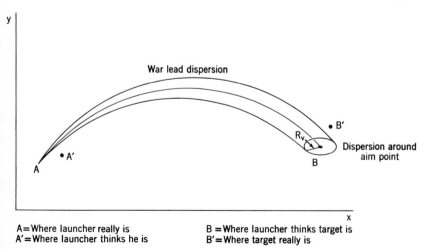

A = Where launcher really is   B = Where launcher thinks target is
A′ = Where launcher thinks he is   B′ = Where target really is

Figure 6-22b.   Sources of warhead inaccuracy.

The first step involved is gathering whatever data are available involving the phenomenon under consideration; in this case, the dispersion from the target at which the warhead was aimed. This set of data samples, comparing an "actual ground zero" (AGZ) obtained with an intended "designated ground zero" (DGZ), would be obtained, for example, from the test firing results of the current United States ICBM system. When plotted, this might resemble the data shown in Figure 6-23a. The modeling problem then, is: How can the locations of these impact points be used to describe the phenomenon quantitatively?

The first problem which the analyst faces is that, in attempting to analyze a limited number of samples (which may be the only data he has), he is implicitly making the assumption that future firings will follow the pattern of past firings (as long as the missile system is unchanged). The problem of determining the amount of error made in the process of dealing with a limited set of data samples is discussed in Chapter 8.

There are several ways used to describe (or structure) the data samples shown in Figure 6-23a. One way is to find that circle whose center is located at target B and whose circumference encloses one-half the sample points, as shown in Figure 6-23b. The radius of this circle is the most likely estimate of the circular probable error (called the CEP). Unfortunately, knowing the value of the CEP is not sufficient to calculate what percentage of these firings would have landed within the radius of vulnerability as previously shown in Figure 5-3. What the analyst needs in order to describe the function quantitatively is the frequency of hits which fall within varying distances of the target B. One such modeling of this information is shown in Figure 6-23c where frequency contours of these firing results are plotted. These contours use circles as lines of constant frequency, a further elaboration of Figure 6-23b; that is, the 10% contour is that circle which contains

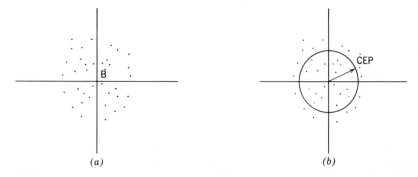

(a)   (b)

Figure 6-23. (*a*) Data samples of warhead dispersion; (*b*) determining the circular probable error.

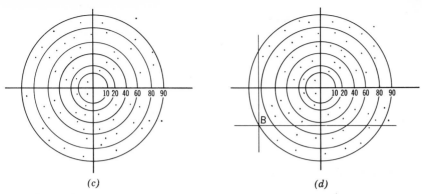

(c)   *(d)*

Figure 6-23. (c) Constant frequency contours; (d) structuring data which contain an offset.

10% of the test points, the 20% contour contains 20% of the samples, etc. By definition, the CEP is identical to the 50% contour. Such a model would be called the "sample cumulative probability function."

Sometimes the test results might not be centered around the DGZ at the origin but contain an offset bias as shown in Figure 6-23d. Since such a fixed bias could be removed from the system (as a rifleman does with his windage adjustment), the analyst could do so in his analysis, thereby translating the data back to that shown originally in Figure 6-23c.

A more systematic way of obtaining a probability distribution from the original data available (Figure 6-23a) is to replot the same data as a sample cumulative distribution over values of x, as shown in Figure 6-23e, and over values of y. These data can be used to predict what an infinite number of samples might look like. This is done in order to deal more effectively with the problem of sampling errors. For example, if one used the actual sample data shown in Figure 6-23e, one might erroneously conclude that there was no likelihood of a firing landing at a distance greater than the furthest points shown (x_1 and x_2). However, now the analyst is confronted with the problem

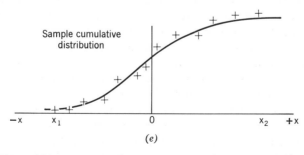

Figure 6-23. (e) Curve fitting a sample cumulative distribution.

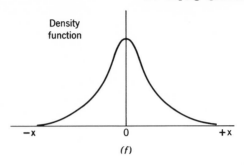

Figure 6-23. (*f*) Probability distribution of X component of dispersion.

of choosing some analytical expression which appears to fit the data samples shown in Figure 6-23e. In general, the following considerations are taken into account.

1. *Closeness of fit.* Obviously, the analytical expression should closely fit the data samples. However, if there are only a few samples and a great deal of dispersion, a close fit may not be possible.

2. *The employment of other related data.* Another way of effectively obtaining more information, rather than through the expense of firing more test missiles, is to analyze the test firings results of other related missiles (such as a previous version of the missile under study) or to use analytical studies made of the phenomenon. This information may provide further insights with regard to the type of distribution to use. It may even provide information useful for extrapolating past results to predict future performance more accurately.

3. *Ease of using the analytical expression.* The expression should be simple to manipulate so that the analytical effort involved in applying the expression to the problem will not be excessive.

With these constraints in mind, a suitable probability distribution may be constructed with the assistance of an experienced statistician. For weapon systems, a normal distribution is generally assumed for both x and y, as shown in Figure 23e, which then enables the analyst to construct the appropriate probability density function, as shown in Figure 6-23f. If the variance (or standard deviation) of the normal distribution for x is equal to that of the normal distribution for y, circular contours can properly be constructed, as was done in Figure 6-23c. If not, elliptical contours could be constructed, as shown in Figure 6-23g. In general, circular contours are used (σ_x is made equal to σ_y) for simplicity. If this is the case, a probability distribution using polar coordinates may be easier to use. Hence a sample cumulative distribution of the test data available as a function of range from the DGZ could be constructed as shown in Figure 6-23h, and a smooth curve fitted to this data. For weapon systems, a Rayleigh distribution is generally assumed for the range distribution, yielding a probability density function, as shown in Figure 6-23i. Thus the probability of target destruction may be found by integrating the probability density function shown in Figure 6-23i between:

$$R = 0 \text{ and } R = R_v.$$

where R_v is the vulnerable radius of the target.

Incidentally, as would be expected, a uniform probability distribution is generally assumed for the angular distribution of the warhead dispersion, as shown in Figure 6-23j.

It should be emphasized that we have assumed independence between the two variables involved in describing the distribution from which we are assuming the sample data was obtained. A correlation analysis could be made to determine if such an assumption is appropriate.

Errors in Target and Launch Position

As indicated previously, additional errors in accurately determining the missile launch position or the exact position of the target (points A and B in Figure 6-22b) will result in a reduction of the probability of target destruction. These errors may be factored into the analysis by considering the revised problem as illustrated in Figure 6-24. The probability contours indicate that there is some uncertainty associated with the location of the points A and B. In general, the launch point A is known more accurately than the target point B, since A was probably measured accurately prior to launch. These

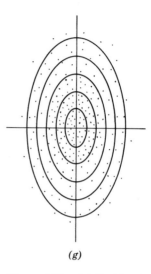

(g)

Figure 6-23. (*g*) Constructing elliptical contours of dispersion.

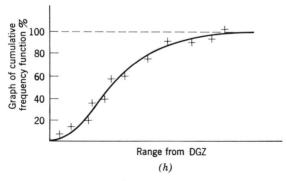

Range from DGZ

(h)

Figure 6-23. (*h*) Cumulative distribution of dispersion in range.

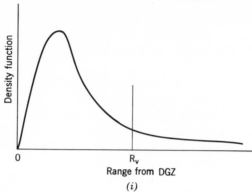

Figure 6-23. (*i*) Density function of dispersion in range.

contours may again be converted into some analytical functions, showing the probability density functions in range and angle. These uncertainties in A and B may then be added to the warhead dispersion (assuming that all three are independent of one another) by adding the variances of all three probability distributions. This new probability distribution of the total warhead dispersion (shown in Figure 6-25a) is then used to determine the probability of target destruction, and is equal to the amount of probability contained by the circle of radius, R_v. This results in a binomial distribution, as shown in the frequency function for targets destroyed, Figure 6-25b.

The analyst could also construct two other functions from the total warhead dispersion function. The first, shown in Figure 6-26a, is the probability of weapon kill as a function of radius of target vulnerability, assuming a constant (total) warhead dispersion. This function will be of help in calculating W_k for different type targets. The second function, shown in Figure 6-26b, is W_k as a function of target dispersion for a given R_v. This will be of help in considering other system improvements affecting system accuracy.

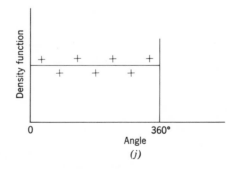

Figure 6-23. (*j*) Density function of angular dispersion.

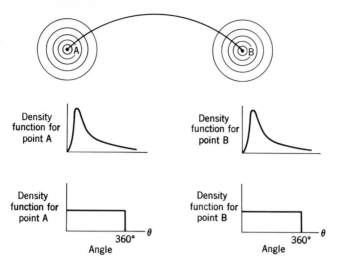

Figure 6-24. Probability distributions for uncertainties in launch and target positions.

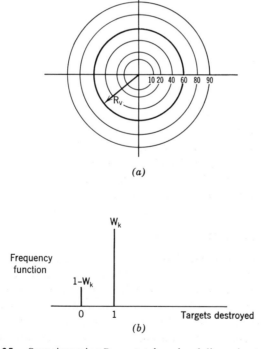

Figure 6-25. Superimposing R_Y on total warhead dispersion to find W_k.

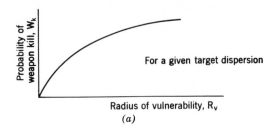

Radius of vulnerability, R_v

(a)

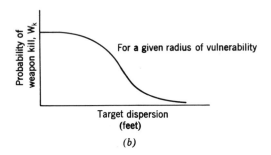

Target dispersion
(feet)

(b)

Figure 6-26. (a) Probability of weapon kill as a function of target vulnerability; (b) probability of weapon kill as a function of total dispersion.

Bonus Kills

Occasionally targets may be so closely located that their areas of vulnerability overlap, as shown in targets B and C of Figure 6-27. Thus, both targets will be destroyed if the warhead lands in the area of overlap, Area II. This is called a "bonus kill" and must be included in a systems analysis. Obviously, the systems planner must find some way of eliminating or reducing areas of overlap to make such targets less attractive to an enemy. This can be done by additional target hardening or relocation.

Reliable Warhead Detonation

Given that the warhead has landed within the radius of vulnerability, what is the probability that it will be properly detonated? This conditional probability may be obtained in the same fashion that the airborne reliability was obtained. As before, the analyst gathers available data pertinent to the existing system and determines whether he should consider the most likely estimate of the conditional probability (i.e., the ratio of number of successes to total number of tries), or whether he wishes to consider the

Figure 6-27. Bonus kills for closely located targets.

probability to be a function of flight time (as in the case of a bomber). If the latter is the case, an exponential probability density function may be used:

$$R_2 = e^{-t_2 / \text{MTBF}_2},$$

where R_2 is the conditional reliability,
t_2 is the flight time variable.

MTBF_2 is the Mean Time Between Failures (of the pertinent subsystems involved). Thus, if the arming, fuzing, and warhead are the only subsystems involved, they may be omitted from the first reliability calculation and considered here alone. In this case, t_2 would be the total flight time involved.

Is Target at Preplanned Position?

There is always the danger that while the detonation may have taken place, the target is no longer at the location. An example of this would be a

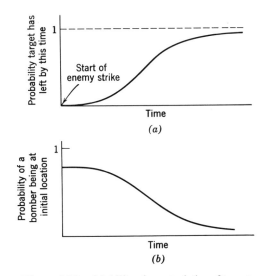

Figure 6-28. Mobility characteristics of target.

missile silo in which the missile had already been launched. This factor will be considered analytically by creating a probability distribution which indicates the mobility characteristics for each target class, as shown in Figure 6-28. Since certain targets are very mobile, an aggressor, in planning a large first strike, might attempt to fire many of his missiles initially to avoid the expected retaliatory strike.

There are two ways to include the factor of a mobile target into the analysis. The first way is to exercise the model by means of simulation (such as Monte Carlo simulation, described in Chapter 7) and keep track of the location of each specific target and warhead as a function of time. Thus, if the warhead lands at a position after the target was moved, it destroys an empty area. A second method is to create a transfer function which describes the probability of the target's still being at the initial location as a function of time, as might be described in Figure 6-28a, where the time lags are due to the launch rate achievable, missile availability, and the enemy withholding policy. The case of a bomber is more difficult to describe quantitatively. Figure 6-28b shows the probability that a bomber will be at its initial base. However, bombers may be orbited before the conflict starts, although they may later return to a base.

7

Model Exercise

Model exercise is the process of combining numerical input data with the transfer functions making up a model, in order to obtain numerical results. Before discussing the various ways that these numerical results can be obtained, let us first review what we have covered on the process of evaluation. Chapter 6 was devoted to describing an operations analysis process such as might be used to determine the various submodel transfer functions making up the operational flow model and the data needed to quantify the model. The output of this phase is the series of events making up the Red versus Blue and Blue versus Red models and the set of algorithms (i.e., equations or relationships) which can be used to estimate both the number of successes and the time required to perform each of the (random) events making up the entire process, as shown in Figure 7-1. (In the general case it is a series-parallel network). These estimates are based on the various performance characteristics assumed for the Blue force and the Red force as well as the rest of the environment involved, which are in turn based on whatever data

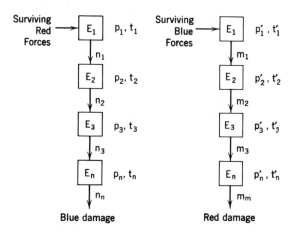

Figure 7-1. Operational flow models as a series of cascaded processes.

are available to the analyst, such as test data, extrapolation methods and the considered or intuitive judgment of experienced personnel. Based on these estimates, a most likely value of each input characteristic plus some range of uncertainty can be obtained. The discussion which follows deals with combining the data; the most likely value of the data will be used. Treatment of the uncertainties in the numerical values of the data will be discussed in Chapter 8.

Prior to his combining the data in any way, the analyst should plan to review the operational data he has gathered and the quantitative operational flow model he has constructed with the pertinent decision-makers and operational personnel. In this way, the data, scenarios, and force management policies to be used can be validated before further effort which is dependent upon these data is expended.

Methods of Exercising a Model

Combining the probabilistic functions such as those previously described may be simple to do in principle, but in practice may be laborious, complicated, time consuming, or even impossible to perform, particularly where dependencies among the functions are involved. Hence quite often some form of approximation must be made to satisfy the analytical resource limitations. Such approximations might include:

1. Choosing probability functions which only approximate the available data, but are simpler to work with.

2. Assuming independent events, where dependencies are apparent.

3. Employing an "expected value model" when dealing with random processes (i.e., assuming the average result will occur as in a deterministic fashion), rather than dealing with the wide range of possible results which could occur due to the randomness of the process.

4. Using a simulation method to combine the separate transfer functions.

Each of these methods provides some approximation to the solution. However, since the decision-maker should have some knowledge of the tradeoff between the analytical approximation made and the cost of the analysis, he should be familiar with the different approaches possible so that he can approve that method which provides the best compromise between degree of approximation and cost.

We shall first describe how a wargaming exercise would be performed for this problem, using Monte Carlo simulation, a form of simulation most appropriate to complex problems involving random processes and interactions, such as the case being considered. Then we shall describe how various analytical approaches can be used to determine the effectiveness of each of the given systems. In all cases, the same submodels and data discussed previously are used as the common starting point.

MONTE CARLO SIMULATION

To illustrate the over-all process of Monte Carlo simulation, Kahn and Mann consider how one could calculate the probability of success in playing a card game such as solitaire.* Since the probability of success may be defined as the ratio of the number of successful outcomes to the total number of possible trials, one method would attempt to calculate each of these two values which make up the ratio. For example, the total number of hands that are possible in solitaire is equal to the total number of possible random sets of 52 cards (i.e., $52 \times 51 \times 50 \ldots 1$, or 52 factorial). The number of possible winning hands is more complicated to determine, involving a very complex, time consuming process of calculation.

Game No.	Result	Total No. Games Won	$P = \dfrac{\text{Games Won}}{\text{Total Games Played}}$
1	Lost	0	0.0
2	Won	1	0.50
3	Lost	1	0.33
4	Won	2	0.50
5	Won	3	0.60
6	Lost	3	0.50
7	Lost	3	0.43
8	Won	4	0.50
9	Won	5	0.56
10	Lost	5	0.50
11	Won	6	0.55
12	Won	7	0.58
13	Lost	7	0.54
14	Won	8	0.57
15	Won	9	0.60
16	Lost	9	0.56
17	Won	10	0.59
18	Lost	10	0.56
19	Lost	10	0.53
20	Lost	10	0.50
21	Won	11	0.52
22	Won	12	0.55
23	Lost	12	0.52
24	Lost	12	0.50

Figure 7-2a. Tabulation of simulation results.

* H. Kahn and I. Mann (1957).

There is another way of determining the probability of success at the game of solitaire. This method requires that a number of games be played and that the results of each game be tallied as won or lost, as shown in Figure 7-2a (using data from actual trials). The percentage of games won is then plotted as a function of the number of games played to obtain a function such as that illustrated in Figure 7-2b. Note that for the first few games, the most likely estimate of the probability of success oscillates rather wildly, but after playing a certain number, it gradually stabilizes.

Such an example demonstrates the simulation process: (a) actually playing a game with all the rules involved; (b) keeping track of the number of wins and the number of losses (or the degree of winning) for each of the runs; (c) estimating the probability of success, based on the limited number of runs. In addition, as discussed in Chapter 8, by assuming what type of probability distribution is apparently involved (e.g., assuming a binomial distribution for the solitaire example), a statistician can provide a numerical estimate of the error involved in estimating the probability of success.

The Strategic War Simulation

Moving from card games to wargames, we shall now show how Monte Carlo simulation can be applied to strategic force operational flow models. Prior to the start of the game, each team separately lays out the different weapon systems of its constant cost force structure onto some geographical map. This deployment is based on each team's decision as to how it wishes to

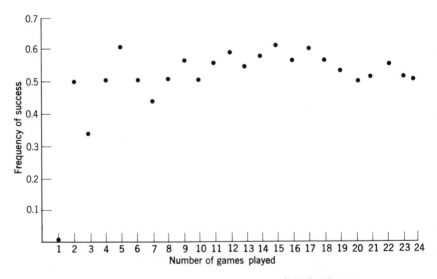

Figure 7-2b.   Estimating the probability of success by sample trials.

allocate its forces at various geographical base locations. Each team may also have available various alternative war plan options which the team feels it would like to use (which may be based on alternative responses of the other team). This is generally helpful in cutting down the time lag required for force management decisions, an important factor in actual operations. The Red team is given a force whose size and performance characteristics are based on the intelligence estimates of the Red System. For now, assume that the Red team forces entail the most likely value of these intelligence estimates.

The main elements which are to be simulated in the exercise have been illustrated in Figure 6-2. For each side, these are the forces, the command, control and communications system (consisting of information sensors, communications links, and a command and staff to control the forces), and industrial centers which, together with the opposing forces, make up the opposing targets. Note that the information sensors may provide information to the commander regarding status (e.g., damage) to his own targets or opposing targets.

To simulate these elements, three functional groups are created to operate the games: the Blue force commander, the Red force commander, and an umpire who will simulate all of the activities involved in Figure 6-3 (with the exception of the decision-making function of each force commander). Since each force commander is connected to his forces and the rest of the external environment through his sensor and communications systems, it will be the umpire's function to provide each isolated force commander (and his staff) with the same information inputs (in terms of accuracy and time lags) which the real (or proposed) sensor and communications system would provide him. In addition, the umpire's task will be to move each of the forces in accordance with the commands given and the numerical values of the system performance characteristics involved in the operational flow models developed in detail from Figures 5-9 and 5-10. He performs this role by first accumulating all information regarding the assumed initial conditions and the various transfer functions of the models, and then presents each side with the set of prewar conditions as delineated in the scenario to be used. As each team issues orders to its appropriate forces, it is the umpire who will receive each command and move the appropriate force elements called for by the command. In so doing, he must keep track of the time when the decision is made so that he can calculate the time taken to complete the event. He must also determine if each operational event was performed successfully or not, based on the appropriate transfer functions previously described. Thus, the umpire will keep track of all events such as the location of all forces at any time, the destruction of targets, etc. Each of the three groups is separated from one another although provision must be made for communication between the two teams if the game should include such means of communications (e.g., a hot line or diplomatic channel). This can again be achieved through the umpire. Obviously, the umpire will need staff assistance if the large amount of data involved is to be handled manually. Later we shall discuss the role which a computer can play in assisting the umpire. Let us now describe the entire gaming process rather than which functions will be performed manually or by a computer.

As indicated previously, the transfer functions to be used include the probability of success of performing each of the random events. This is where the term "Monte Carlo simulation" arises. In this form of simulation, the analyst wishes to follow (and tabulate) the success (or failure) of each weapon against each target. However, because of the nondeterministic characteristic of most (or all) of the events, the analyst needs a way of applying the probabilistic distribution associated with the likelihood of a successful performance of the event to each of the occasions when the event will be performed. As will be seen, the principle of the Monte Carlo roulette wheel can be used to solve this problem. In initially setting up the exercise, the purposes of the exercise should be recognized so that a decision can be made on whether or not to play the game in "real time." "Real time" means performing each event, such as missile travel to target, in the same time that it would take in a real operation. The main advantage of a real time exercise as opposed to an accelerated exercise is that it permits the actual, limited amount of time to be given to the force commanders so they may decide which forces to move as a function of the information they are receiving. The exercise may also be played in accelerated time which is done when all decisions have been programmed ahead of time and the exercise may be performed quite rapidly. Lastly, the exercise may be run more slowly than actual time. This is generally done if the exercise is to be run manually or might be done if research is to be done on the decision-making possibilities.

MONTE CARLO PROCEDURE

Strategic Alert Moves Before Strike

Assume that the scenario which has been chosen for analysis begins with a Red counterforce surprise attack against the Blue side. Hence, the first model to be used for this attack is the Red versus Blue operational flow model of Figure 5-10, using the initial forces in place as previously described. Note that since Blue does not strike first, all of the initial Red forces are available to enter the model at the point shown as "Surviving Red Forces."

The Red side, in beginning its operations, has the option of dispersing some of its forces, such as bombers or ships, in anticipation of a Blue counterforce counterstrike. However, in so doing, the Red side runs the risk of having its force movements observed by elements of the Blue side's strategic intelligence system, as indicated in the model. Such Blue strategic indicators could lead to a Blue reaction, bringing its forces to a higher level of strategic alert. Thus higher alert level might include dispersal of bombers, alerting personnel at station, checkout of weapon systems, etc. Thus, Red must take this possibility into account in whatever strategic moves it takes initially.

Thus, if any abnormal moves are made by any Red forces, the pertinent strategic intelligence information is transmitted to the Blue force commander in accordance with the appropriate information accuracies and time lags of the Blue intelligence system. Note that since Blue does not strike first, all of the initial Red forces enter the model at the point shown as "enemy surviving forces."

Red Launches Weapons

The Red force commander now decides which missiles are to be launched in the first strike against the Blue targets, using the launch weapons submodel in Figure 6-1. This force management decision is based on some previously generated attack plan. The Red commander now issues commands to launch specific weapons to fire at specific targets.

It is important to emphasize that in Monte Carlo simulation, each weapon unit must be identified by a particular number, since the procedure requires that the path of each weapon unit be tracked by the umpire in order to determine if it performed its assigned job or not. Thus, suppose a number of missiles, including missile 23, for example, have been ordered to get ready to fire at certain targets. Assuming this command was received at time zero, the umpire must now decide when missile 23 will be available for firing. He determines this by using the Red force's launch missiles transfer function showing the probability of a missile being ready for launching by a certain time, as shown in Figure 7-3.* Here the umpire has assigned a set of numbers from 0 to 99 to this cumulative distribution. Thus, the numbers of 0 through 94 represent the particular missile being ready to fire by time t_1. Numbers 95 through 99 indicate that the missile is not ready for launch, and represent the total time, including repair, it would take before the missile is ready. In other words, the 100 numbers are distributed in accordance with the cumulative probability distribution. Obviously, greater accuracies could be achieved by using 1000 numbers instead of 100.

* Recall that this transfer function for the Red missiles is based on some extrapolation of American missile firing capability in conjunction with intelligence estimates of the Red missile system characteristics.

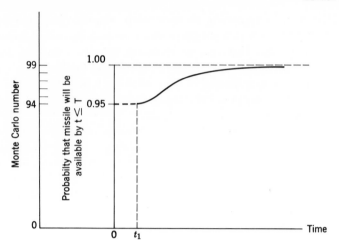

Figure 7-3. Determining time missile is ready for launch.

Thus, a roulette wheel having 100 numbers from 0 through 99 could be used to decide what specifically happened to missile 23 by merely spinning the wheel and comparing the resulting number with the cumulative distribution. This would be the function of the umpire.

In actuality, a roulette wheel is not used; either a table of random numbers such as that published by the RAND Corporation or a random number generator is used. The umpire then informs the Red team if missile 23 is ready at time t_1, or the time it is expected to be ready (as would be transmitted to the command center by the missile force). If the time lag is too excessive (it is possible that the random drawing resulted in a time of thirty minutes until it was ready for launch), the Red commander may decide to launch a different missile at this particular target and still have missile 23 available for later action. The sequence of times taken to complete each of these events is kept track of by the umpire.

Note that the same approach may be used for deployment of other weapons, such as bombers or submarines. The time of launch is recorded by the umpire so that he can predict the time when subsequent events will occur.

In similar fashion, when the Red commander orders missile launch, the umpire makes a random draw for each missile and by comparing this with the cumulative distribution for launch success (as shown in Figure 7-4 where $P_1 = 0.97$), he can determine the failure of each launch (drawing 0 through 2) or success (drawing 3 through 99).

Warheads Arriving at Defense Zone

Given that a weapon has been successfully launched, the umpire must next determine if it has reliably arrived at the defense zone. One way is to determine the reliability of the flight by determining the total flight time (when failure could occur) and take a Monte Carlo draw based on this reliability. A second way is to divide the total time into small intervals and make a draw for each of these time intervals (by using the appropriate higher reliability for the smaller time).

Tactical Early Warning by Blue Forces

The next event to be examined is the time when the Red forces enter the volumetric coverage pattern of Blue early warning sensors, and are detected. An example of such a sensor system is the "ballistic missile early warning system" (BMEWS). This system generates a signal which is transmitted to the Blue force commander when one or more missiles pass through a radar beam arranged in a particular volumetric configuration. By following the trajectory of each missile successfully launched, the umpire can determine both the time in which each missile intersects the radar beam as well as the radar range at which surveillance occurs. The umpire can then determine the value of the probability of detection at this range from the detection transfer function of Figure 6-17; then he can determine if detection actually did take place by performing a Monte Carlo draw. Note that the umpire may wish to perform a series of Monte Carlo draws, each corresponding to a range r_1, r_2,....r_n, since the sensor is observing a moving target. These ranges might be separated greatly in the case of a sensor whose beam or coverage pattern may be scanning and, hence, intersecting the target only intermittently. Every time that the sensor is observing the target, the umpire makes a Monte Carlo random number draw at that time to determine if detection has actually occurred. The umpire continues to

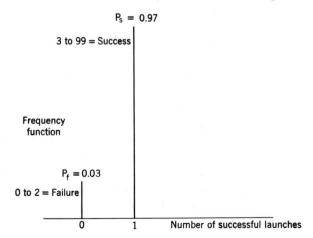

Figure 7-4. Determining if missile has launched successfully.

make these random draws until detection has occurred or until the missile is out of the detection coverage pattern.

This early warning information may then be used by the Blue team for alert and dispersal of American forces and/or population, if such a system has been included in the Blue force structure. Thus, the umpire transmits the early warning information to the Blue commander at the prescribed time lag Δt after detection has taken place.

Destruction by Blue Defenses

Given that particular Red warheads have been delivered to the vicinity of the Blue defenses, the umpire must now determine which warheads survive the Blue defenses and, hence, can be delivered to target. This is done by determining the probability of missile survival which is a function of the number of incoming targets simultaneously in the defense zone. The umpire then uses Monte Carlo draws to determine which warheads survived, as is done for a binomial distribution.

The Red forces may use decoys as penetration aids to help saturate the defenses, and the number of decoys accompanying the true targets in the defense zone must be included in determining the appropriate defense kill probability to be used by the umpire. If the defense forces have a target discrimination capability, they may be able to distinguish these penetration aids from the true warhead targets. This would be done with some probability of success and a Monte Carlo draw for each decoy would determine the number of decoys which fooled the defenses and caused a defense missile to be fired. If electronic countermeasures such as jammers or chaff were used, the reduced defense effectiveness transfer functions shown in Figure 6-21b would be used before making the Monte Carlo draw. In similar fashion, if any defense system element had been destroyed by a previous wave attack, the reduced defense system effectiveness would also have to be used by the umpire.

The last point to be included is the use of a second round of fire against a target which has survived the first round. In general the defense policy employed when firing a missile against an incoming target might be to fire the first missile and if the target survives, fire the second, and to continue this until the target has been destroyed or it leaves the lethal defense zone. Hence when the umpire informs the Blue commander that a target has survived, the commander may direct that a second round be fired. If so, a Monte Carlo draw would again be made to determine if the target was destroyed by the second round. This process could continue until the target leaves the defense zone.

Warhead Explosion and Target Damage

Given that a number of warheads have been delivered to their targets and have survived the Blue defenses, the umpire must next determine which Blue targets have been destroyed by these warheads. The umpire does this by first determining if the warhead has satisfactorily landed within the target radius of vulnerability by computing the probability of its doing so, based on the total warhead dispersion involved as described in Chapter 6, and then making a Monte Carlo draw based on the cumulative distribution. The umpire next makes a second draw, using the reliability distribution to see if the warhead detonated. Lastly, the umpire must determine if there still is a target at the impact point. This is why the umpire must keep track of the position of all targets (including weapons) as a function of time. For, if the Red weapon fires at a Blue weapon which has already been launched, the only destruction will be to real estate, not to a target. If an airfield is hit by a warhead, the umpire will destroy all aircraft still at the airfield and located within the vulnerable radius.

Bonus Kills

Occasionally targets may be so closely located that their areas of vulnerability overlap, as was shown in Figure 6-27. Thus, each target will be destroyed if the warhead lands in Area II, the area of overlap. Such a "bonus kill" can be included in the systems analysis by modifying the preceding approach. Instead of using a binomial probability distribution for determining if the warhead landed within the target radius of vulnerability, Monte Carlo draws are taken to determine the actual ground zero where the target landed, with respect to the designated ground zero. This may be done by using the cumulative probability distributions of the polar coordinate density functions of Figures 6-23i and j, where a Rayleigh distribution is assumed for the total uncertainty in range, and a uniform distribution is assumed for the uncertainty in angle (0 to 360°). Thus, a Monte Carlo draw from each distribution would provide the exact AGZ. By seeing if the value of R is less than or equal to R_v, the umpire can determine if target B was destroyed. By seeing if the AGZ is also within the area of vulnerability of target C the umpire can determine if target C was also destroyed.

Blue Weapons Employment

Based on the Red first strike scenario assumed, Blue forces activities would probably proceed as follows, using the transfer functions indicated. Strategic intelligence information and tactical early warning information are provided the Blue force commander by the umpire in accordance with the Blue sensor transfer functions and at the appropriate time it would be available. The Blue commander would place certain Blue forces on alert and disperse others such as airborne forces. These would be moved by the umpire in accordance with the appropriate times required. Following the first wave of Red warheads and subsequent Blue target destruction, damage assessment information would be provided the Blue commander (by the umpire simulating whatever sensors are available in the Blue system). An appropriate Blue weapon launch would then be commanded in accordance with an appropriate war plan and these weapons would destroy targets in accordance with the appropriate Blue versus Red system transfer functions. These would be manipulated by the umpire's taking a series of Monte Carlo draws.

The exercise would then continue allowing each commander to move his forces (through the umpire) at whatever time he chooses. Again, each commander would be provided information regarding his own forces and enemy forces (by the umpire), using the transfer functions (accuracy, time lags) inherent in each of the possible information inputs. The game would be continued until some time when both sides would agree to halt operations.

Analyzing Results

Following completion of the game, the outcome is compiled by the umpire in accordance with each of the two measures of the problem (i.e., destruction of Red and Blue population and industrial capacity). In addition, it may be useful to keep track of these outcomes as a function of time for more detailed analysis. Other information might also be tabulated, such as the number of Red and Blue weapons available in inventory as a function of time, or any other measure useful for later analysis.

The reader will, no doubt, realize that the results from one such exercise may not be very significant in predicting what might be expected in an actual nuclear war. One reason is that pure randomness, as represented by the many Monte Carlo draws required, will influence the results. However, while the results of one wargame exercise may not be an accurate indication of the actual outcome, the composite pattern of results of many runs of the same game may have great significance, as in the solitaire card game example. Thus, in Monte Carlo simulation, the same wargame, using the same scenario, must be played a number of times, using a new series of Monte Carlo draws. The final numerical results of these runs can then be compiled in the form of a sample cumulative distribution of Red (or Blue) destruction as shown in Figure 7-5.

Three points are significant in this collection of sample data: the minimum and maximum values obtained, as well as the average value. It should be realized, however, that there is some chance that a lower minimum and a higher maximum could be reached if more runs were made. This yields the

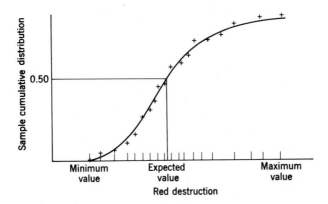

Figure 7-5. Sample results of war game.

conclusion that some appropriate probability function, based primarily on judgment and a general understanding of the real phenomena under study, might be fitted to the sample data points obtained, as shown in Figure 7-5.* This function provides a means for obtaining the probability that at least a given amount of Red destruction (e.g., one-third of the total population destroyed, the measure of assured destruction) would occur. Thus the probability function of Figure 7-5 could be used as the basis of the measure of effectiveness for the force structure which was exercised. A good example of this probabilistic reporting of information is the way that the weather bureau currently indicates the probability of snow. Thus, if they report that there is a 40 % probability of snow, you (the decision-maker) may use this information to determine whether or not you want to wear boots. The systems analyst functions in a similar manner in that he provides the decision-maker with structured information which shows the expectation that a particular job or mission will be successfully completed. Thus, maximum information is provided if the amount of the job is kept as a variable as shown in Figure 7-5.

USEFULNESS OF COMPUTERS IN MONTE CARLO SIMULATION

While it is possible to perform a wargaming exercise in a manual fashion, it may take considerable time, not only of the participant, but particularly of the umpire, who must find a way of rapidly retrieving and using the various transfer functions in performing the necessary calculations and random Monte Carlo draws. Since many runs of the same game must be played over

---

* A process meeting these conditions is called a "Bernoulli process with known parameters". See R. Schlaifer, (1959), Chapter 10.

and over to obtain statistically significant results,* this brings to mind the possibility of using a computer for performing the more routine umpire functions (i.e., storage and retrieval of transfer function calculation, and Monte Carlo draws). To do this, the analyst must make explicit all of the transfer functions of the operational flow model and the systems performance characteristics, including their time dependencies, and program a computer to store these. Under this arrangement, the Red and Blue commanders would communicate their commands either directly to a computer † or through a computer operator.

One of the main reasons for playing the game more slowly than real time when a computer is used is to do research on human functions such as decision-making. For example, an analyst may wish to uncover and experiment with different force management decision rules and policies and uncover the effect of these rules on the outcome of the game. As part of this type of research, the game could be played and replayed from a given activity point in the game, each time using a different decision-making policy and seeing how the final scores vary. In this case, the computer could record the game condition at this "junction point" so that the same set of conditions could be reproduced at the start of each replay.

If the logic of the decision-making functions can also be made explicit, this can also be programmed into the computer, and hence the entire exercise can run much faster than real time without human interaction. In the more complex situations, however, particularly those which involve a high degree of complex decision-making which depends on human evaluation of data and where the logic is not completely explicit or consistent, it may not be possible to do this without assuming some arbitary decision rules. However, it is a good rule of thumb in wargaming to identify the routine parts of the problem and program these separately onto a computer. In so doing, the computer can make many more runs in the time available, improving the statistical confidence in the results as well as reducing costs.

In the case of those wargames which do have a number of decision-making steps requiring human actions as part of the system operation, it is still possible to program the explicit portions of the game onto a computer and play the game in "modified real time." Under such an arrangement, the computer would go through the explicit steps more rapidly than real time, keeping track of proper time sequences until the time for human decision-

---

* The relationship between the accuracy of these results and the number of Monte Carlo runs made is discussed in Chapter 8.
† The development of time-shared computers and remote consoles which permit easy communication in English command-oriented language makes the possibility both feasible and economical in terms of minimizing the actual computer time used, since the greatest part of the time used is in decision-making, not computer calculations.

making. At this point the computer would stop and wait for the human interaction. When this step is completed, the computer would then continue with the game until the next decision point when it must stop and wait for the next human interaction. While such an approach offers the advantages of reducing the time requirements for a given run, it may produce errors in the game if the players have not been given proper time to make the necessary observations of data before making their decisions.

Last, the wargame may be played in "real time" which is the most accurate way of obtaining system evaluation results, particularly in testing the abilities of the human decision-makers involved.

In summary, the Monte Carlo simulation offers an accurate way of evaluating the effectiveness of complex systems whose activities involve many random processes with probability distributions which cannot readily be combined analytically. Note that when Monte Carlo simulation is used, the probability distribution may be any function desired and need not be simplified to the common, more easily manipulatable functions such as the normal distribution. Further, it permits the inclusion of performance characteristics whose numerical values are dependent on some other factor, as, for example, the probability of kill of a defense system whose value varies as a function of the number of incoming targets. It also can take into account various interactions of moves where these interactions may cause a change in the value of the performance characteristic. For example, one force management policy might be to fire the first salvo of missiles at the air defense components or the anti-ballistic missile (ABM) radars, for example, thus causing a reduction in the effectiveness of these defense systems to subsequent moves if these components were destroyed. Simulation offers a relatively simple way of including this type of activity which involves interdependent probabilities.

It should be stressed that the role of the computer is merely to keep track of data, combine the mathematical transfer functions and logic of the situation and determine the quantitative results for each of the runs, since computers make no decisions which have not been programmed into them by a human being. The run results are then structured statistically and examined by the decision-maker.

MODEL EXERCISE: ANALYTICAL METHODS

While Monte Carlo simulation may be an accurate way of combining the gathered data in the operational flow model, it does, in general, require the use of a computer for making the sufficiently large number of runs needed for statistically significant results. If the systems analyst does not have sufficient time and funds to perform such a simulation, he must find some way of combining the separate transfer functions to obtain the probabilistic results. We shall now discuss various approaches for combining probabilistic transfer functions involved in competitive models of this type, indicating different degrees of problem simplification which can be made, the degree of accuracy achievable, and the degree of analytical effort required.

Simplified Approach

We shall start our review of the analytical techniques available with the simplest method of combining data, based on making a number of assumptions as follows:

1. Examine only one pure weapon system at a time (such as each of the missile alternatives under study) and avoid examination of a mixed offensive force (including bombers and submarine launched ballistic missile systems and any intraforce interactions such as the effect of launching missiles first on the opposing air defense elements, then launching the bombers). The specific output of such an approach might be stated to the decision-maker as follows: "If you were to spend a fixed sum of money on either of these three missile systems, here are the results you could expect from each." This approach, however, still requires that the higher level (OSD) analyst combine the preferred missile system with the other force elements (as well as any other proposed changes), taking into account their interrelationships. This basically moves the wargame exercise to the higher level headquarters, and has merit since one organization is doing this rather than several.

2. Assume a homogeneous target set (i.e., all targets being of equal importance).

3. Avoid time dependencies (i.e., assume a mass wave attack during a small time period).

4. Concentrate on trying to evaluate *relative differences* among systems rather than trying to find exactly how well each system performs.

5. Assume that each of the random processes involved in the total operation is independent of the others, and that each consists of a series of independent trials, with a given numerical value of probability of success for each trial. Hence a binomial distribution could be used for each of the processes. This assumption removes the difficulty of computation associated with a dependent process such as the defense system whose kill probability, p_k, varies as a function of number of warheads simultaneously in the defended area. In this case some value of p_k will be chosen, based on the expected number of warheads in the defended area, for example.

A simplified scenario for this problem, illustrated in Figure 7-6, would be as follows:

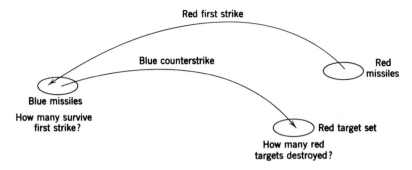

Figure 7-6. Simplified scenario.

1. The Red forces fire a first strike missile attack against each of the three Blue missile systems under evaluation.
2. Certain of the Blue missiles survive the Red attack.
3. The surviving Blue missiles are fired at the Red targets.

The evaluation problem is then to determine the number of targets destroyed in this process for each of the three cases. While this simplified scenario may differ from what the real operation would be like, the answers obtained would provide some relative measure of effectiveness for the three systems being evaluated.

Gathering Data

The probabilities to be used in this simplified example are shown in Figure 7-7. The first third of the table structures the Red versus Blue attack, using the transfer functions of the Red versus Blue operational flow model of Figure 5-10. Note that the analyst can make an estimate of the relative performance (conditional probability) for each of the pertinent activities. For example, it was assumed that for the time period of the mass attack, 95 % of all Red missiles would be available and would be launched successfully (i.e., p = 0.95). It was further assumed that for the number of Red warheads which are expected to be in the Blue ABM defense zone simultaneously, the probability of survival of each warhead will be 0.6. Next, consider the probability of landing inside the vulnerable target area for each of the three systems. If it were assumed that the Red missile has a probability of success of 0.7 against the hardened silo, the analyst could calculate the probability of success of the Red missile against the superhardened silo, based on its decreased vulnerability. (It was assumed the reduced probability of weapon kill is 0.3). In a similar fashion, it was estimated that the probability of success for the mobile ICBM would be 0.1 because of the uncertainties of keeping track of these moving missiles. In similar fashion, it was assumed that

	Hardened Silo Characteristics	Superhardened Silo Characteristics	Mobile ICBM Systems
I. RED ATTACK:			
Launch Weapons	0.95	0.95	0.95
Dependable Travel (D)	0.90	0.90	0.90
Survive Blue ABM	0.60	0.60	0.60
Land in Vulnerable Area	0.70	0.30	0.10
Target at Location	1.00	1.00	1.00 (included in above)
Warhead Detonation	1.00	1.00	1.00
Probability Blue Target Destroyed	0.36	0.15	0.05
Probability Blue Target Survived	0.64	0.85	0.95
II. BLUE ATTACK:			
Launch Weapons	0.95	0.95	0.80
Dependable Travel (D)	0.90	0.90	0.70
Survive Red ABM	0.60	0.60	0.60
Land in Vulnerable Area	0.70	0.70	0.50
Target at Location	1.00	1.00	1.00
Probability Red Target destroyed	0.36	0.36	0.17
III. CALCULATIONS:			
Probability Blue Target Survived and Red Target Destroyed	0.23	0.31	0.16
Number of Blue Missiles (for fixed cost)	1200	1100	1000
RESULTS:			
Total Number Red Targets Destroyed Probability Distribution	Binomial:1200 targets, each with a probability of 0.23	Binomial:1100 targets, each with a probability of 0.31	Binomial:1000 targets, each with a probability of 0.16
Expected Value	276 targets	341 targets	160 targets
Relative Effectiveness	81%	100%	47%

Figure 7-7. Numerical data for strategic planning cost-effectiveness analysis. (Activity probabilities and number of missiles)

each of the fixed ICBM targets would be at their fixed locations (p = 1.0) for a surprise attack. This same probability was used for the mobile system since it was assumed that the reduced uncertainty of target location was included in the probability, p = 0.1, assumed for landing in the target's vulnerable radius.

The second third of Figure 7-7 contains the probabilities associated with the Blue versus Red operational flow model of Figure 5-9. Again, an estimate of the relative performance characteristics among each of the systems is determined as shown. Now, however, it is assumed that the surviving Blue missiles will fire at fixed (countervalue) targets (i.e., the probability of the target being at its location is unity).

The next step in the analysis is to determine the number of missiles which can be procured for an equal cost configuration of each of the three systems under study. This will permit the decision-maker to select on the basis of highest effectiveness. In this example it was assumed that for the mobile ICBM system, 1000 missiles would be removed from the silos and placed on mobile carrier-launchers. This results in some incremental costs being required for required missile modifications trucks, launchers, additional personnel, etc., to obtain the mobile system. Total system life costs are to be considered in this analysis, including any costs for research, development, test, and engineering (RDT & E), procurement of quantity units, and operation and maintenance (O & M) for some years of operation (generally five years).*

Using this same incremental cost figure, the analyst can determine how many superhardened missile silos can be constructed. Assume that this calculation indicates that not only can all 1000 superhardened silos be constructed but some additional funds still remain. The system planner could use the remaining funds to buy more ICBM's, with their superhardened silos. Let us assume that the calculations indicate that a total of 1100 superhardened missiles and silos can be obtained with the same incremental funds that would be expended for the mobile system. Again, using the same incremental cost figure, the analyst could determine how many additional missiles and standard hardened silos could be purchased for the same funds. Assume that this results in an additional 200 missiles for the current system. Thus, a total of 1200, 1100, and 1000 missiles would be available for a constant cost configuration of the three systems, as shown in Figure 7-7.

The analyst must now determine the number of Red missiles that would be fired against the Blue missiles. There are two ways to approach this problem of enemy uncertainty. The simplest way is to assume that the Red force

---

* A detailed discussion of these costs is contained in Chapter 9.

will have the same number of missiles as the Blue force. This assumption may be reasonable since the opposing sides may tend to base their armament on the number of equivalent type weapon units which the other side has. An alternative assumption would be that the Red side would spend an amount of funds equivalent to that which the Blue force is spending for each of the three systems. Thus, the assumption could be made that 1200 Red missiles of the type described in Table 7-7 were available. This presents a more difficult but possible calculation since this would involve allocating more than one Red missile to 100 or 200 of the Blue missile targets, depending on which Blue system is being evaluated. For simplicity of calculation, we shall assume that the Red force fires one missile at each Blue missile.

Combining Probability Distributions

By assuming independence of events, the probability of success of the entire process can be calculated as the product of the separate conditional probabilities. Thus the probability distribution of a Blue target surviving and destroying a Red target may be calculated as a binomial distribution as in Part III of Figure 7-7. Thus the distribution of the total number of Red targets destroyed would be as indicated at the bottom of Figure 7-7. For this case of the hardened silo system, the number of Red targets destroyed varies from 0 to 1200, and the distribution is binomial with a probability of success of 0.23. Thus F(n), the probability of n targets being destroyed is

$$F_n = C_n^{1200} \quad (p)^n (q)^{1200-n}$$

$$= \frac{1200!}{n!(1200-n)!} \quad (.23)^n (.77)^{1200-n}$$

Since for large n the binomial distribution tends to approach a normal distribution, such an approximation can be made to reduce the calculations involved.*

Again, as in the Monte Carlo simulation form of model exercise, two types of numerical values may be extracted from the probability distribution obtained. The first is the expected value, which in the case of the three missile systems being compared, is the expected number of targets destroyed. This is merely the product of the number of missiles and the total probability of kill. The main advantage of using the expected value approach is its simplicity of calculation, which is generally good enough for a cost-effectiveness analysis that is primarily intended to focus on the relative differences among alternatives. Also note that merely constructing the comparative structure of Figure 7-7, showing the individual conditional

---

* See Mosteller, et al. (1961).

probabilities and the simple multiplications in arriving at an expected value solution, results in the following information:

1. A relative ranking of the three alternatives based on the expected value measure of results.

2. The particular parts of the system which cause a specific system to be deficient. For example, while the main advantage of the mobile ICBM system under study was its greater survivability against attack, its main weakness was its greater expected equipment malfunction which produced lower availability and reliability. In addition, its lower accuracy guidance system resulted in a smaller percentage of missiles landing inside the vulnerable target area. The system was more expensive and hence the total number of available missiles (1000) was lower than the other alternatives. By noting this relative performance of the various systems, the systems planner may be able to construct other system possibilities which overcome these deficiencies. This, of course, is the role of systems planning, discussed further in Chapter 8.

The main deficiency of the expected value calculation is that it gives only an *average* result of what would occur over many trials and not a range of values. This is not much of a problem if there are a large number of trials (e.g., weapons) involved. But even were this true, "good luck" or "bad luck" may play some role and the extent should be quantified. The expected value approach would certainly not accurately predict the result of a small number of trials such as in predicting the number of successes of ten Apollo space vehicle launches. Hence the second numerical result obtainable from the probability distribution is the probability of at least D targets being destroyed as was described for the Monte Carlo simulation results. This probability of performing a given job (such as destroying X per cent of the targets) could be used as one of the system selection criteria (instead of the expected value of system effectiveness).

IMPROVING THE SIMPLIFIED APPROACH

As indicated in the introduction to the analytical approach to model exercise, various simplifications to the transfer functions (probabilistic distributions) can be made to simplify the task of model exercise. Each of these assumptions is now examined to show how the model would be exercised if the restrictive assumption were removed and the model made more nearly similar to the actual operation.

Salvo Attacks

One assumption made in the previous analysis is that there is a large number of Red targets to be fired upon and that the number of targets is the same as the number of Blue missiles (i.e., 1000, 1100 or 1200 targets, depending on the Blue missile system being evaluated). This is a particularly poor assumption when the analyst notes the low total kill probabilities involved for each system (i.e., 0.16, 0.31 or 0.23). A more realistic assumption which could be made is that the number of Red targets is much less than the number of Blue missiles procured, so that several Blue missiles are initially available to be fired at each target in order to increase the kill probabilities involved. For example, assume that there are 400 Red targets involved, each having an equal worth. Thus the problem is to take the data originally structured in Figure 7-7, allocate the missiles in each case to the 400 Red targets, and calculate the expected number of Red targets destroyed. As shown in Figure 7-8, the 1200 Blue missiles would be allocated three missiles to each of the 400 Red targets. In this fashion, the total probability of killing the target is then increased because a salvo of three missiles is fired. We shall now calculate the increased salvo probability of kill. For this, consider the hardened missile system in which three missiles, each having a kill probability of 0.23, are targeted at each of the 400 different targets. The missiles may be fired simultaneously or one at a time, but all three are to be fired. However, we shall not permit the force commander to know if the first or second missile destroyed the target (thus saving one or two missiles by firing at an already destroyed target). Now we calculate the probability of the target being destroyed by one or more missiles. There are three ways of calculating this probability.

The first method makes direct use of the frequency function. That is, the total probability of success is the sum of the individual probabilities of success for each of these three possibilities (i.e., 1, 2, or 3 successful missiles);

$$P_k = p_1 + p_2 + p_3 = .41 + .12 + .01 = .54.$$

A second way of determining the probability of success of the salvo is to find the complement of the probability of total failure (i.e., no missile destroying the target).

$$P_k = 1 - p_0 = 1 - (.77)^3 = .54.$$

A third way is the incremental approach which was discussed in Chapter 3 and is used when dealing with incremental allocations later. We shall calculate the total probability of success P_i and probability of failure Q_i as a function of the ith shot fired. Firing the first shot:

$$P_1 = p_k = 0.23; Q_1 = 1 - P_1 = 0.77.$$

1200 missiles, $p_k = 0.23$
Assign 3 missiles
to 400 targets
$P_3 = 1 - 4(0.77)^3 = 0.54$

1100 missiles, $p_k = 0.31$
Assign 3 missiles
to 300 targets
$P_3 = 1 - (0.69)^3 = 0.67$
Assign 2 missiles
to 100 targets
$P_2 = 1 - (0.69)^2 = 0.52$

1000 missiles, $p_k = 0.16$
Assign 3 missiles
to 200 targets
$P_3 = 1 - (0.84)^3 = 0.41$
Assign 2 missiles
to 200 targets
$P_2 = 1 - (0.84)^2 = 0.29$

Expected
Number of
Red Targets
Destroyed

$\overline{T.D.} = 400(0.54) = 216$

$\overline{T.D.} = 300(0.67) + 100(0.52)$
$= 201 + 52$
$= 253$

$\overline{T.D.} = 200(.41) + 200(0.29)$
$= 82 + 58$
$= 140$

Relative
Effectiveness

82%

100%

55%

Figure 7-8. Expected value calculations for a salvo attack.

When the second shot is fired, the incremental gain is that amount of failure which has been converted to success.

$$\Delta P_2 = Q_1 p_k = (0.77)(0.23) = 0.177.$$

Thus $\quad P_2 = P_1 + \Delta P_2 = 0.23 + 0.177 = 0.407,$

$$Q_2 = 1 - 0.407 = 0.593.$$

Similarly, on the third shot, the incremental gain is

Thus $\quad \begin{aligned}\Delta P_3 &= Q_2 p_k = (0.593)(0.23) = 0.137.\\ P_3 &= P_2 + \Delta P_3 = 0.407 + 0.137 = 0.54.\end{aligned}$

Note that while all three solutions give the same answer, in general, the second approach is the easiest to calculate; but the third approach is the one most used for systematic allocations of resources.

We shall now apply the above technique described by allocating the total missiles available for each of the missile systems to the 400 targets. The hardened missile silo system has 1200 missiles available for the 400 targets, resulting in an assignment of three missiles to each target, with a three-missile salvo probability of kill of 0.54.

For the superhardened missile silo system, the 1100 missiles would be allocated as follows: three missiles to each of 300 Red targets, and two missiles to each of the last 100 targets. The kill probabilities obtained for each of these salvos are increased from a single shot kill probability of 0.31 to a salvo kill probability of 0.67 and 0.52 respectively, as shown in Figure 7-8.

For the mobile missile system of 1000 missiles, three missiles would be allocated to each of 200 of the Red targets and two missiles to each of 200 Red targets. Here the kill probabilities for each of these salvos would be increased from a single shot kill probability of 0.16 to a salvo kill probability of 0.41 and 0.29 respectively.

The final calculation of the expected number of targets destroyed is the product of the total number of targets and the probability of kill of its salvo. Notice that the relative ratios of the effectiveness among systems has changed somewhat between Figures 7-7 and 7-8 due to the nonlinear way that the kill probabilities change when a salvo of missiles are used. The spread would have been even less if the single missile kill probabilities had been greater.

It should be emphasized that the previous calculation is only an approximation to the expected value. The same problem will now be analyzed by obtaining the frequency function of the number of targets destroyed, as a further example of the application of probability theory. In this case, the analyst might wish to determine the probability of destroying 100 or more of

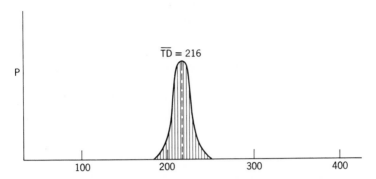

Figure 7-9. Probability density distribution for hardened missile system.

the 400 Red targets, as an example. By assuming a binomial probability distribution for each of the event probabilities, it is a straightforward matter to calculate the frequency function for the hardened silos (using the normal approximation) as shown in Figure 7-9. The result is a discrete function contained by the envelope shown. By using this frequency function, the analyst could then determine the probability of destroying 100 or more targets by a summing process.

This procedure is a bit more difficult to do for the other two systems since they involve combining two separate frequency functions, one having a kill probability from using a three-missile salvo per target and one from a two-missile salvo per target. For example, in the superhardened missile system the analyst would obtain the two frequency functions as shown in Figure 7-10. Thus, to obtain the total probability density distribution function, the

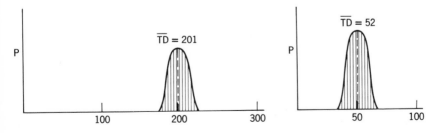

Figure 7-10. Probability density distribution for superhardened missile system.

analyst would have to use these two functions and sum the probabilities for all possible ways of obtaining a given number of targets destroyed. The way this is done may be seen by examining the following related but simpler problem.

Consider two dice, one of which is a six-sided die having number 0, 1, 2, 3, 4, or 5 on each of the six sides. The second die has four sides having 0, 1, 2 or 3 on each side. To calculate the probability of obtaining the sum of five or more when tossing the two dice, we shall construct the complete frequency function as obtained from structuring a logic table, such as the one shown in Figure 7-11, which considers all possibilities. Since, in this simple case, one can assume a multinomial distribution for each die (a probability of $\frac{1}{6}$ for each number on the six-sided die and $\frac{1}{4}$ for each number on the four-sided die) the total probability for any sum is $\frac{1}{24}$th the number of possibilities which exist for that sum. Thus, the frequency function may be obtained for any given sum, as shown in Figure 7-12. From this it can be seen that the probability that the sum will be five or more is 10/24.

Note that the same approach can be used for the previous missile case. The analyst could construct the logic table for each possibility of target destroyed, and determine the probability of each event as the product of the

		Die 2 Possibilities			
	Die 2				
	Die 1	0	1	2	3
	0	0	1	2	3
	1	1	2	3	4
Die 1	2	2	3	4	5
Possibilities	3	3	4	5	6
	4	4	5	6	7
	5	5	6	7	8

Sum	Probability
0	1/24
1	2/24
2	3/24
3	4/24
4	4/24
5	4/24
6	3/24
7	2/24
8	1/24
Σ	24/24

Figure 7-11. Logic table.

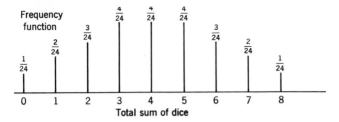

Figure 7-12.   Probabilistic results of dice throws.

separate probabilities involved. Summing each of the total probabilities for each value of target destroyed would yield the frequency function and then the cumulative distribution.

It should also be mentioned that calculating the expected value of the frequency function is the only correct way of obtaining this value (i.e., the expected method previously described is only an approximate method when interdependencies are involved). As was mentioned, however, the former method is easier and generally accurate enough to be a relative measure among competing systems.

Nonhomogeneous Targets

We shall now re-examine the simplifying assumption that there is a homogeneous target set (i.e., all targets are of equal worth). In the actual operations this is not true, and because some targets are of greater worth than others, more missiles should be assigned to those targets of higher worth. This particular situation is covered by the following problem:

Problem As Given

You are a systems analyst and have been asked to develop a method for properly assigning available missiles to targets. There is a group of 500 homogeneous missiles to be assigned to a total of 170 targets of interest. Sixty of the targets are "soft"; hence for each missile $p_k = 0.9$. One hundred and ten targets are hardened; for each missile $p_k = 0.4$. The worth of each target is indicated in Figure 7-13.

Target Worth	No. of Soft Targets $(p_k = 0.9)$	No. of Hard Targets $(p_k = 0.4)$
100	10	20
60	30	40
40	20	50
	60 total	110 total

Figure 7-13.   Target set.

Two proposals have been made for allocating the 500 missiles to the 170 targets. The first proposal would allocate missiles to targets solely on the basis of the relative worth of targets. This proposal would allocate five missiles to each 100 point target, three missiles to each 60 point target, and two missiles to each 40 point target.

The second proposal would also take into account the kill probabilities of the targets and would allocate missiles to each of the six target classes in proportion to the following figure of merit:

$$FM = \frac{WN}{P_k},$$

where $FM =$ figure of merit,
        $W =$ target worth,
        $P_k =$ missile kill probability for this target class,
        $N =$ number of targets in this class.

1. Determine the missile to target allocations for each of the two suggested allocation procedures.

2. Which is the better allocation procedure?

3. Is this the optimal manner of assigning the 500 missiles to the target set given above?

4. Make explicit the optimal procedure you would use to assign any given number of missiles (up to 600) to the target set.

5. Using this optimal procedure plot the effectiveness you would accrue as a function of the total number of missiles available (0 to 600 missiles).

Solution to the Problem

This problem represents an example of a classic resource allocation problem. Given a fixed set of resources (i.e., 500 missiles) what is the best way to allocate these resources among different jobs in order to maximize the objective? In this case, the effectiveness measure of the objective to be maximized could be defined as the *worth* of the expected number of targets destroyed, since the set of targets have different worths. While there is a desire to assign missiles to the higher worth targets, it soon becomes apparent that as the number of missiles assigned to a given target increases, a state of diminishing returns is reached so that it becomes inefficient to continue to assign a large number of missiles to even the higher worth targets. We shall now examine the various proposed ways of allocating the 500 missiles and compare the results obtained in terms of the effectiveness measure (i.e., expected worth of targets destroyed which is the same as the worth of expected targets destroyed).

Using the salvo kill probabilities for different numbers of missiles in the

A. Soft Targets ($P_k = 0.9$)

P_k (5 missiles) $= 1 - (.1)^5 = 1 - .00001 = .99999$
P_k (4 missiles) $= 1 - (.1)^4 = 1 - .0001 = .9999$
P_k (3 missiles) $= 1 - (.1)^3 = 1 - .001 = .999$
P_k (2 missiles) $= 1 - (.1)^2 = 1 - .01 = .99$

B. Hard Targets ($P_k = 0.4$)

P_k (7 missiles) $= 1 - (.6)^7 = 1 - .028 = .971$
P_k (6 missiles) $= 1 - (.6)^6 = 1 - .047 = .952$
P_k (5 missiles) $= 1 - (.6)^5 = 1 - .0778 = .9222$
P_k (4 missiles) $= 1 - (.6)^4 = 1 - .13 = .870$
P_k (3 missiles) $= 1 - (.6)^3 = 1 - .216 = .784$
P_k (2 missiles) $= 1 - (.6)^2 = 1 - .36 = .64$

Figure 7-14. Salvo kill probabilities.

salvo as shown in Figure 7-14, the missile allocation and results obtained from the first proposed method can be found as shown in Figure 7-15. The method of allocating missiles and the results obtained from the second proposed method are shown in Figures 7-16 and 7-17. Note that the second method yields a higher effectiveness than the first method. However, is there another method which will yield an even higher effectiveness?

To answer this question, the analyst can employ the concept of maximizing the marginal effectiveness, which is extremely useful for all problems of resource allocation. This concept is actually a very simple rule. It states that resources should always be allocated so that if they had been added sequentially, each addition would provide the maximum increase in effectiveness (or return, or benefits) per unit resource used. In this problem the marginal effectiveness is measured by the increase in worth of the expected targets destroyed. The unit of resource is the missile. The application of the concept of maximizing the marginal effectiveness when assigning resources is described in the following.

Consider the structure of Figure 7-18 which indicates the incremental effectiveness obtained by the assignment of the nth missile to each target in

	Target Worth	Missiles Assigned	Number of Targets	P_k	Expected Worth
SOFT	100	5	10	0.99999	1000
	60	3	30	0.999	1800
	40	2	20	0.99	792
HARD	100	5	20	0.922	1844
	60	3	40	0.784	1880
	40	2	50	0.64	1280
				TOTAL EXPECTED WORTH:	8596

Figure 7-15. Allocations and results—Procedure 1.

$$FM_1 = \frac{WN}{P_k} = \frac{(100)(10)}{.9} = 1111$$

$$FM_2 \qquad = \frac{(60)(30)}{.9} = 2000$$

$$FM_3 \qquad = \frac{(40)(20)}{.9} = 890$$

$$FM_4 \qquad = \frac{(100)(20)}{.4} = 5000$$

$$\frac{(60)(40)}{.4} = 6000$$

$$\frac{(40)(50)}{.4} = 5000$$

TOTAL FM $= 20001$

$$M_1 = \frac{1111}{20000} \times 500 = 27.8 \text{ or } \frac{27.8}{10} = 2.78 \text{ missiles/target}$$

$$M_2 = \frac{2000}{20000} \times 500 = 50 \text{ or } \frac{50}{30} = 1.67 \text{ missiles/target}$$

$$M_3 = \frac{890}{20000} \times 500 = 22.2 \text{ or } \frac{22.2}{20} = 1.11 \text{ missiles/target}$$

$$M_4 = \frac{5000}{20000} \times 500 = 125 \text{ or } \frac{125}{20} = 6.25 \text{ missiles/target}$$

$$M_5 = \frac{6000}{20000} \times 500 = 150 \text{ or } \frac{150}{40} = 3.75 \text{ missiles/target}$$

$$M_6 = \frac{5000}{20000} \quad 500 = 125 \text{ or } \frac{125}{50} = 2.5 \text{ missiles/target}$$

$$\overline{500} \text{ Missiles Total}$$

Figure 7-16.   Calculations of figures of merit and allocations—Procedure 2.

each of the six target classes. Such a compilation will be the basis for applying the key decision rule to be followed in allocating missiles to these targets, which is, "Always assign the next missile to that target which will yield
the highest marginal effectiveness of all of the assignment choices available." Thus while there are six possible choices involved in the first allocation decision (i.e., assign the first missile to any of the six target classes), the

Target Worth	Missiles Assigned (per target)	Number of Targets	Total Missiles Assigned	P_k	Expected Worth
SOFT 100	2	2	4	0.99	198
100	3	8	24	0.999	799
60	1	10	10	0.9	540
60	2	20	40	0.99	1188
40	1	18	18	0.9	648
40	2	2	4	0.99	79
HARD 100	6	15	90	0.952	1430
100	7	5	35	0.971	486
60	3	10	30	0.784	470
60	4	30	120	0.870	1567
40	2	25	50	0.64	640
40	3	25	75	0.784	784
		TOTAL MISSILES ASSIGNED:	500	TOTAL EXPECTED WORTH:	8829

Figure 7-17.   Allocations and results—Procedure 2.

highest marginal effectiveness is obtained by assigning the first missile to the first target class, yielding an expected worth of 90.* This first decision is noted by the circled number one in Figure 7-18. Actually, the same decision can apply to all targets in a given class. Hence decision one will consist of allocating the first *ten* missiles to the ten targets in Target Class 1. Decision two has as its six possibilities allocating a second missile to a target in the previously assigned Target Class 1 (having an expected worth of 9), or, allocating the first missile to a target in any of the remaining five target classes. Target Class 2 has the highest marginal effectiveness (equal to 54) and hence thirty missiles are allocated to these targets under Decision 2. This procedure can be continued as long as there are additional missiles to be allocated. The results of the procedure are tabulated in Figure 7-19a, showing the following:

1. The marginal worth associated with each decision (always decreasing).
2. The number of missiles accompanying that decision (i.e., the number in the target class).
3. The incremental effectiveness obtained with each decision.
4. The subtotal of the missiles assigned.

* Expected worth = (residual worth) (probability of kill of the next missile).

Target Class	Target Worth	No. Targets	1st Missile Assigned	2nd Missile Assigned	3rd Missile Assigned	4th Missile Assigned	5th Missile Assigned	6th Missile Assigned
1	100	10 Soft	$100(0.9)=90$ (1)	$10(0.9)=9$ (11)	$1(0.9)=0.9$			
2	60	30 Soft	$60(0.9)=54$ (2)	$6(0.9)=5.4$ (15)	$0.6(0.9)=0.54$			
3	40	20 Soft	$40(0.9)=36$ (4)	$4(0.9)=3.6$ (18)	$0.4(0.9)=0.36$			
4	100	20 Hard	$100(0.4)=40$ (3)	$60(0.4)=24$ (6)	$36(0.4)=14.4$ (9)	$21.6(0.4)=8.64$ (13)	$13(0.4)=5.2$ (17)	$7.8(0.4)=3.12$
5	60	40 Hard	$60(0.4)=24$ (5)	$36(0.4)=14.4$ (8)	$21.6(0.4)=8.64$ (12)	$13(0.4)=5.2$ (16)	$7.8(0.4)=3.12$ (20)	
6	40	50 Hard	$40(0.4)=16$ (7)	$24(0.4)=9.6$ (10)	$14.4(0.4)=5.76$ (14)	$8.6(0.4)=3.44$ (19)	$5.2(0.4)=2.08$	

Figure 7-18. Force allocation problem. Marginal effectiveness (Expected target worth) versus number of missiles assigned.

Decision Number	Marginal Worth	Missiles Assigned	Subtotal Missiles Assignment	Incremental Effectiveness	Subtotal Effectiveness
1	90	10	10	900	900
2	54	30	40	1620	2520
3	40	20	60	800	3320
4	36	20	80	720	4040
5	24	40	120	960	5000
6	24	20	140	480	5480
7	16	50	190	800	6280
8	14.4	40	230	576	6856
9	14.4	20	250	288	7144
10	9.6	50	300	480	7624
11	9	10	310	90	7714
12	8.6	40	350	344	8058
13	8.6	20	370	172	8230
14	5.8	50	420	290	8520
15	5.4	30	450	162	8682
16	5.2	40	490	208	8890
17	5.2	20	510	104	8994
18	3.6	20	530	72	9066
19	3.4	50	580	170	9236
20	3.1	40	620	124	9360

Figure 7-19a.   Force allocation procedure.

These results are plotted in Figure 7-19b which shows that the decisions are made on a basis of unequal blocks of missiles. Note how the rule of diminishing returns applies, since the slope of each allocation is always decreasing.

Problem Conclusions

A tabulation of the effectiveness resulting from each of the three allocation policies is shown in Figure 7-20. This indicates that Policy 2 is only 1 % less than optimum and, since it is so much easier to implement, might be the allocation policy used for missile assignment purposes. Note that the decision-maker would not know this if the analyst had not compared the two suggested policies with the optimal.

In conclusion, it should be stated that the concept of allocating resources on the basis of maximizing marginal effectiveness is one of the most important concepts in systems planning and has many applications. Another application of this concept is contained in Chapter 11.

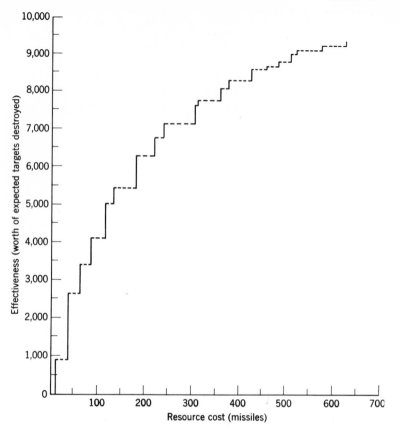

Figure 7-19b.   Demonstrating the principle of maximizing the marginal effectiveness.

CONDITIONAL PROBABILITIES AS A FUNCTION OF ANOTHER VARIABLE

In the simplified approach to model exercise, each of the transfer functions making up the operational flow models of Figures 5-9 and 5-10 were treated as an independent binomial probability distribution. We shall now discuss how to deal with those distributions whose probability of success is a function of some other variable.

Nonstationary Targets

One such transfer function is that related to the question of whether the target is still at the targeted location. Fixed targets such as industrial com-

	E (Worth of Expected Targets Destroyed	% of Optimum
Policy 1	8596	96.1
Policy 2	8829	98.8
Policy 3 (Optimum)	8942	100

Figure 7-20. Comparison of results of different allocation policies.

plexes, or fixed weapon installations such as air bases or a fixed surveillance radar, will always be at the target location. However, many of the offensive counterforce targets (such as aircraft and missiles) are mobile, and may have left the targeted location by the time the warhead arrives. Many times this time dependency must be explored. For this the analyst can use the transfer function of Figure 7-21a which, for a given type of weapon, such as a bomber, relates the probability of the target being at the location as a function of time. This figure indicates that even at the start of a surprise attack, a target such as an aircraft may not be at its airbase, but may be on airborne alert. Then after the initial attack, the remaining weapons are launched from their original location. Thus weapons may be moved to some orbiting location in the case of aircraft, or launched to target in the case of a fixed missile whose vulnerability can always be assumed to be in jeopardy.

This transfer function of Figure 7-21a may be used in the analysis by deciding what total time will have elapsed until the first offensive weapon reaches target, and use that value of the conditional probability for the binomial distribution which represents the probability of the target being at its location.

Defense Effectiveness

Another dependent probability distribution to be considered is the probability of surviving the enemy defense system. Recall that the probability of

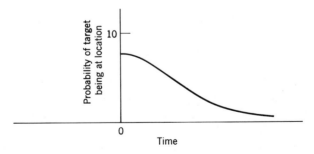

Figure 7-21a. Target mobility considerations for bombers.

target survival increases with the number of targets simultaneously penetrating the defense zone, as shown in Figure 7-21b. Since the probability of success of this stage is now a function of the input, the analysis is complicated somewhat. This dependency may be handled in the following fashion.

Figure 7-21b. Probability of surviving defense system.

Consider the same two-stage random process that was discussed previously, but not let p_2, the probability of success of the second stage, be a function of the number of incoming objects, as shown in Figure 7-22a. Assume for this example that three objects enter the first stage; x objects survive the first stage (and enter the second stage where there is a probability of success p_2 (a function of x) that the object also survived the second stage. Our task is to find the frequency function of the entire process, F(y). As shown in Figure 7-22a, there are four possible values (i.e., 0, 1, 2, and 3) for both x and y. Note, however, in Figure 7-22a, that if none of the three missiles was successful in the first stage (x =0), there is no opportunity to advance to the second stage. The frequency distribution for three objects (three independent trials) passing through the first stage whose probability of success is $p_1 = 0.7$ is a binomial distribution having the four possible values as shown in Figure 7-22b. Now we must determine the frequency distribution for each of these four possibilities as they pass through the second stage where p_2 is some function of x, as shown in Figure 7-22c. Let $p_2(x)$ be equal to 0.5, 0.6 and 0.7 for values of x of 1, 2, and 3 respectively. Note that p_2 could represent the survival probability of a missile as it passes through the defense system (the survival probability increasing as the number of targets entering the defense zone increases). Note also that only if x >0 does a missile go through the second stage, as shown in Figures 7-22a and c. Figure 7-22c also indicates all of the possible values of y which could result from each of the possible values of x. Thus, by summing all of the probabilities which are associated with each of the four values of y, the total frequency function desired may be obtained.

This task may be systematically attacked by constructing a logic table, as shown in Figure 7-23, which considers all combinations of values of x and y which might occur. The starting point of this table is the frequency function of x (i.e., the four values of x and the probability of each occurring). The second step is to rule out all conditional values of y which cannot occur. For example, if x =0, y cannot be greater than zero. If x = 1, y cannot be greater than 1, etc. The third step is to determine the probability of the joint event, which is the product of two probabilities as shown below:

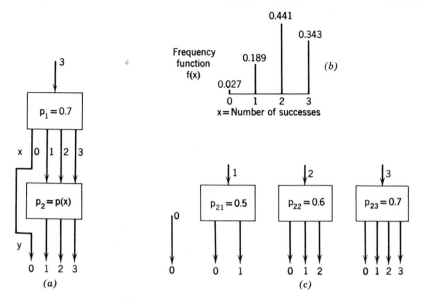

Figure 7-22. Handling interdependent events.

$$P(x,y) = [P(x)] \ [P(y|x)],$$
where $P(x,y) =$ probability that x and y occur,
$$P(x) = \text{probability that x occurs.}$$

$P(y|x)$ equals the probability that y occurs, given that x has occurred. This is defined as a conditional probability and is used when y is dependent upon the state of x.

This product of probabilities, which represents the joint probability of each combination of x and y, is entered into the logic table of Figure 7-23. Note that while p_2 is a function of x, it is assumed that there is independence of each of the three trials of the missiles as they pass through each stage; hence, the frequency function of y (for each value of x) is also a binomial distribution.

It should be evident from the discussion of the logic table approach versus the use of some average value of the probability that the latter is a simpler approach. However, only with completely independent processes is it true that

$$\bar{y} = np_1 p_2 = \bar{x} \ \bar{p}_2.$$

If there is not complete independence among the various functions this equation is not true and a logic table approach must be used for the correct distribution and expected value of the result. However, for most cost-

Values of y					F(y)
3				$(0.343)(1)(0.7)^3(0.3)^0$ $= 0.118$	0.118
2			$(0.441)(1)(0.6)^2(0.4)^0$ $= 0.159$	$(0.343)(3)(0.7)^2(0.3)^1$ $= 0.151$	0.310
1		$(0.189)(1)(0.5)^1(0.5)^0$ $= 0.0945$	$(0.441)(2)(0.6)^1(0.4)^1$ $= 0.212$	$(0.343)(3)(0.7)^1(0.3)^2$ $= 0.065$	0.372
0	0.027	$(0.189)(1)(0.5)^0(0.5)^1$ $= 0.0945$	$(0.441)(1)(0.6)^0(0.4)^2$ $= 0.071$	$(0.343)(1)(0.7)^0(0.3)^3$ $= 0.009$	0.202
Frequency Function in x	0.027	0.189	0.441	0.343	
Values of x	0	1	2	3	

Figure 7-23. Logic table for handling interdependent events.

effectiveness analyses where a relative comparison of systems is the main consideration, the above approximation is generally adequate.

Finally, if the true frequency function is desired, and the logic table approach is used, this technique must be continued for the remaining stages to the final result. However, if all of the remaining stages are composed of binomial distributions, these may all be combined into one stage (whose probability of success is that of the product of all remaining stages) and only one more logic table is required.

Subsequent Defense Rounds Fired at Surviving Targets

If the lethal defense zone is large enough, there may be sufficient time to fire a second or subsequent round against any incoming target which survives the first defense round fired. This procedure may be analyzed as a multistage process, as shown in Figure 7-24, for a three-stage system (where

there is sufficient time to fire three rounds). Note that this is not the case of firing a salvo of three rounds at each incoming target to increase the total probability of kill. The process to be analyzed might be called a "look, shoot, look" system in which only surviving targets are refired upon. This is obviously a more efficient policy since it avoids overkill, but is only effective if there is sufficient time to permit sequential firings.

This case may be analyzed in the same way as previously described (i.e., by structuring a logic table for each stage of fire, determining the resulting frequency function, and using this as the input to the next logic table until all the stages are covered). Of course, the frequency functions resulting from each stage will permit the analyst to better predict how many rounds are expected to be fired for any given incoming raid.

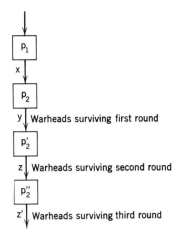

Figure 7-24. Multi-round defense system.

CHOOSING THE APPROPRIATE SELECTION CRITERION

Consider the problem of choosing between two equal cost systems (whose effectiveness is illustrated in probabilistic form in Figure 7-25).* As indi-

* A more detailed discussion of this topic is presented in E. S. Quade and W. I. Boucher (eds.) (1965), Chapter 4 by L. D. Altaway, and Chapter 5 by A. Madansky.

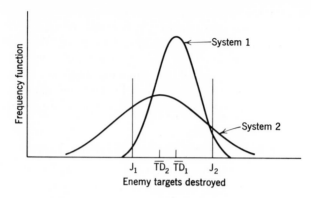

Figure 7-25.   Interpreting probabilistic results.

cated earlier, one way of comparing the two systems is on the basis of expected value of effectiveness $(\overline{TD})$. In this case, System 1 has a slight edge.
The disadvantage of using the expected value as the measure is that it is
merely the average of all the values that would be obtained if the system
were to perform the job a large number of times; and this is not true in the
present problem. Note that System 2 has much more uncertainty connected
with its operation and this should be stressed to the decision-maker. Thus, a
second measure of the system is the probability of performing a job to at
least a given level. By integrating the area under each of the probability frequency functions of Figure 7-25, the analyst can obtain the probability that
the system will perform to at least a given level. For example, if it is important that at least the level J_1 be attained, System 1 will provide a higher
probability of attaining J_1 than System 2. On the other hand, if J_2 is the important level and anything less than this is considered worthless, the
decision-maker may wish to procure System 2.

While this second approach may be a more sophisticated one for purposes
of choice between two (or more) systems than the first, there may be some
problems in which neither of these approaches would be a suitable method
of selection to the decision-maker. Hence there is a third way of evaluating
the responses of each system which might be considered. This method uses a
value scale called "utility," which is an attempt to translate the probabilistic
results previously obtained into a higher level measure based on the intuitive preference of a decision-maker for different possible results. This concept, which we shall now summarize,* is quite often useful in aiding an individual to deal with problems of risk. Consider, as an example, a "worth
function" which represents the feelings of an individual who currently has

---

* For a more detailed discussion of utility theory, see R. Schlaifer, (1959), Chapter 2.

$20,000 in assets with respect to winning or losing various sums of money. Such a function might be pictured as shown in Figure 7-26. Notice that in the region of the origin, the curve is fairly linear, meaning that in the individual's mind, either winning or losing relatively small sums of money (say up to a total of $100) at a gambling table might give him the same sort of gain or loss with respect to how it would affect his life or way of living. Winning larger and larger sums of money causes his utility function (happiness) to go up, but in a diminishing fashion.

On the other hand, losing $200 might make him despondent for a month (due to a nagging wife, for example). And as losses increase, the adjustment to his life becomes more and more drastic as he loses all of his assets and goes deeper and deeper into debt. (In fact, greater and greater gambling losses might actually put his life in jeopardy were he to borrow from loan sharks to repay his debts.)

While it is difficult to obtain a precise curve for any individual, this example illustrates the following interesting points: first, it might be possible to construct a transfer function, such as illustrated in Figure 7-26, which relates some measurable level of effectiveness to a higher-level measure of worth to an individual or an organization. Such a function would illustrate that even though on an expected value basis an individual might be better off to self-insure (he saves, at least, the administrative costs of the insur-

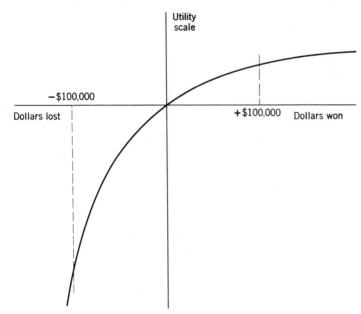

Figure 7-26. Decision-maker's utility function.

ance company), people are more content to pay a higher amount in small sums of money each year rather than run the risk of having unexpected bad luck and having to face a very large loss at any one time. The concept of obtaining and using some quantitative utility function is an attempt to bring such feelings of the decision-maker into the analysis quantitatively.

Madansky also discusses how utility theory might be applied as an attempt to quantify McNamara's stated feelings regarding the worth (or utility) of strategic offensive systems, whose primary function is deterrence.* According to Secretary McNamara's stated logic, if the United States has a deterrent capability (as measured by its capability of providing assured destruction to a potential enemy even if attacked first), it is of high worth to this country. On the other hand, if the United States forces would not provide an assured destruction capability, we would be in danger of being attacked first and such a system is of no worth to the United States.

A utility function which quantifies these statements is shown in Figure 7-27. Here the worth of a force which provides anything less than assured destruction is considered to be zero. The worth of the force then suddenly climbs rapidly once it has achieved deterrence, gradually tapering off as it provides greater capability than assured destruction, D_0.

By means of this quantitative function, the analyst can translate the probabilistic aspects of the effectiveness of each of the two weapon systems illustrated in Figure 7-27, by summing the product of the frequency function and utility at each level of target destruction, thus yielding the expected utility of each system. Such a measure could then be used as the basis of system selection. Notice that even though each system had the same expected value for its effectiveness, it can be seen that the utility of system A is greater than that of system B. This is obvious since the utility function is zero up to D_0; thus, a larger portion of system B's distribution function has zero worth than for system A.

Many times an analyst will attempt to construct not only the American utility function but a utility function for a potential enemy. Such a function would be based on our belief of what the country's decision-makers feel, and would be limited to whatever information is available. By this means, we would be able to obtain a more meaningful measure of effectiveness. This will be found useful in Chapter 8.

An example of the usefulness of this might be the case where China rather than the Soviet Union might be the potential enemy. In this case, some deci-

* Also see "The Deterrence and Strategy of Total War, 1959–1961: A Method of Analysis" (The RAND Corporation, RM-2301, April 1959) for a major analysis of utility theory as applied to the problems of American deterrence and the choice of strategy should deterrence fail and the Soviet Union initiates a premeditated preventive war.

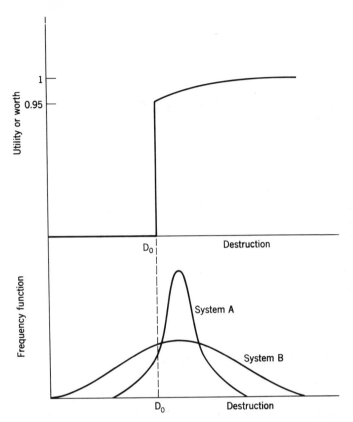

Figure 7-27. Converting probabilistic results to utility.

sion-makers may argue China might not be deterred merely because we have an assured destruction capability of destroying one quarter of the industrial capacity of China. In other words, the utility function as pictured in Figure 7-27 may be shifted to some higher value or it may be an entirely different function. Knowing what it may be can aid in evaluating the analytical results obtained.

8

Coping with Uncertainties

Chapter 7 indicated various ways of combining a series of random processes so as to determine the over-all effectiveness of a system operating in a given environment. However, many times when the analysis is reviewed by the hierarchy of decision-makers involved or by others who have an interest in the problem, the results (and the analysis) may not be believed and, hence, rejected by some of them. Thus, the analytical effort may have been wasted. We shall now explore some of the reasons why people reject analytical results with the objective of recognizing pitfalls which may occur, thus avoiding them in future studies.

Why People Reject Analytical Results

It should be recognized there may be many reasons why people reviewing the analytical results may choose to disagree with them. Some of these reasons are stated; others not. For example, if the results of the analysis are used as reasons to support a course of action which may adversely affect the security, power, or prestige of an individual or his organization, it is only natural that an individual might wish to oppose this recommended course of action on some basis. Thus, objections to an analysis must be anticipated.

However, while the analyst may (and should) be aware of these political considerations, he can really only defend the credibility of his analysis on explicitly stated grounds. In fact, he should insist on being told the reasons why the analysis is not accepted by an individual, since only by knowing these reasons can the analyst do anything to resolve the differences. Thus, we shall focus on some of the reasons generally stated for disagreeing with an analysis.

Sometimes the reasons given can be directly related to a particular part of the analysis, such as the following:

1. Disagreement with the logic or assumptions used in the analysis. Thus, there may be disagreement with the models employed, particularly

190

the scenario(s) employed or the operational flow model and the transfer functions generated which quantify the scenario(s).

2. Disagreement with the numerical values of force size or performance characteristics of the United States and/or the competitive (enemy) systems involved.

3. Disagreement with the form of model exercise used.

The preceding disagreements are generally based on either omissions ("You did not include the following important factors or scenarios"), or differences of opinions ("I believe the value of this performance characteristic should have been this, not the one you used"). Many times the analyst encounters a more serious situation where important individuals express their objections in terms of their experience, judgment, or intuition, often in a language which is different from that which the analyst has used. Hence such objections may not be simple to handle. The analyst, however, cannot afford to ignore such comments or objectives for two good reasons. First, ignoring these factors tends to lower (perhaps to zero) the credibility of the entire analysis, since doubts are raised which, if left unresolved, may jeopardize acceptance of the entire analysis. Second, the doubters may, in fact, be correct. The important point to be made here is that the methods described thus far are based on a logical structuring of an explicit scenario of operation, containing a series of operational events and a set of performance characteristics which numerically describe the performance of both the United States and enemy systems involved. Any of the various disagreements in the analysis cited can, in fact, significantly alter the results (i.e., systems effectiveness) obtained, and reverse the recommended selection, as made apparent by Figure 8-1. Hence it is necessary that the analyst "go the

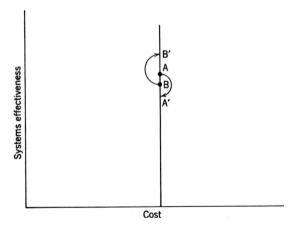

Figure 8-1. Uncertainty in systems effectiveness.

extra mile" by attempting to translate the objectives in terms of the models and the numerical values he has used.

Pedagogically, the problem may be approached in terms of establishing proper two-way communication in making explicit what the relevant issues are. Thus objections such as "Based on my experience I do not believe:

1. The war would be fought that way;
2. The system would be used in that fashion;
3. The system reliability will be that high"

can be used to initiate a dialogue around the point of:

1. How do *you* believe the war might be fought?
2. How do *you* believe the system would be used?
3. What numerical value of reliability would *you* expect?"

Of course, any data available to either side of the issue would be quite useful in such discussions.

Recall that the data to be supplied the decision-maker consists of the effectiveness of the various systems considered (A and B), each having been configured for equal cost. Various uncertainties introduced in the analysis will result in the effectiveness of each system being higher or lower than was originally calculated. Thus, points A and B might actually turn out to be A' and B' because certain performance characteristics actually have different values (e.g., an incorrect selection is being recommended).

Systems analysts have an expression regarding the accuracy of their analytical results: "garbage in—garbage out," meaning that the final results obtained will only be as accurate as the accuracy of the input qualities.

As was mentioned before, the analyst should review the scenarios, operational flow model, and numerical data to be used, as well as the method of exercise to be used, before performing the model exercise. However, even at this point it will be recognized that various uncertainties in performance characteristics and enemy capabilities cannot be avoided because of data limitations, and must therefore be coped with. This chapter describes methods for doing so.

TYPES OF UNCERTAINTIES

Different analyst-authors have attempted to classify various types of uncertainties in different ways. We shall use the following nomenclature: "Statistical uncertainty" will include the uncertainty in the final outcome of a nondeterministic or random process involved, given that the probabilistic parameters of the performance characteristics are precisely known. Thus, statistical uncertainty, sometimes called "risk," involves the element of chance inherent in every random process. Methods of dealing with randomness or statistical uncertainty in which the effectiveness of the system can only be predicted statistically (if one were absolutely certain of the probability distributions of the various events) have already been discussed in Chapter 7.

"Real uncertainty," on the other hand, involves the uncertainty in system effectiveness caused by not knowing the exact probability distributions of all of the processes involved, the capability of a potential enemy, or how this capability will be used. Some authors use the term "uncertainty" (as opposed to risk) to describe this phenomenon.

Problems involved in making an accurate total cost estimate of a system which has never been built before will be discussed in Chapter 9.

Analyzing Competitive Uncertainties

Here the key question to be addressed is, "What are the various strategies and tactics that the competitor (enemy) can employ, at what strength, and what impact will these have upon the effectiveness of each of the system alternatives being analyzed?" Perhaps the most important aspect of dealing with competitive uncertainty is recognizing that it is an important problem. This may seem obvious, but many systems analyses and systems planning efforts are deficient because it is assumed that the competitor will act in only one assumed way.

The first rule for dealing with competitive situations is to recognize what the competitor's system possibilities are at the future operational time being examined. Here the analyst may list the systems the competitor may have in his inventory at the future time, as well as alternative ways he might use them. This information generally comes from intelligence sources, such as the new threat estimate which initiated this strategic planning study.

Given that the competitor may have certain systems in his inventory, the analyst must also cope with the additional strategic uncertainty of how the competitor might use these weapon system units (i.e., the different scenarios which might occur). For example, what type of attack might be planned? Will low altitude or high altitude bomber penetration be involved? How might the enemy allocate his aircraft among different bases?

Having listed the different types of enemy systems and their possible scenarios involved, the analyst might next structure these into a matrix composed of all reasonable competitor possibilities which might be effective against the various American system alternatives under consideration, as shown in Figure 8-2. Thus we must consider how well each of the possible American system alternatives under study would perform against each of the competitive possibilities. One area of uncertainty which becomes appar-

Soviet Possibilities / American Possibilities	System A′	System B′	System C′		System J′
SYSTEM A					
SYSTEM B					
SYSTEM C					
SYSTEM I					

Figure 8-2.   The competitive matrix.

ent is how many weapon system units should the American force structure contain and how many weapon units should the competitor's force contain? The best way of coping with this problem is to compose all American force structures for the same cost. Similarly, all competitor systems should be constructed at equal cost, although the competitor costs may be different from the American costs. Later the analyst will vary the costs of the American systems and/or the competitor systems and see how the results vary as each of the costs (or the cost ratio) varies.

Before determining the effectiveness of each of the possibilities uncovered (i.e., each of the intersections of the competitive matrix), the analyst should consider any new competitive systems or strategies which could be introduced in response to any of the American alternatives under examination, once they have been detected by enemy intelligence. For example, an enemy might observe that it is difficult to accurately fire on the mobile ICBM system due to a lack of accurate target location information, and hence, might wish to consider some alternative strategy. One such alternative is the so-called "decapitation attack" in which the fixed command centers such as the national command authority or the headquarters of the Strategic Air Command would be fired upon. If these command centers were destroyed, positive target assignment might be withheld from the missile force, and hence such an enemy policy must be considered. This contingency and the systems planning effort associated with it is considered in detail in Chapter 12.

Thus planning in a competitive environment may be looked upon as a feedback control system. We observe the competitor to some degree using available intelligence information sensors, and the inferred information enables us to configure systems to best cope with our limited view of his systems. He does the same, using his intelligence sensors. Thus each competitor attempts to be adaptive and endeavors to uncover the weak points of the opposing system and exploit these through his own developments. The problem is further complicated by the lead times involved in developing new systems and the uncertainties involved in how precisely and timely one side perceives the other side's position. Thus, as in any competitive game, what our competitor does will most likely be a function of what we do.

The competitive matrix must next be evaluated by determining the effectiveness of each American system versus each of the possible competitive systems or strategies involved, using a given cost level for each side. As was indicated in Chapter 7, the effectiveness measures will be those used previously (i.e., Blue target damage as well as Red target damage). If applicable, these effectiveness measures could be translated instead into United States and competitor utilities by using some derived utility transfer function for each side.

It is probably evident to the reader that examining all of the competitive possibilities involved can be quite time and resource consuming. Hence, in practice, some preliminary filtering is done to rule out obviously less effective possibilities on each side, by conducting some preliminary analyses of the entire matrix, making suitable assumptions to simplify the model exercise, or by means of considered judgment. Following this preliminary filtering, a more detailed analysis can be made to determine the effectiveness of each of the remaining systems when operating under each contingency in an attempt to determine the preferred system. Several examples of this process follow.

The first example of the principles discussed is the case of the Maginot Line erected by the French before World War II. One can initiate a lively debate among military strategists by considering the wisdom of the expenditures for this defense approach. Some would say that this defense line was never used to stop an anticipated German advance and, hence, it was a poor investment. This view can be further reinforced by the statement that building this defense line provided a false sense of security to France, since many thought that it would prevent a German invasion of France. If the French had built their Maginot Line without any holes, and used it properly to enable a relatively small number of soldiers to defend a large frontier, they could have concentrated troops where they needed them—to conduct offensive operations. It is one of the purposes of a good defense to enable one to pursue offensive operations when and where needed. A good defense not only prevents the enemy from destroying one's offensive capability, it causes him to divert and use up large resources in his offense. This presumably weakens his defense.

Kahn and Mann * state that a completely different opinion can also be presented; that is, that the Maginot Line was a good idea, but it was not implemented properly. We shall examine this example in greater detail and see how it relates to the systems analysis principles of dealing with enemy uncertainty. Briefly, the reasons for building the Maginot Line can be seen from Figure 8-3. Before the Line was erected, the perimeter shown offered what seemed to be the easiest and, hence, most likely penetration path that a German attack on France would take. However, once this fortified defense line was built, other attack routes to France were now easier and less costly in German resources. In fact, the actual attack route used by the Germans completely avoided the line and went around through Holland and Belgium, completely bypassing the Maginot Line. These different attack routes are what we have called "scenarios." Some advocates of the use of subjective probabilities, in attempting to quantify what a competitor might do, would have originally placed a high probability on the attack crossing the locations covered by the Maginot Line and would have built the Line on some "most likely value" basis. While this may be a reasonable approach in dealing with random events, it is quite fallacious in dealing with competitive uncertainty since the competitor's main objective is to seek methods of frustrating his opponent. Thus, once the Line had been built, the subjective probabilities originally estimated would now change, since this was no longer the easiest

---

* H. Kahn and I. Mann (1957).

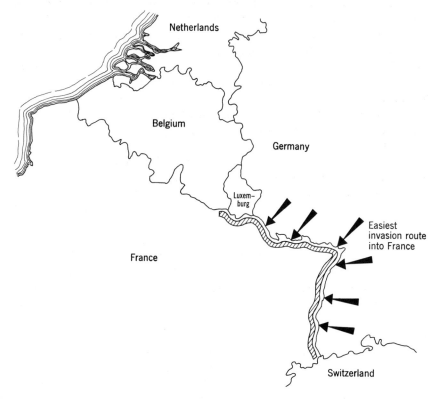

Figure 8-3.   Invasion routes into France.

invasion route. This gets back to the feedback system nature of the problem, in dealing with competitive behavior.

The proper concept of an optimal defense is a *balanced* defense. To design this, consider a cylinder which completely surrounds the objective to be defended by 360° and rising vertically through space and down through the earth. We shall conduct the analysis by pivoting on constant cost (although a similar analysis could be conducted by pivoting on constant effectiveness). Thus, we initially assume a total resource expenditure for the offensive force and one for the defensive force. (These may be changed for subsequent analyses.) Next, examine all ways that the offensive forces can penetrate the undefended (or currently defended) cylinder, and the total effectiveness obtained by the offensive force for each penetration route (scenario). Defense expenditures are next added sequentially to counter the various possible penetration routes, always adding just enough resources to equalize the effectiveness among alternative penetration routes. Once defense resources are completely allocated, the penetration tactics can be re-examined to see if any additional scenarios should be added for consideration. If so, the defense system may need modification, always subtracting resources from one portion to add to another. The final result is characterized by the term "balanced" defense system which is one that will provide the same level of effectiveness (e.g., targets destroyed), regardless of what scenario the offensive force uses.

This is another example of the use of maximizing the marginal effectiveness when each increment of defense capability was added in turn, and is similar in concept to the process of assigning missiles to a target set on an incremental effectiveness basis, as described in Chapter 7. This is obviously the same way that additional defense resources would be added to increase the defense effectiveness further.

Another example of this balanced approach to system design is illustrated by the incremental approach used in the construction of the ballistic missile early warning system (BMEWS). At the time that this system was being considered, a ballistic missile attack from the north was the most likely attack path (i.e., would provide the Soviet Union with the highest effectiveness for a given missile cost level). Thus, northern volumetric coverage was provided. Later, as the potential threat from submarine-launched ballistic missiles (SLBM) began to increase, it was decided to use an additional increment of expenditures to add an SLBM surveillance network to the continental defense early warning system, as shown in Figure 8-4, since a surprise attack was now feasible from these directions. Even with such an addition, note that the southern route is still open, and the question remains of whether a BMEWS system should be constructed for a missile attack from the south.

To answer this question, the analyst might play the role of a potential enemy and calculate the damage to a given number of American targets, using the southern route as compared with any other route. Using the southern route attack permits a surprise attack, perhaps catching additional bombers on the ground. However, since the southern route is much longer, the propulsion requirements, as well as the accuracy requirements, are much greater than using shorter missile trajectories, resulting in a higher cost for each of these longer trajectory missiles. Thus by comparing the effectiveness (targets destroyed) gained by purchasing these more expensive missiles with the effectiveness which could be obtained by saturating the area with more Soviet missiles or SLBM's using the northern or sea-launched attacks for the same cost, it can be determined that the scenario of an attack from the south is dominated by other scenarios, and hence not practical from the Soviet point of view (assuming a given state of technology resulting in a given set of performance and cost characteristics.

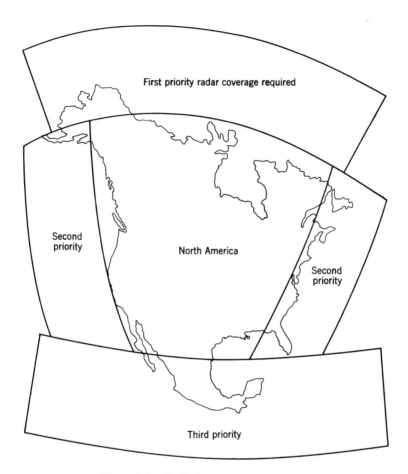

Figure 8-4. Ballistic early warning needs.

Another important principle can be noted from each of these two examples. Even if a system is not used (in anger), it may still be worth buying, since it does accomplish its function by forcing the competitor to consider other ways of accomplishing his mission (if he is not prepared to pay the cost of direct confrontation with the particular system under consideration). Thus a system such as BMEWS, or the Atlas ICBM program, may be looked upon as "fire insurance," which most people are willing to pay even though the fire is unlikely to occur.

FINDING THE PREFERRED SYSTEM

Determining which of the system alternatives analyzed should be chosen is a difficult task, mainly because of the large number of options open to an adaptive and creative competitor. However, a choice must be made (even if that choice is the "do nothing" option). T. C. Schelling * and Kahn and Mann discuss some of the key principles involved which should be considered in making such a choice. These principles are now summarized, with particular emphasis on relating these principles to the analysis performed thus far in the ICBM problem under study.

Determine Effectiveness of Each Alternative for Each Contingency

The heart of the selection process is the completion of the competitive matrix of Figure 8-2 which indicates the effectiveness resulting if a given American system alternative were employed against a competitive system / contingency, using such measures as destruction of forces, population, and industrial capacity on each side. The multidimensional measure of effectiveness could then be summed in terms of one utility measure as viewed by each side. Based on this, each side could construct a "preference matrix" in terms of a ranking of each of the system alternative–environment possibilities as viewed by each competitor. (See Figure 8-5a for the American possibilities and Figure 8-5b for a similar portrayal indicating the American belief of how the Red forces might rank each possible alternative.) Note that the best possibility for the United States (signified by 1) may be the worst possibility for the Red force (signified by 9). However, both sides may mutually prefer one possibility over another. This might be the case where an American victory is obtained (in both cases), but with lower losses on both sides than higher losses. Notice also that no one system dominates the others under all competitive conditions.

Who Moves First?

Principles of game theory may be applied to the competitive intricacies of Figures 8-5a and 8-5b to aid in making a proper system selection. One key factor to be considered, however, is which side moves first. We shall now consider the results obtainable, depending on which side moves first. Consider first a situation where the American side is forced to make a system procurement decision first (i.e., before the Red side) because of a longer procurement lead time requirement. An example of this would be an Amer-

* See E. S. Quade (1964), Chapter 10.

	Red Force System Possibilities		
American System Possibilities	A′	B′	C′
A	3	1	9
B	7	8	4
C	2	5	6

Figure 8-5a. American ranking of possible outcomes.

	Red Force System Possibilities		
American System Possibilities	A′	B′	C′
A	6	9	1
B	4	2	8
C	7	5	3

Figure 8-5b. Red forces ranking of possible outcomes.

ican decision to deploy a particular defense system at specified locations from among many alternative locations considered. If we can assume further that Red intelligence will detect this decision in sufficient time to plan a countermove, our selection might be made as follows:

Based on the Red rankings of each of the possible strategy combinations, it is possible to indicate which Red strategy we would expect to be employed for each of the three possible first moves by the United States. For example, from Figure 8-5b, if the United States chooses A, Red would choose C′ since we believe this combination provides the highest utility to Red. Similar expected Red responses could be determined for American moves of B or C, and these are shown in Figure 8-6a. Based on this information, the best outcome that the United States can achieve is its sixth choice, by choosing System C, since any other move would result in a lower ranking result for the United States (and higher ranking for the Red side).

However, notice what happens if, as in the example of a Red offensive system procurement having a longer lead time, Red is forced to move first, and the American intelligence system detects this move. In this case (assuming that the Red side goes through the same type of analysis as described above and arrives at the same results, as shown in Figure 8-6b), the Red side would choose System C , since any other choice would permit the United States to choose a higher ranking outcome. Thus if the United States can move second, it can, in general, choose a system which will provide a superior outcome, as compared with moving first. Thus, one rule of systems planning in a competitive environment is to make the competitor move first.

American First Move	Red Response	Results Red Ranking	American Ranking
A	C′	1	9
B	B′	2	8
C	C′	3	6

Figure 8-6a. Red system selection if American force moves first/Red force moves second.

Red First Move	A′	B′	C′
American Response	C	A	B
Results			
American Ranking	2	1	4
Red Ranking	7	9	8

Figure 8-6b. American System selection if Red force moves first/American force moves second.

One way in which this may be done is to have many development programs underway which are continued up to, but not including, the stage of procurement so that when the competition does enter procurement (i.e., moves first), the United States is in a technological position to counter with the most appropriate response. This strategy, unfortunately, is opposed by some (generally those whose development program does not go immediately into procurement) as being wasteful, for two reasons: first, money is spent on many developments, but the results may not be used (in procurements); second, by waiting for a competitive move and then responding, we need to undertake "crash programs" involving massive efforts, using premium cost overtime, which is not very efficient. A good example of such crash efforts would include the crash developments of the Atlas ICBM program and BMEWS, both initiated in 1954 when intelligence reports confirmed an accelerated Soviet ballistic missile program.

These arguments would normally have merit until we fully understand the alternatives to the strategy. Why procure an expensive BMEWS system if the Soviet Union is not planning to build up a ballistic missile capability? Why procure a large number of ballistic missiles when we had a bomber superiority at the time? What could be criticized was the small-sized development effort in the ballistic missile system area, which a management strategy emphasizing developments might have overcome. Thus, it may be better to spend relatively smaller amounts of resources and not use all of them, than go into costly procurements too early and find that the competitor has out-maneuvered the system, such that it contributes little to over-all effectiveness.

Designing the Preferred System

We shall now discuss a systematic method for arriving at a preferred system choice. The original completed competitive matrix which shows the system effectiveness (or utility) for each possible competitive situation can also be used to determine the best we could do if we had complete foresight regarding what the competitor would do. That is, for each one of his alternatives we could not only specify our system choice, but could also determine the maximum utility we could hope to attain for each competitive contingency. While we cannot hope to procure all of the systems which make up the optimal set, this information is quite useful for systems planning purposes, since it indicates the best we can do. Thus, as Kahn and Mann indicate, what we are seeking is a new system which can do almost as good (90 to 95% as much as the maximum attainable by each of the separate systems) over the entire spectrum of possible contingencies. Configuring this new system, which is *insensitive* to the different contingencies involved, is the goal of sound systems planning and the real challenge of systems analysis.

The first method of finding the preferred system to be discussed is that of system improvement. If following the evaluations, a particular system is found to be superior for most contingencies, but weak for a few, an attempt should be made to improve the system deficiencies in some way for the limited cases. The operational flow model is an excellent vehicle for attempting to improve the system effectiveness since this model shows in which stages of the over-all process a system is deficient (i.e., where the "bottlenecks" are). For example, an examination of Figure 7-7, which indicates the numerical values of the probabilities used in the operational flow model, reveals that while the mobile ICBM system was superior in survivability to this particular type of enemy attack, it was lowest in over-all effectiveness primarily due to low reliability, low accuracy in hitting the target, and smallest number of missiles available. This information is useful to the systems planner since it pinpoints possible opportunities of improvement. For example, why is the reliability low? Is there any one particular component which is mainly responsible for equipment malfunctions? How much would it cost to either replace, improve, or provide some backup to the responsible component which might raise the over-all reliability and, hence, the effectiveness of this system.

In a similar fashion, the causes of target misses could be explored. Perhaps the main reason is the inaccuracy in knowing the missile launch position as compared with a fixed silo launched missile in which the launch position has been carefully determined. If so the system planner might propose a strategy of carefully surveying several thousand alternative launch sites, marking each with some stake. Then when the alert command was given, the missile launch crew could leave its random position and move to one of the nearest sites and prepare for launch. This mixed strategy provides a high degree of survivability (in proportion to the number of redundant surveyed sites employed) as well as obtaining an accuracy of locating the launch position which could approach that of a fixed site.

Both of the above examples demonstrate the key underlying principle of systems planning; that is, seeking ways of improving system performance (and, hence, effectiveness) and comparing the improvement in effectiveness to the incremental increase in system cost, to determine whether this marginal effectiveness is high enough (compared with other alternatives) to warrant implementation. This is the theme of Part V.

Another way of improving the system is the cost reduction approach. In this case, focus is directed to the cost model which contains the costs of each system element as well as the operation and maintenance costs of the system. This is discussed in greater depth in Chapter 9. Analysis of the cost model may direct attention to those areas of high cost so that alternative ways of reducing costs through techniques such as value analysis, which basically examines the function of an element or operation and explores alternative ways of accomplishing the same function or operation at lower cost. Obviously, if the cost can be reduced, more missile units could be procured and the total effectiveness increased.

Creating New System Concepts

The operational flow model may also be used as a means of creating other system concepts in arriving at the preferred system under various conditions of competitive uncertainty. After these other system concepts are created, they will then be evaluated in terms of not only their technical feasibility, but also their costs, as compared with the costs of other system concepts.

We shall now describe a systematic process which will aid us in finding alternate solutions for accomplishing the same system objective. One way of doing this is to focus on each of the specific performance characteristics in the operational flow model, determine the generic function to be satisfied, and then find other specific ways of achieving this function (perhaps at lower costs). For example, the two main alternatives with which the problem is concerned are hardening the missile silos and the mobile Minuteman system. Note that both of these are specific means for decreasing target vulnerability: one through hardening the target (i.e., the silo), and the second by denying the enemy information about the target's exact position. However, there are many other ways for decreasing missile vulnerability, some of which are shown in Figure 8-7. For example, given an enemy warhead with a given lethality, one might reduce missile vulnerability by moving the missile around randomly in many ways. This could be done through a road mobile system, a railroad system, a submarine system, a warhead in a random earth orbit, or many other ways.

It is also possible to move only a portion of the missile. For example, move only the warhead in a random fashion having redundancy at different fixed launching sites, each containing missile boosters.

A. Missile vulnerability (see enemy damage submodel).
Given enemy warhead lethality, have to reduce missile vulnerability.
1. Move around randomly (road mobile ICBM system, railroad car system).
2. Move only portion (missile, not launcher) around randomly and have redundancy in launching sites.
3. Hide (camouflage, Polaris submarine, dark side of the moon).
4. Deceive (shell game).
5. Difficulty in reaching (orbiting bomb which changes orbit each cycle).
6. Decrease target vulnerability (hardening).
B. Number—buy more/larger quantity/redundancy.
C. Early warning of attack (would permit dispersement of the force before they are hit by enemy weapons).
Improve volumetric coverage.
Range⸺⟶increased early warning time.
Angle⸺⟶no "open windows."
2. Improve credibility of warning.
Highly credible warning device would permit more of force to leave sites.
Summation of outputs of different type sensors might reduce uncertainty of information.
D. Recallability of Weapon.
(Weapons could leave site and return if false alarm).
E. Improved defense forces.
Destroy enemy warheads at different time phases (prelaunch, midcourse, terminal).
1. Antisubmarine warfare systems (ASW).
2. Antiballistic missile systems.
a) Ground based.
b) Air based.
c) Space based.
d) Sea based.
e) Combinations.
3. Jamming missiles.
F. Improved force management system.
Improved weapon assignment policy (discussed in Chapter 12).
G. Reconnaissance Systems for detecting mobile targets.
1. Airborne systems.
2. Space systems.
3. Other.

Figure 8-7. Offensive force improvements model.

It is also possible to hide the missiles through camouflage or by using a vehicle such as a Polaris submarine, or locating the missiles behind the moon, for example. It is also possible to decrease target vulnerability through hardening as was one of the specific cases. Another approach to increase effectiveness is merely by buying a larger number of missiles.

One can now look at the list of system performance characteristics, as shown in Figure 8-7, and concentrate on each of these characteristics, trying to find alternate ways of accomplishing this. One example would be the improved defense forces or increasing the reliability or supportability of the equipments, and, of course, increasing the timeliness, accuracy of the command, control, and communications systems. Thus by constructing a structure such as in Figure 8-7, which dissects the problem into smaller parts, one is able to examine the generic performance characteristics required, the function to be performed, and find specific other ways of performing or implementing this function.

As will be seen in subsequent chapters, one must then determine not only if each of the ways is technically feasible, but how much it costs to implement this approach. By focusing on the deficiencies of any particular system under given scenarios, the planner may be able to find alternative ways of accomplishing the function which circumvents these deficiencies, and see if the resulting marginal effectiveness of the improvement is high enough to include the improvement.

Build Flexibility Into System

Another way of coping with future uncertainties is to design the system to include certain flexible features. For example, a computer can be designed with planned expansion of other compatible components in mind. Mobile defense (or other) systems may be considered rather than fixed installations, so that if a future need occurs at some other location, it may be met. An example of a fixed system built with a lack of flexibility was the gun emplacements intended to defend Singapore harbor. Since an attack by sea was the one expected, gun installations which could fire at sea only were constructed. As it occurred, the Japanese did not attack by sea but took a land route, rearward of the guns. Since these could not be rotated to the rear they were never used in combat.

There is generally some cost associated with building flexibility into the design. In general, however, the cost is slight, especially when compared with the possibility of having an obsolete system available at some future date, because the environment has changed. Thus, designing in flexibility may be a worthwhile investment. An application of the flexible design principle is contained in the bridge problem of Chapter 11. A flexible design may also assist us to force the competitor to move first, since we now have the capability to respond to alternative competitive strategies.

Multipurpose systems, such as a general purpose computer may also be examples of built-in flexibility. In this case, one system is designed to accomplish a number of related missions, and hence offers a cost saving as compared with the purchase of a number of systems, each designed to perform one mission.

Design Mixed Systems

Another way of obtaining over-all flexibility to cope with a set of different contingencies is to purchase a force which is composed of a mix of a number of different type systems. For this reason the United States has purchased some ICBM's, as well as some submarine launched ballistic missiles, as well as maintaining a bomber force. While each is an offensive weapon, a mix of weapons avoids the disadvantages of "putting all of your eggs in one basket." Again, the benefit comes at some extra cost.

The procurement of refueling tankers (at the expense of more bombers) permits longer range flights, thus adding much more flexibility to the penetration path finally chosen. Thus, a potential enemy is forced to maintain a fairly uniform air defense system to defend the desired targets, rather than concentrating the defense forces (which could be done if he knew the penetration path to be taken if the flight distance were more restricted). In this fashion we are causing him to adapt his system in a way that is beneficial to us.*

How Much Will Competitor Spend / How Much Should We Spend?

The analyst must consider the uncertainty in how much the competitor will spend for his opposing system. Obviously, the more Red ICBM's procured (and used in an initial attack), the lower the effectiveness the Blue system will have. This is also true in nondefense problems. For example, the higher the advertising budget of a competitor, the lower our percentage of total market sales will tend to be. Of course, we can counteract this by increasing our own advertising. Hence there is some relationship between what we should spend and that spent by our competitors in a given area, such as advertising, manufacturing, or offensive/defensive systems.

With this in mind, the problem of system selection might be reduced to the following two key questions:

1. Is the system concept effective enough to warrant the purchase of *any* system units, regardless of what the competitor may do? Namely, are there more efficient ways of coping with the range of competitive possibilities?

2. Given that the system concept is sufficiently efficient, how many system units should be procured?

* See E. S. Quade (1964), Chapter 3, for a more detailed discussion of this point.

The heart of the rationale which McNamara stated that he used for judging the worth of the NIKE-X antiballistic missile system (and for rejecting its procurement) offers a good example of an analytical principle for answering the first question; that is, do we wish to leave the development phase and procure the system in *any* amount? McNamara indicated that the cost of such a defense system to protect the United States against a Soviet massive attack would be approximately forty billion dollars. Thus, if such a system were installed, the effectiveness measure, as measured by United States population surviving, would increase from some level E_1 to some higher level E_2 at the cost of forty billion dollars, as shown in Figure 8-8. But, note that if the United States effectiveness measure were raised by this procure-

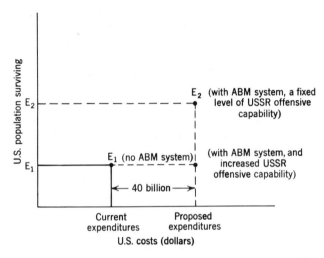

Figure 8-8. Effectiveness resulting from different force expenditures.

ment, the Soviet Union could negate this increase in American effectiveness by increasing the number of their offensive missiles, missile warheads, and penetration aids to saturate the defense and return it to the original level (also shown in Figure 8-8). The key question then is, "How much would it cost the Soviet Union to negate the increase in United States effectiveness?" Analysts have determined these costs to be approximately five billion dollars. Thus, a cost of forty billion dollars on the part of the United States could be negated by a five billion dollar cost by the Soviet Union. McNamara has defined such a situation as a "poor buy" for the United States. A "good buy" would be that system which would cause the competitor to expend a larger cost than our own to negate it (if he chooses to do so). Note that we never know whether he will expend these funds or not. Our only concern is what it would cost if he did decide to negate our increase in effectiveness.

Note that this technique involves analyzing the costs to each side while pivoting on a level of constant effectiveness (i.e., the job to be done).

We shall now consider the second question: If we decide to procure a given system, how much expenditure (number of system units) should be undertaken in the face of uncertainty in competitive expenditures? One way of including this type of problem in the analysis is to show system effectiveness for various levels of competitive expenditures. In Chapter 12, for example, the effectiveness of an improved strategic force management system is displayed as a function of the size of enemy attack. This is called a para-

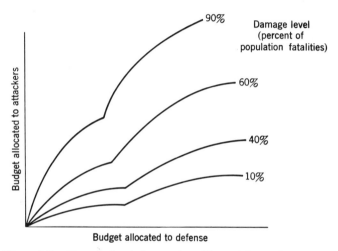

Figure 8-9. Damage levels for varying offense-defense allocations.

metric analysis and permits the decision-maker to see how sensitive the proposed system gains are to the size of competitive action.

In the more practical two-sided case, where there is uncertainty about how much each side will spend, parametric analyses for both types of expenditures may be performed. Such results are described by McMahan and Taylor * for the general war case of an attacker most efficiently spending variable amounts versus a defense efficiently spending variable amounts (illustrated in Figure 8-9). This figure indicates to the decision-maker how much population damage he will absorb as a function of his defense budget, for different levels of attacker offense budget. The decision-maker can then

* See Stephen Enke (Ed.), (1967), Chapter 7.

use this information as an aid in making his defense allocation. In such a two-way strategy, it is a good idea to intensify our intelligence system in an attempt to monitor the progress of a competitor and to see how many units the enemy is procuring and deploying before taking our own action on a large procurement. This is part of the "let the competitor move first" strategy discussed earlier.

UNCERTAINTY IN ESTIMATING NUMERICAL VALUES OF SYSTEM PERFORMANCE CHARACTERISTICS

We shall now discuss how to deal with the uncertainty in estimating the numerical values of the performance characteristics of our own and competitive systems.

As he acquires input data for the model, the systems analyst must recognize that one of the fundamental causes of possible error in predicting system performance characteristics involves the limitation in the amount of available data (i.e., the more limited the number of data samples available, the greater the degree of inaccuracy). Consider the case of the performance characteristics of an American system which may be currently operating, such as the hardened ICBM system. Here the performance estimates can be based largely on past test or operating data. Take, as an example, the problem of determining the probability of success of a particular missile system. Assume that eight out of ten test firings have flown successfully; how well can the analyst estimate the probability of missile success for additional firings, based solely on the above data? If the analyst were asked for one number which would estimate the total probability of missile success, he would probably use the most likely estimate ($\bar{p} = 0.8$). Unfortunately, the maximum likelihood estimate for a small number of data samples will almost certainly differ from the true value of the quantity being estimated. Hence the analyst should provide some indication of the accuracy of his estimate, providing this estimate in the form of a band of uncertainty, called a confidence interval, provides such an indication.* For example, Clopper and Pearson have published the chart shown in Figure A-13 in Appendix I which provides confidence limits for a confidence coefficient of 0.95 and a binomial probability distribution. Thus for any value of the most likely estimate of the probability, and the number of trials on which it is based (e.g., eight successes obtained out of ten trials result in $\bar{p} = 0.8$), the analyst can estimate with ninety-five per cent correctness that the true probability will lie between 0.43 and 0.98. That is, over the long run of such estimates,

---

* See Appendix I for a more detailed discussion of confidence limits.

the estimator using this rule would be correct ninety-five per cent of the time. In addition, such an approach is only mathematically valid if the probability distribution is known (or assumed), such as the case of the binomial distribution described in the above example.

Suppose more information were available. For instance, if 16 successes were obtained from 20 trials, the true probability can be estimated to lie between 0.56 and 0.95. If 80 successes were obtained from 100 trials, the true probability can be estimated to lie between 0.70 and 0.88. Again, it is emphasized that we can never guarantee any of these statements to be true—only that ninety-five per cent of them are true.

Subjective Estimates of Performance Characteristics

Evaluation of new systems which have not been built before leads to a second source of uncertainty (technological or development uncertainty) in estimating system performance characteristics. Again the analyst should use whatever data he can readily obtain, which might include the following:

1. Performance data of laboratory or development prototypes of any system components available.

2. Performance data of any tests of systems or system components which have some relationship to the system under evaluation. For example, if the characteristics of the mobile missile were to be predicted, data of the performance characteristics of a fixed missile would be useful for extrapolation purposes.

3. The use of expert opinion. In this case, operational personnel or system designers who have experience with this type of system would be shown whatever data is available, and asked for their opinions of how the data could be extrapolated. This subjective estimation of performance characteristics is particularly appropriate if no other data are available. Generally the technique of comparative analysis should be used with some related type system (whose performance characteristics are known to some degree) as the basis of comparison.

Estimation of the competitor's system performance characteristics is an even more difficult task. Data on factors such as the size of his force and its performance characteristics are varied and much more limited. The analyst would still apply the same technique in determining the estimate, but he would rely more heavily on expert opinion in such cases because he would be concerned with increasing the accuracy of what can be inferred from whatever limited samples are available. Even so, the analyst would realize that one of the chief difficulties is determining the degree of uncertainty of the estimate. Consider the following example.

The systems analyst receives an estimate of the enemy's system performance characteristics, which includes the expected value he might assume, as well as a range of values which bracket the expected value. He may then ask several experts to examine the original intelligence data independently and provide him with the results of their inferences, including the best estimate as well as optimistic and pessimistic estimates. From these subjective opinions, the analyst might select a majority mean estimate as well as an upper and lower limit for the optimistic and pessimistic estimates. Dalkey, Helmer, Meier, Gordon, Campbell and Brown discuss methods for dealing with group opinions.

Coping with These Uncertainties

Although it may seem obvious, perhaps the most important principle in coping with uncertainties of information is to make maximum use of all data which can be made available within the time and resource limitations of the analysis. This includes operation or R & D test data, theoretical extrapolations or theory that can be used to derive operational test data from related systems to the current system under analysis, and subjective opinions based on the judgment of experienced personnel.

Given that all of these have been accumulated, the analyst should next determine the range of numerical estimates of performance. In the case of uncertainties caused by limited test data, as in the preceding example of performance testing, the analyst can find the confidence limits of performance. In the case of estimates based on the judgment of operational personnel or decision-makers, one can accumulate these beliefs and show the range.

The third step in coping with uncertainties is to test the sensitivity of the characteristic whose value is uncertain (i.e., to find out how important even the range of values will be on the final effectiveness result). One form of this is called a dominance or *a fortiori* analysis; here, an attempt is made to show that a particular system is still preferred in spite of disagreement on some numerical value performance characteristic since even if the worst case for performance were true, the system would still be the preferred one.

One example of this analysis is that of a tactical fighter. During the analysis, the aircraft turn-around time (the time taken between that when a fighter lands and when it becomes refueled, rearmed and ready to go on the next mission) was assumed to be a certain number of hours. One of the decision-makers, upon seeing the analysis, insisted that the turn-around time, based on his experience, would be closer to three times as long. This new figure was then inserted into the analysis and the model re-exercised to note the difference in effectiveness. It turned out that the effectiveness (i.e., the number of targets destroyed) was within one per cent of what had been calculated previously. Thus, the results were basically insensitive to even this large difference in the value of the performance characteristic.

In this type of analysis, the value of the computer becomes apparent. The computer can be used to make additional runs in which the range of values of the uncertain performance characteristic considered important is inserted to determine how the effectiveness would vary with the change in the value of the characteristic. Even the randomness described previously from Monte Carlo simulation, for example, can be removed from these additional runs. This is done by using the same random draws previously made in the simulation, while inserting the new value of the performance characteristic when rerunning the game. Identical random draws can be achieved in a computer through the use of a random generator which, in fact, uses the same set of "random" numbers.

A variation of the *a fortiori* analysis is the sensitivity analysis in which the system effectiveness is calculated as a function of various numerical values of the characteristic. Including this as part of the analysis permits the decision-maker to see how sensitive the final results are to the particular characteristic under question. Again, the value of using a computer for making this repetitive form of calculation is apparent.

If it were found that the range of uncertainty in the performance characteristic were important to the system decision (i.e., this uncertainty could change the selection from one system to another) this uncertainty would then have to be resolved. Thus the final procurement decision would have to be delayed until more data could be obtained, since the uncertainty could not be reduced without obtaining more information. Many times parallel development programs are initiated and continued until additional laboratory or other test data are obtained which can narrow down the uncertainties of the critical characteristics.

Uncertainties in the values of enemy system performance characteristics can be treated in several ways. One way is to conduct an analysis over the entire range of uncertainty for certain key characteristics, to see if the decision (as to which system is superior) would change for different values of the parameter. If the selection is insensitive to the value of the parameter, there is no problem. However, if the uncertainty range is large enough so that it affects the system selection, the American system acquisition decision can be delayed while more data about the characteristics are gathered through intelligence channels. If the risks of delay are high, it is possible to start the acquisition of a parallel American system while delaying setting the maximum number of units to be procured until more intelligence information is obtained.

A second and easier way to treat uncertainties is to "lump together" all of the enemy uncertainties into one uncertainty of enemy expenditures, since these are essentially equivalent. Methods of treating this uncertainty are illustrated in Chapter 12.

KEY PRINCIPLES OF CASE

We shall now summarize the key principles used in analyzing this problem.

1. There is a formal problem-solving process which can be followed in complex problems such as the example presented. The steps in this process were listed in Table 2-1 and the models involved in this process were shown in Figure 4-8a.

2. The system evaluation process involves determining the operational effectiveness of each of the systems by comparing the objectives of the operation and their measures with the performance characteristics of each system as it operates in its environment. This may be done for those operations which involve a series of steps in the operational flow model; these steps consist of the series of events and the transfer functions which combine the system performance characteristics with each of these events.

3. It is important to validate each of the models used with the decision-makers, the operational personnel, and the system designers before proceeding further in the evaluation.

4. There is a need to obtain various data concerning system performance and operational environment, as well as the competitive environment, in order to quantify and exercise the operational flow model. These data may be obtained through operational exercises of current equipment as well as the use of theory for extrapolating the data or expert opinion to estimate the performance characteristics of future systems.

5. There are two ways of exercising the operational flow model: the analytical method and the use of simulation. Each of these methods has its own degree of complexity and accuracy of result. This should be borne in mind and emphasized to the decision-maker to aid him in his allocation of systems analysis efforts.

6. To aid the decision-maker in choosing among system alternatives two ways of structuring the information are found to be helpful. The first is to design each of the system alternatives on an equal cost basis. Then the system which yields the highest effectiveness may be chosen as the preferred system. The second approach is to design each of the systems for an equal effectiveness basis, particularly if a critical level of effectiveness is required (such as to achieve "assured destruction"), and then choose that system which requires the least cost.

7. Various ways for coping with uncertainty of data have been discussed. These involved determining how much uncertainty exists in the data, and conducting sensitivity analyses which indicate how much of an impact this

uncertainty of data has on the decision-making process (i.e., what the influence on effectiveness or cost may be). If it is determined that the uncertainty has a large effect on the decision, several alternatives should be continued while seeking to obtain more data.

Strategic uncertainties can also be dealt with by determining the various possibilities of competitive or environmental actions. The final system chosen should be one which performs almost as well under the varying conditions as the best of all possible systems could. This "preferred system" is determined through systems planning by focusing on the weaknesses of each of the key system contenders.

8. In addition to the quantitative outputs previously described, it is good practice to assemble all qualitative considerations which could not be quantified in the analysis into one section so that the decision-maker can also focus upon these in reviewing the analysis as a further aid in making his decision.

IV

PRINCIPLES OF
RESOURCE ANALYSIS

9

Resource Analysis

The analyses in the preceding chapters focused on the problem of determining how well different system alternatives would accomplish an operational objective as measured by some effectiveness measure. In general, each alternative system configuration will provide a particular level of effectiveness (different from its system competitors). Also, each system alternative will require the expenditure of a different set of resources which will have to be procured before the system can be procured, phased into operation, operated and maintained over its system life. To decide which of several different system alternatives should be selected, the system planner can pivot on a constant level of effectiveness as follows:

1. Choose an arbitrary level of effectiveness which appears to be appropriate to the problem.
2. Determine the number of complete system units of each alternate system type required to meet this given level of effectiveness.
3. Choose that system which requires the smallest amount of resources in meeting the effectiveness level.

Alternatively he can pivot on a level of constant cost, as was done in the strategic force planning problem. This requires that the systems planner be able to estimate the amount of scarce resources required for each alternative. This chapter focuses on that task which will involve such problems as ways of translating scarce resources into common denominator(s) of cost, determining which costs should be included in the analysis, and how to combine costs which occur at different times over the system life. Methods of coping with cost uncertainties are also discussed, including techniques for translating performance uncertainties into cost uncertainties.

It should be noted that the fields of cost analysis and estimation of system cost are specialized areas in themselves and require the contribution of a trained cost analyst. It is not intended that this chapter provide wide coverage of these fields. Rather it is the objective to describe some of the economic aspects of the problem which a systems analyst should consider while performing a systems analysis, the various key principles involved, and how they are applied.*

PROBLEMS INVOLVED IN ESTIMATING COSTS OF REQUIRED RE- SOURCES

Several problems arise if either of the system selection criteria defined is to be used. The first problem is in being able to compare one total set of resources required for the first system against a different set of resources required for the second system. For example, in comparing the superhard- ened missile silo system against the mobile missile system, how does one compare the large amount of construction material required for the first sys- tem with the large amount of vehicles required for the second? For long range systems planning problems of this type, dollar or monetary costs are a reasonably good common denominator for measuring diverse resources. However, even here difficulties will arise in treating certain types of re- sources required, such as lives lost in combat. Difficulties also arise in using dollar costs to represent resources when dealing with shorter range planning problems, where time dependencies are present in obtaining these re- sources. For example, in the post-World War II years when fissionable ma- terial for nuclear bombs was in very short supply, the design of the strategic force had to strongly consider this scarcity, since quantities of this material, over a limit which varied with time, were not available at *any* cost (consider- ing the production facilities fixed). The difficulties involved in properly dealing with such nonmonetary costs is discussed later in this chapter, but until then we shall use dollar costs as the common measure of resource re- quirements.

To obtain such dollar costs, the analyst must identify all of the costs in- volved in adding the proposed system to the organization. Since a complex system may contain a large number of operating systems as well as support system elements, a systematic procedure must be created so that no cost ele- ment is omitted. Here the analyst can create several cost models, each ad- dressed to a different phase of the system life cycle [i.e., research, develop- ment, test and engineering (RDT&E), investment, and operations and main- tenance (O&M)].

The investment cost model consists of a series of arithmetic expressions for determining the total cost of procuring all of the system units (including

* For the reader interested in more detailed information, a rather complete bibliog- raphy of this field, as compiled by Dr. Martin Jones of The MITRE Corporation, is included.

replacements for units which are destroyed, such as in training or in combat). The operations and maintenance cost model predicts the cost for operating the system and replacing equipment malfunctions while keeping it at a specified level of effectiveness during the entire system operating life. The RDT&E cost model predicts the cost of engineering efforts needed to be performed before the system can be manufactured. Many times this cost is included as a prorated part of the selling price of the system. In the defense industry, however, the government generally purchases the RDT&E for a system as a separate item. Hence, RDT&E costs must be included separately in the tabulation of total system cost.

Many times the useful life of two systems being evaluated will differ, and this difference must be taken into account in the analyses. Even if the useful life were the same for both systems, the "cost streams" (i.e., the expenditures required over time) may be different, even though the total cost is the same, as in Figure 9-1a. Thus, there is a problem of how the time preference of expenditures should be taken into account.

Another problem is how to consider those elements of an existing system which could be used as part of the improved system. In some cases these elements can be used for another purpose. In other cases, they are of no use at all. Thus, the systems analysis problem concerns itself with whether to include the worth of these elements in the analysis and, if so, how to deter-

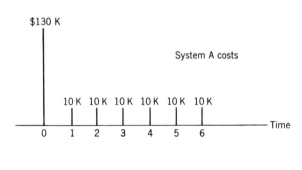

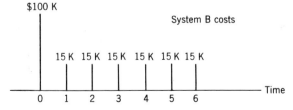

Figure 9-1a.   Different cost streams.

mine the correct worth (i.e., is it original cost, depreciated cost or book value, or salvage value?).

Besides being of use to a decision-maker or budget director, cost information is of great importance to the systems planner. If the planner is configuring constant cost systems, he needs to know how total costs increase as additional units are added to a system configuration. In addition, the systems planner must make a number of performance tradeoff analyses to arrive at the preferred system, and these require performance versus cost information. Chapter 13 discusses the tradeoff analyses needed to determine whether a small number of sophisticated systems will do the job at lower cost than a large number of simpler systems.

UNITS OF COST

Before proceeding with a discussion on how to estimate the true total cost of a system, we shall first discuss the meaning of the term "system cost" and the units by which it is to be measured. Critics of cost-effectiveness analysis as practiced by the Department of Defense have stated that the systems analyst is only concerned with monetary costs (dollar costs) of the system and ignores other important factors such as lives lost or casualties. If this is the case, it is a criticism of inadequate analysis rather than a deficiency of the systems analysis approach, for proper systems analysis will indicate to the decision-maker *all* scarce resources which are required to implement a system alternative. The difficulty comes in properly identifying and finding appropriate units of the many different elements required for system implementation. We shall now demonstrate how this can be done in a fashion meaningful to a decision-maker, the real test of the credibility of the analysis.

What Are the Real Resources?

The real resources which any system requires for full implementation include the following.

People. Various different skills might be required for a wide variety of activities. Such activities would include system and component development and testing, production, operation, and maintenance.

Natural Resources. Examples of natural resources include metals to be used in the production of new system components which are required.

Facilities. Various capital facilities and equipment would be needed for manufacturing, storage, transportation, etc.

Obviously, if these limited resources are used during the time that one system alternative is being implemented, they are unavailable at the same time for other alternatives which would satisfy other needs. One way of expressing these limitations is to translate the amount of resource used into its monetary equivalent (dollars required) to obtain the resource over the total time it is needed. Measuring resources in dollars offers the main advantage of combining different types of resources by providing a common denominator for expressing the worth of a resource. A second advantage of using dollars to represent the resources involved in a system or collection of systems is that budgets, such as the United States budget submitted to Congress, generally use this as a measure.

Problems in Using Monetary Cost

Several problems are involved when dollars are used as the only measure of cost in evaluating alternate systems. Several of these problems are now discussed.

Loss of Resources. In performing an analysis of many systems, it can be anticipated that there will be some attrition of system elements such as hardware, facilities, or even people, because of equipment malfunction, accidents, or combat attrition. This is particularly true in the case of defense systems where losses in training and combat are expected and must be included in the planning. Such predicted losses may be included in the total system cost estimate by calculating the dollar cost of replacing parts or equipment. While the the dollar cost of replacing people can be estimated by considering such items as hospitalization insurance, pensions and replacement training, such costs might omit the cost of time delays in obtaining system replacements. Spare parts kept in storage (at a cost) to anticipate such needs would reduce, but not eliminate, such delays. Casualties, particularly loss of life, present a separate problem. It is impossible to replace a lost leg or life as can be done with a malfunctioned machine, and such system costs have an impact on the morale of an organization or entire nation. Thus it is necessary to measure the predicted lives lost and casualties incurred as a separate system cost, in addition to monetary costs, when this is a differentiating cost among alternative systems.

Maximum Resource Limitation. Sometimes representing a particular scarce resource by a per unit dollar cost is an inadequate measure, since it omits the maximum amount of this resource which can be made available. For example, in post-World War II years, fissionable material for nuclear devices was in particularly short supply. Thus, systems requiring such material were given close attention and system tradeoffs were conducted to obtain systems which provided a given level of effectiveness by using acceptably small amounts of fissionable material. A dollar cost could be used to describe the cost of this material by performing the following calculations. Assume that the total output of fissionable material was 5,000 pounds per year and the total cost of operation was one hundred million dollars per year. Thus, the cost of fissionable material was $20,000 per pound.* The use of such a cost factor alone would hide the fact that there were only 5,000 pounds of material available per year. Thus the systems analyst, aware of such shortages, needs to call the decision-maker's attention to the need for a certain required amount of this resource as a separately identified system cost.

Time Dependency. As indicated in Chapter 2, cost and effectiveness represent two of the three key dimensions of a systems analysis. Time (the third key dimension) is an important factor in the analysis in several ways (e.g., the time when the system will become operational). In the analysis of Part III, it was assumed that all system alternatives had to be available at a time which was related to the enemy's buildup of missiles over time (shown in Figure 5-2).

The process of changing an existing system within an organization may be looked upon as translating a given set of resources. On the one hand, an organization can phase out or eliminate certain systems, equipments or facilities, using whatever salvage value is obtained for other purposes. On the other hand, certain resources can be expended to produce other resources. For example, a farmer can expend energy for plowing, fertilizing, and seeding to obtain fruit and vegetables. Likewise, manpower and laboratory facilities can be expended in a development program which will provide new technological elements, such as a high power laser or a microwave tube which can be used as part of a new system. If the nuclear material production rate mentioned earlier were deemed insufficient, additional facilities could be built, and trained people hired to enlarge the production capability. All of these translations, however, generally require the ingredient of time.

In some cases, such as farming, the time requirements may be independent of the physical resources required. In other cases the required resources are related to the time requirements, and the tradeoffs between time and resource costs should be considered explicitly. Consider as an example that the surgical technique of transplanting human kidneys was perfected and an additional 10,000 surgeons were needed in the United States to perform such surgery for these individuals per year who otherwise would die if

---

* As long as no more than 5000 pounds per year are needed. Any further needs have essentially infinite cost if the nuclear production facilities are considered fixed.

they did not receive such an operation. Various plans can be formulated to build up to the additional 10,000 level. For example, by using the existing training facilities, the needed number of physicians can be reached in t_1 years (or $t_2 < t_1$ if overtime and summer training is utilized) as shown in Figure 9-1b. By building larger facilities and hiring additional teachers, physicians can be trained in less time but at a greater monetary cost. Finally there is some minimum time t_3, below which it is impossible to do the job, using the operational concepts which have been identified by the planner. By examining the cost in time and dollars to build up the other system elements required (such as hospitals, nurses, etc.) a tradeoff between total dollar cost and total time required for the various alternative plans can be illustrated as shown in Figure 9-1b. Such a function is needed to choose the proper plan which will be based on total dollars and number of needless deaths which would take place during the system buildup.

The same sort of transfer function holds true in transporting supplies from Boston to Los Angeles, for example. It is possible to use railroad transportation for a given time and dollar cost. Jet air transportation decreases time, but may raise cost. Thus this concept of examining various ways of performing a job and keeping separate time and dollar costs is a good concept when no critical operational time constraint can actually be specified. Of course, only by conducting a system effectiveness evaluation can the analyst determine the impact which time has on system effectiveness (and consequently, dollar costs). Such an analysis would quantitatively determine the proper time-dollar operating point. If this analysis is not conducted, the decision-maker is forced to use his intuitive judgment to select the final operating point. Occasionally one system will dominate all others by having the lowest cost in both measures. For example, with today's state of technology, the use of electronic data processing equipment to perform routine tasks such as payroll and record keeping operations is lower in both time and total dollar costs than a manual method of performing these jobs.

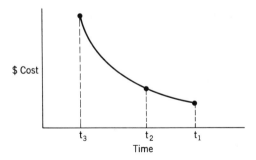

Figure 9-1b. Trade-off of time versus dollars.

Key Principles in Handling Incommensurable Resources

The following list summarizes the preceding discussions for coping with required incommensurable resources.

1. Define the job to be done and the level (or alternate levels) of effectiveness to be attained by each of the competing systems.

2. Determine the time when each system alternative will become available operationally.

3. Design all systems to provide the same level of effectiveness.

4. Determine the different costs of each system in terms of dollars, lives lost or casualties, as well as any other scarce resources required to perform the job.

5. Examine the costs to see if any required resources exceed the constraints of the problem. If so, attempt to redesign the system to reduce the particular critical cost, even at the result of raising other costs.

6. Present the decision-maker with the final, multidimensional costs of each acceptable system alternative, and its time of availability. The decision-maker then chooses on the basis of his set of preferential values.

DEVELOPMENT OF KEY PRINCIPLES OF COST ANALYSIS

Having discussed the problems of costing the nonmonetary resources required in implementing a system, we shall now consider the problems previously delineated in estimating the total dollar cost of a system. These key principles used in cost analysis are now developed by means of several illustrative problems. The first of these problems is a classical example of engineering economy * and illustrates the following principles:

1. How to combine costs which occur at different times over the system life cycle.

2. How to evaluate and select a preferred system from alternatives which differ in both cost and useful life.

SELECTING A BRIDGE

A new bridge is to be built across a river which is currently serviced by a ferry. Two alternative bridge designs are being considered. Each bridge will handle the same maximum traffic capacity; however, Bridge A is stronger and, hence, is expected to have a longer life than Bridge B. In addition,

* For a more detailed description of these principles see E. L. Grant (1950).

Table 9-1.

Cost Elements	Bridge A	Bridge B
Initial Cost	$3 M	$2 M
Maintenance	$10K/year	$30K/year
(Operation same for both)		
Expected Life	30 years	15 years
Salvage Value	$100 K	$50 K

Bridge A is constructed of a new steel material which develops an oxide coating and does not require painting, whereas the maintenance expenses for Bridge B are higher, primarily due to the need for repainting. The expected costs for each bridge, as well as its expected operational life, are shown in Table 9-1. These costs include the costs of original construction and maintenance, as well as the salvage value. The operating costs (e.g., toll collectors) are the same and hence have not been included in the analysis.

Systems Analysis

Since both bridge alternatives are designed for equal effectiveness (i.e., equal traffic capacity) the problem narrows down to determining what the total lifetime costs will be for each of the two alternatives. There are two problems which must be taken into account in performing the cost analysis. The first is that each of the systems has a different system life (30 years for Bridge A versus 15 years for Bridge B). This may be handled by taking the lowest common multiple, 30 years, and accumulating the total costs which would occur over the 30-year period. These costs to be considered for each of the two system alternatives are plotted as a function of time in Figure 9-2. Each of the two sets of cost is called the "cost stream" which occurs over the total system life. In this example the investment cost of the bridge is paid at time zero, the start of operations, followed by each of the yearly maintenance costs (end of year payments are assumed in this analysis). At the end of 15 years, a new Bridge B is built at a cost of two million dollars minus the fifty thousand dollar salvage credit.* Salvage credit for Bridge A occurs at the end of 30 years.

---

* Note that the following assumptions have been made to define a simple problem which will be useful in describing certain economic principles:

1. Each bridge has a different design life. This assumption may not always be true when proper maintenance provides an indefinite system life. It is introduced only to show how to cope with systems which do have different "wear out" characteristics.

2. Traffic capacity is the same at the end of 15 years. Techniques for handling traffic which will probably increase over time are described in Chapter 11.

3. Construction costs of the second Bridge B will be the same as initial costs. Techniques for including inflation effects for both construction and maintenance costs use the same method to be described.

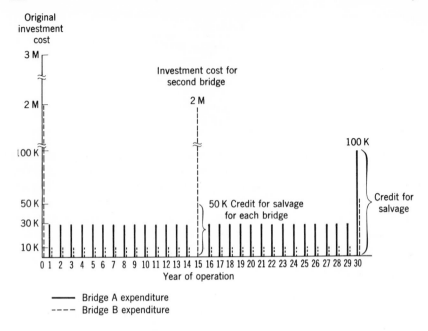

Figure 9-2.   Cost stream of expenditures.

The second problem the analyst has is that he cannot merely sum the total costs over the 30 years in a direct arithmetic fashion. One reason for this is that the state government must solicit bonds to raise the initial costs of whichever alternative is chosen; hence, interest on the bonds must be paid annually to the bond holders. One way of taking this interest cost, or future enhanced value of money, into account is by determining the "present worth" of each of these future expenditures. This is done by reflecting each of the expense terms back to some common reference time such as the zero reference, by means of some compound interest function, and accumulating each of the reflected cost elements. For example, as shown in Figure 9-2, while two million dollars is needed for Bridge B, 15 years from the initial starting time, this sum is equivalent to the smaller sum of money needed at time zero, based on some effective interest rate which the organization ascribes to the worth of money. This interest rate could be the same as the normal cost of money (i.e., the same interest rate which the organization would have to pay on its bonds). There are other reasons why the interest rate to be used in the analysis might be even higher. These reasons are discussed later in this chapter. For this problem, however, we shall assume that the bond interest rate will be used throughout. The formula for determining the present worth of $2 million to be spent 15 years from today is given below:

$$P = S(1 + i)^n = S(SPPW),$$

where P = present worth,
S = future worth,
i = interest rate,
n = number of years in the future.

SPPW is the "single payment present worth" factor, sometimes called the "amount of annuity." Values for this factor which are pertinent to this problem are given below.*
SPPW = 0.4810 for i = 5% and n = 15 years.
SPPW = 0.2314 for i = 5% and n = 30 years.

While the same approach could be performed for each of the smaller periodic payments, an easier approach has been formulated. The formula for the present worth of an entire uniform series of periodic payments is as follows:

$$P = R(USPW)_{i,n} = R \; \frac{\left[1 - (1 + i)^{-n}\right]}{i} \, ,$$

where P = present worth,
R = future payments,
i = interest rate,
n = years in the future.

USPW is the "uniform series present worth" factor, sometimes called the "present value of annuity". Values for this factor which are pertinent to this problem are given below: †
USPW = 10.38 for i = 5% and n = 15 years.††
USPW = 15.37 for i = 5% and n = 30 years.††

The present worth calculations for each of the two bridge alternatives are given in Table 9-2. Note that the more expensive investment, Bridge A, is still slightly less expensive in total cost than Bridge B, but not by as much as if arithmetic additions of cost would have been taken, omitting the interest rate.

* These values are obtained from standard interest tables which provide values of P for different values of i and n. Such information is contained in C. D. Hodgman, "Mathematical Tables from Handbook of Chemistry and Physics," Chemical Rubber Publishing Co., or Grant (1950).
† From Hodgman "Mathematical Tables from Handbook of Chemistry and Physics."
†† Note that these factors would have been 15 and 30 respectively had not interest been taken into account.

Table 9-2. *Present Worth Calculations*

Initial Cost	BRIDGE A (thousands) $3000	BRIDGE B $2000
	2000(.4810) =	962
Maintenance Costs	10(15.37) = 153.7	30(15.37) = 461.1
Total Expenses	$3153.7	$3423.1
Salvage Income	100(.2314) = −23.1	50(.4810) = 24.05
		50(.2314) = 11.57
		35.62 = −35.6
Total Cost	$3130.6	$3387.5

We shall now summarize the principles involved in the preceding example. A systems planner is always faced with the problem of deciding how much to spend for the initial investment costs of a system. If he builds more quality into the system (at greater initial cost), he may benefit by having the system last longer before it needs replacement; or related to this, he may reduce the maintenance costs needed to keep the system operating.

The role of the systems planner, who is concerned with total system life, is to design a system for minimum total cost. To calculate this total cost, he must sum all the costs which occur over time by reflecting them to some common reference time, such as the starting time. Thus, the costs for various systems would be compared on the basis of present worth.

COPING WITH UNCERTAINTIES

Several other factors which involve aspects of uncertainty must be taken into account in the evaluation before a final selection can be made, since these factors reflect into the cost calculation. The previous calculations were based on an implicit assumption that the operationally useful life of the bridge would be equal to its life to wear out. There are other factors, however, which may reduce the life of each bridge.

One factor to consider is the uncertainty of future traffic; another is the bridge's resistance to hazards, such as flood or fire, which may differ for each bridge. Each of these factors will now be discussed.

Probability of Bridge Destruction

Even with careful designing there is some probability that either of the bridges may be destroyed by some hazard, such as a flood. One way of coping with this factor is to purchase an insurance policy whose premium will be based on the probability of the bridge's being destroyed (or damaged) by flood during a given period of time. Another way of taking the possibility of damage into account is to self-insure by including in the yearly cost the expected cost of replacing the bridge, based on the probability of destruction and the replacement cost of each bridge.

The most accurate estimate of the probability of destruction can be obtained by using statistical data relating to the particular bridge's construction. If only a small amount of data is available, considered judgment could be used, based on some relative or comparative basis. Such an estimate of the probability of this occurrence was made for each bridge, taking into account that Bridge A has been built to be stronger than Bridge B. The figures used are:

$$P_{dA} = 0.002,$$
$$P_{dB} = 0.005 \text{ per year.}$$

Thus, we could obtain an expected value of the loss of the bridge and include this in the cost calculation, as indicated in Table 9-3. On the basis of this calculation, Bridge A's cost advantage is even greater over the assumed 30-year life.

Uncertainty in System Demand

The previous problem of examining bridge life was mainly inserted to show how to deal with systems having different operational life spans and how to combine system costs which may occur at different time periods. However, one of the most important factors which affect system life is the future system demand function which, in the context of our bridge problem, is the traffic expected to use the bridge. In the previous section we assumed that we knew what the traffic would be over time and chose a bridge size which could suitably meet the traffic density. Suppose, however, that the traffic demand over time grows at a larger rate than is anticipated, such that the maximum capacity of the 30-year bridge is, for example, exceeded after only 20 years. In this case, the bridge may become obsolete and some other way of handling the increased traffic must be found. Hence the total costs may become much greater than originally predicted, and perhaps another alternative would have required lower costs. On the other hand, the traffic may grow at a slower rate than expected. In this case the bridge size, or ca-

Table 9-3.

	BRIDGE A	BRIDGE B
Previous Total Cost	3130.56	3387.48
Expected Destruction	(.002)(3000 − 100)(14.37)	(.005)(2000 − 50)(15.37)
	= 83.40	= 149.80
Total Cost	3213.96	3537.28

pacity, would have been designed at too high a value and, hence the costs would be higher than they need have been. In fact, the decreased toll collections may now be insufficient to pay off the bonds. Thus some systematic way must be found to cope with an uncertainty in the system demand function (i.e., expected traffic), so that the system capacity may be properly set. This part of the problem is discussed in the bridge problem of Chapter 11.

SUMMARY OF MAIN PRINCIPLES OF COSTING

*Sum the total cost of the entire system over the total expected system life.** It is in the implementation of this simple principle that many problems arise. The purpose of this section is to discuss a number of these problems which, unless the systems analyst is aware, may lead to the following sources of error in predicting system cost:

Make Certain that the Three Main Cost Elements are Included

It is possible to tabulate all of the system cost elements which occur as a function of time, and structure these into three main classes as shown in Figure 9-3. The three main cost elements are those costs involved with:

1. Research, Development, Test, and Evaluation (if applicable);
2. Investment of equipment elements;
3. Operation and maintenance.

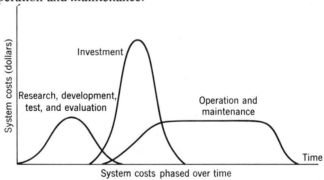

System costs phased over time

Figure 9-3. System life cycle costs.

* The systems analyst wishes to calculate and display for the decision-maker both the total system cost as well as the "cost streams" as shown in Figure 9-3. The advantage of displaying cost streams is to aid the decision-maker in obtaining a stable budget over time. Thus, a given system may be lowest in total system cost but it may inject a very sharp high cost at a given time which may be inappropriate for an over-all stable budget. This factor must be taken into account in considering total budgets of several systems.

Too often only the investment costs of the original prime equipments are considered. Frequently, research and development are needed before the prime equipments can be manufactured. These RDT&E costs must be included as part of total system costs.

Many times the operating and maintenance costs are equal to, or several times greater than, the original investment costs, and thus may comprise a substantial proportion of the total system costs. Consumers, many times, tend to ignore operating and maintenance costs, which for equipments that may last many years may be as great or greater than the cost of the original investment. For example, in purchasing a television receiver consumers tend to concentrate on the initial purchase or investment cost and ignore maintenance costs. Surveys have shown that there may be large differences in reliability, and hence in the maintenance cost of a TV receiver. For example, while the investment cost of a completely transistorized receiver may be greater than a vacuum tube receiver, maintenance costs should be less. Hence, both costs should be considered in the decision-making process when considering various TV receivers. The same considerations are true when choosing between an electric and gas clothes dryer. While the electric dryer may have a lower initial investment cost (perhaps $40 lower, and perhaps $20 lower in installation), the electric dryer normally costs approximately ten cents per load to operate whereas the gas dryer costs approximately four cents per load. Thus, even without considering the maintenance cost, which in general is quite low for this type of appliance, one could calculate the number of loads required over the total system life in which the gas dryer would, in fact, be less expensive than the electric dryer. This particular example, however, leaves out one factor of uncertainty: the number of moves which would be required over the total system life of the appliance. Every time the gas dryer is moved to a new house, a new twenty-dollar installation cost may be required which is not needed with the electric dryer. In fact, the owner may move to a house which does not have gas service at all; thus, he might be forced to sell the dryer and purchase another one.

A third example is gas versus oil heating. Normally an oil heater is more expensive to buy than a gas heater, although in a new house a builder will often supply either one upon the owner's option. Oil heat is approximately 20 to 30 per cent lower in fuel costs for the same heat output. An oil heater, however, requires annual maintenance which is generally covered by a service contract which may cost $15 to $30 per year. Again, the tradeoff between investment cost and operating cost can be made.

Uncertainty in Predicting System Life

If the analyst were to calculate the cost of operating and maintaining the system, he must find some way of predicting the total useful life during which the system will be in use. In general, system planners who are advocating a system change tend to predict longer useful lives for their systems than can be permitted in the analysis, particularly when the RDT&E and/ or investment costs are high, since extending the total operating life will obviously tend to lower the average total cost per year of service life. Recall that this was the case for the 30-year bridge versus the 15-year bridge. Unquestionably, an aircraft carrier or battleship could physically last 20 to 40 years, but can this operating period be used for costing purposes?

While such systems may, in fact, physically last for the longer predicted times, as World War II carriers have lasted, there are many uncertainties which may prevent these systems from being used for this entire time. Some of these reasons are listed below:

Technological Improvements. Technological improvements continue to occur so that the same job may be performed at lower operating or maintenance cost using new technological elements. The vast improvements in the field of computers serve as a good example of this point.

Competitive Improvements. A competitor may obtain a new technological element and employ this in his system thus rendering our system obsolete. Some defense examples of this are as follows:

1. Missiles and high-speed aircraft helped to render the battleship obsolete since it is too costly to properly defend battleships against air attack, for example. Hence, the Navy continued in the development of the aircraft carrier and Polaris (the missile-carrying submarine) as the most efficient offensive weapon.

2. Surface-to-air and high-speed interceptors rendered slower, high altitude bombers obsolete as primary offensive weapons for general war purposes.

3. There is always the possibility of a new breakthrough in antisubmarine warfare detection equipment which may render the submarine obsolete.

4. Industrial examples would include a competitor installing new automated manufacturing equipment which would permit the competitor to reduce his manufacturing or operating costs and thereby reduce his selling cost. Such a situation would force us to find a way of similarly reducing our manufacturing costs in order to meet the new selling price.

Changes in Demand. The predicted demand for our product or service may change. Thus, specialized manufacturing equipment which was installed specifically for a particular type product may be rendered worthless if the demand for this product is radically reduced or eliminated. A related example occurs when new highways are constructed, by-passing the restaurants and gas stations on the old highway. The worth of these investments may now become greatly reduced.

Uncertainty in Predicting Malfunctions. As equipment ages in its later years, equipment malfunctions may become larger or more frequent than was originally predicted, thus increasing the tendency to retire the equipment sooner than anticipated.

All of these uncertainties result in strong pressure for assuming a conservatively small system operating life when performing a system cost analysis. The Defense Department has been performing cost analyses by assuming five years of operation (although there is some pressure in the Defense Department in recent times to increase this to ten years). In either case, when comparing two systems which may have a different operating life, such as when considering a missile versus an aircraft carrier task force, quite often the analysis is made on a consistent (if arbitrary) basis, using five years of operating life. The decision-maker should be provided with total system cost as well as an indication of the cost streams involved. The latter is useful since it indicates any peaks in the over-all organization budget which may have to be absorbed. This point illustrates a factor which many consumers who purchase on the installment plan tend to ignore. They become so accustomed to cost streams (i.e., monthly payments) that they many times lose sight of the total investment cost involved.

Industry tends to perform cost analyses on the same basis, though sometimes using even shorter life predictions. It is not uncommon to use a system life of three years to factor in the technological obsolescence possibility when reviewing a request for a new capital appropriation. Of course, industrial systems analyses have one advantage over defense type analyses since their main measure of effectiveness is corporate profit. It is possible to develop a model which will relate the total cost elements over the total system life to the expected change in corporate profits over the same system life. Thus, many companies will reject a request for capital appropriations if the system will not "pay for itself", including all cost elements, within a three-year period. Obviously, the analysis of certain longer range capital expenditures such as a new building which could be converted to other uses if particular uncertainties materialized, could assume a longer, more realistic system life. Finally, if there is some uncertainty regarding the useful system life, the analyst can conduct various sensitivity analyses with respect to useful life time to aid in comparing alternate systems and seeing how sensitive time differences are in making the final decision. For example, he may vary system life from five to seven years, and see to what extent this changes the cost ranking of alternative systems, or he may determine the break-even point;

that is, how many years of system life are required when the costs of alternative systems would be equal.

Insertion of Interest Rate. Most engineering economy studies use some interest rate, thus charging for the use of money, when conducting cost analyses. This approach was illustrated earlier when considering the bridge alternative. There is justification for using an interest rate since generally the money will come from a bank or some other lender. If the corporation's own funds are used, they could be invested in another alternative, such as bonds or bank deposits. Thus, regardless of whether the money is loaned or comes from the corporation fund itself, corporate studies should insert some interest factor. The problem is, what interest rate should be used? There is, of course, a lower limit to this factor, since if the money is loaned from a bank the interest rate which the bank charges could be used. On the whole, however, larger "effective interest rates" are generally used in analyses to compensate for the risk to the capital investment, some corporations using an interest factor as high as 20%.

In Hitch and McKean's discussion of this problem,* they rejected the use of an effective interest rate since they did not know what correct value of interest rate to use, and felt that technological and competitive uncertainties could be accounted for by including only five years of operating life in any system proposal. This provides the same effect as using a high interest rate since such a high rate would greatly discount the present worth of cost elements which occur in later years. In addition, this simplifies the cost calculations by eliminating the use of interest tables. However, the decision-maker must still analyze the cost stream of each alternative, since postponing expenditures may be desirable. In more recent years, the Department of Defense has been tending to introduce an interest rate of 5 to 10% in their economic analyses of proposed defense investment projects, such as computers, equipment, and facility modifications which tend to reduce future operating costs.

Properly Handling Existing Equipment

In analyzing a given system need, the analyst should attempt to determine what system elements already exist which can contribute to the system need. Generally, there are some equipments available which can be utilized to some extent within the improved design. Such inherited equipments which are of use in the new system design are called "sunk costs." They are beneficial since they help to reduce new "out-of-the-pocket" costs which would have to be incurred if these equipments were not available. Thus, the system

---

* C. J. Hitch and R. N. McKean (1960).

planner should be alert for any way of using assets already available. One such example of the use of available equipment was the addition of the air-to-surface missiles to the B-52 in order to improve the over-all effectiveness of the bomber force since it would not have to penetrate local surface-to-air defenses. The only "out-of-the-pocket" costs, or additional amount of expenditures required for this proposal, would be for the air-to-surface missiles and their maintenance equipment (plus any modifications which would have to be made to the B-52 aircraft for carrying and releasing the missiles).

Frequently, sunk costs implicitly aid the systems planner, as when a decision has been made to order more quantities of these same equipments. In this case the same engineering drawings, documentation, production tooling and know-how gained are immediately useful for the additional order. Note, however, that if a production line were reopened to accommodate the additional order, the costs of reorganizing the assembly line would have to be borne as part of the procurement.

Occasionally an obsolete system may be modified and used for some other purpose; in such a case it may be possible to "trade in" the old system and thus reduce the net cost of procuring a new system. Other times a system may have a salvage value, as in our sales of obsolescent weapons to friendly, allied nations. Often obsolete industrial equipments may have such low salvage value that it actually costs money to disassemble the old system and cart it away.

Remember that sunk costs are not directly relevant to the systems decision. They enter into the decision implicitly in that presumably the incremental costs involved in the new system would be less because of the sunk costs already involved, as compared with adding a completely new system which does not use those elements whose costs have been previously paid for. Unfortunately, errors are sometimes made in dealing with sunk costs as illustrated by the next example. This example concerns an industrial firm that several years before had purchased, rather than leased, a large electronic computer for their engineering analysis department. For various reasons, the accounting department had decided to depreciate the computer over a period of ten years rather than follow the usual practice of perhaps four years. Obviously, this meant that on paper their depreciation costs were much lower than they would be if the shorter time period had been used. The particular electronic computer was of the vacuum tube variety, having been purchased before solid-state computers were available. Thus, the maintenance expenses were much higher than a currently available solid-state computer, as were the operating costs for the expanding workload, as compared with a faster machine. The question which then faced the company was, "Should we trade in the computer for a new, more reliable and efficient solid-state one, or continue to operate in the present fashion?"

Constructing the cost streams for each of the two alternatives under study would indicate that the analysis becomes a matter of comparing the high initial investment cost plus the low operation and maintenance costs of the new computer (minus the small salvage value of the existing computer), versus the high operation and maintenance costs associated with keeping the existing computer over the next four years (the usual life of the new computer).

The error which some firms make in coping with this problem is their uncertainty of how to treat the difference between the current book value of their equipment, such as the old computer, and its actual smaller value. Many times they add this loss in book value as a cost element associated with making the change to the new computer, and then find they cannot afford to change to the new equipment. It should be emphasized that this loss has already occurred whether they trade in the computer now or not, since when they originally purchased the computer, they assumed a commitment to pay for it and this commitment must be met. The time taken for equipment amortization is a separate issue and depends on many other factors including cash flow, tax laws, the desire for the manager to show a higher profit over the short run, etc. If, in fact, the computer's useful life was overestimated in the original analysis, a new decision must now be made, and a second error should not be permitted to compensate for a past error. Thus, the principle for coping with existing equipment is to cost only those additional elements needed to make up the system. This can be done by explicitly ignoring any inherited resources or equipments (i.e., "sunk costs"). It may also be done by totaling the cost of all resources required and subtracting from this the current worth of the inherited resources if these were to be purchased today. This is the concept of incremental costing which ignores any sunk costs. Presumably, if one has high sunk costs already, it should not take much more to continue the system over the useful system life, so that sunk costs are implicitly factored into the decision.

Good judgment must be used in considering whether sunk costs should be ignored. For example, consider the possibility of a system proposal for a military airborne command post. The airframe recommended in this proposal was the new C-5 aircraft. When the systems planner was questioned as to why the cost of a C-5 aircraft was not included in his proposal, he replied that the Air Force had already made the decision to purchase these aircraft, and hence their costs were irrelevant! Unfortunately, what the systems planner overlooked was that the Air Force decision for purchasing C-5's was for a specific military airlift command need. Since the additional C-5's would be required if the airborne command post proposal were accepted, these costs would have to be included in the system proposal.

Next, examine the same proposal where the planner is now considering the refitting of existing C-130 aircraft for the airborne command post function. He must now include the cost of refitting the aircraft but does not include the original cost of the C-130 as long as the military airlift command had the C-130 aircraft available and had no other plans for their use and was planning to phase them out of operations. If, on the other hand, the command decides that they require additional cargo aircraft to replace the C-130 aircraft to be used, the problem should be regarded as a joint decision problem. That is, there are two jobs involved—one for an airborne command post, and one for airlift needs. The decision should then be made in terms of what equipments are required to perform both of these tasks. Thus, it could be most economic to use the C-130 aircraft for the airborne command post and have as its replacement C-5 aircraft which can haul large loads over large distances less expensively than can C-130 aircraft. Each case should be judged on its own basis and may include the performance of more than one military task.

Include the Indirect "Cost Elements"

There are many indirect supporting system elements that are required in operating the total system. These include manpower and facilities for administration, supply, food services, training, etc. Fortunately, there are many planning factors to aid in determining the manpower and facilities required for these indirect elements as a ratio of the direct system element. This point is discussed later in the chapter along with the costing of the missile systems of the strategic planning problem.

System Attrition

The cost elements included thus far should make provision for system replacement costs due to equipment malfunctions, peacetime training, or operational failures. In time of war, however, defense equipments are subject to attack by an enemy and hence may require additional replacement. This presents a problem of evaluating different systems where one system may have much higher survivability to enemy attack than does a competing system. There are four possible ways of coping with equipment survivability considerations and each of these is listed below.

Use only peacetime costs. This costing procedure assumes that there will be peacetime operation and maintenance costs, including anticipated alerts, practice/training and accidental destruction of equipments; it further assumes that wartime destruction of system equipments is not to be considered. Thus, the only replacement of equipments will be that due to equipment unreliability and accidental destruction.

In this costing approach, the analyst determines the total amount of equipments required, including 90 days of spares, plus the amount required to fill the logistic "pipeline." Such an approach ignores equipment survivability considerations.

Use peacetime costs plus an initial inventory based on equipment destruction anticipated in an assumed initial combat operation. This approach assumes that equipment quantities will be purchased on the basis of not only 90 days of spares due to malfunctions, but also due to the initial destruction to the system during these first 90 days of combat. Although there is some uncertainty regarding the amount of enemy damage that may occur during the first 90 days, this approach does have the advantage of recognizing that some additional stores must be acquired for the wartime contingency basis.

System costing having the same basis as the preceding but including an additional inventory based on enemy destruction until contractor production can achieve some desired level. In this case, the attrition costs are based on a greater time period than 90 days. While this method may introduce more uncertainty since the attrition over this longer period is unknown, it is a more conservative approach since it may take more than 90 days for a contractor to resume production.

Include total war destruction. This method extends the costing period to include the total war. Presumably, this method gives the highest consideration to equipment survivability; however, it also introduces the greatest uncertainty.

The preceding methods are all attempts to bring into the decision-making process the differences of system survivability. They all suffer from the same disadvantage of not knowing exactly how much (if any) system destruction will actually occur. In general, system decisions are based on the criterion which includes combat losses over the first 90 days.

DYNAMIC NATURE OF THE COSTING PROCESS

Systems planning is a dynamic process; decisions are made based on the best information available at the time. As new information becomes available, or as conditions change, another systems analysis may be needed to determine a preferred choice among the new alternatives available at the time, including the alternative of not changing one's present condition. If it turns out that the assumed data such as predicted cost or operating life is incorrect, the only recourse the analyst has is to re-examine new alternatives and try to improve his then current position. Past errors, such as the example of costing the computer on a ten-year basis, must be paid for and attributed to experience.

Cost Estimation Process

This process consists of a method of predicting the future costs of a system. There are two types of cost estimates which can be made. The first type involves system elements which are well defined, such as in short range procurements, particularly dealing with available (off-the-shelf) equipments. Here the analyst should be able to arrive at a very accurate total cost estimate.

On the other hand, in systems planning projects there is a need for estimating the cost of a system which is not completely specified and has not even been completely developed. This is obviously a more difficult problem.

Model of Cost Analysis Process

We shall now describe the logic of the cost estimation process as illustrated in Figure 9-4. The first objective of cost analysis is to determine all applicable cost elements which occur over time. This can be accomplished as follows:

1. Subdivide the system into its component parts down to the appropriate level of detail consistent with the analysis time and data base available.

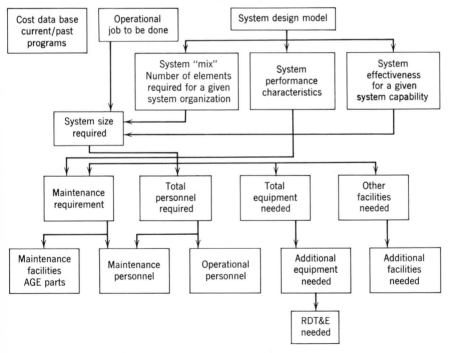

Figure 9-4.   Work process in cost estimation.

2. Further subdivide the components into time-phased procurement functions to be performed, including RDT&E, investment, and operation and maintenance.

Some of this information may be obtained from the system design model which indicates all of the elements making up the total system, including hardware, personnel, and facilities. Other information may be obtained from the operational flow model which indicates the system activities, including operation and maintenance, as well as other supporting activities required.

The next step is to determine the total number of units required. This may be obtained from the system design model as well as tables of organization which show the size relationships existing among elements. If the systems analysis is pivoting on a constant level of effectiveness, the results of the effectiveness evaluation will determine the number of units required to achieve this level. On the other hand, if the systems analysis is pivoting on constant cost, the cost analysis must work in an iterative trial and error basis to develop unit costs in order to see how many units can be purchased for the constant cost level.

Thus, from the total number of units required, the total number of required elements can be obtained. Some of the elements, however, may be inherited from available systems; hence they do not need to be purchased. These elements should be subtracted from the total number of system elements required for purchase.

The next step consists of estimating system element costs, for which an economic data base is needed. Such a data base, analogous to the system performance data base discussed in Chapter 8, dealing with predicting system effectiveness, consists of accumulating all past and current costs involved in acquiring, operating, and maintaining past related systems. An important part of the economic data base consists of those cost extrapolations called "cost-estimating relationships" which can be used to predict future system costs based on past system costs of related systems.

Examples of such cost-estimating relationships might include:

1. The cost of high power radar prime equipment as a function of peak power output and antenna area.

2. Aircraft depot maintenance cost as a function of aircraft gross weight, speed, and activity rate.

3. Ballistic missile booster cost as a function of missile weight, quantity, type of propellant, etc.

4. Various estimating relationships between prime mission equipment, training, logistics, maintenance, etc., in the form of equipments, facilities, and manpower required to perform these functions.

Since these relationships are developed from historical cost data and various system technical and performance characteristics, it is important that such data be stored, and relationships developed, for future application. These cost-estimating relationships should include some estimate of accuracy or uncertainty of the estimate. This would be analogous to the estimate of uncertainty which is connected in the estimate of system performance. Unfortunately, this generally is not done, so that all the systems analyst would receive is an expected value of cost. Two other aspects should also be mentioned:

1. In conducting a cost analysis, the analyst should concentrate more heavily on the costs involved with the newer system elements since, in general, these have the highest degree of uncertainty connected with them.

2. The analyst should also concentrate on determining relative cost differences among alternatives, as opposed to hoping to arrive at a highly accurate absolute value of cost.

Estimating the Cost of the Mobile Missile System

As an example of the cost estimating approach previously summarized, we shall now describe the logic and factors to be considered in estimating the cost of the two proposed missile systems described in the strategic planning case of Chapters 5 through 8. For the superhardened missile system, only new silos are required, since their operation is identical to that of the current hardened Minuteman missile system. Thus, an investment cost and possibly some RDT&E costs will be required to design the new silos. Operations and maintenance costs should be the same as the current hardened system. Since this problem is rather straightforward and simple, we shall concentrate now on a procedure for costing the mobile missile system. Three main inputs are the starting point for this analysis:

1. The description of the operational job in as much quantitative form as possible.

2. The system design model, which indicates all of the system elements, including personnel, hardware, facilities, etc.

3. An economic data base of currently existing information relating to our past and current costs in procuring, operating, and maintaining previous systems which have some relationship to the current missile system.*

We shall now describe in greater detail the cost estimating logic.

* Perhaps the most relevant information pertains to the current operating and maintenance costs of the Minuteman system as well as to the past costs for RDT&E and procurement. This data will provide a means for predicting modification costs to convert the current Minuteman missiles into a mobile system as well as the operating costs of such a system.

The Operational Job

One cannot overstress the importance of the cost analyst's need to understand the operational job as described to him by the operational personnel and system designers. Recall that the cost analyst's function is to translate any additional complexities inherent in the new job into expenditures required to perform the new operation. Many times such translation is hampered by the system designer who, in his zeal to "sell" his proposal, minimizes some of these complexities and fails to provide for the additional performance or personnel required for such complexities. Therefore, the cost analyst, being in possibly a more objective position, should recognize this factor as he performs his cost analysis.

We shall now consider, specifically, the mobile Minuteman system. The following key points should be highlighted:

1. One thousand regular Minuteman missiles are to be used. These will be removed from the silos and mounted on trucks.

2. There will be continuous, 24-hour movement of the missiles by truck vehicles through the less populated parts of the country over roads capable of withstanding the weight involved. Any new real estate required for actually launching the missiles upon command will be indicated.

3. There can be assumed some minimum level of availability as a fixed requirement for this system, with the remaining missiles being under repair.

4. The training of the operating personnel will be accomplished by missile simulation equipments in a school as well as through a planned schedule of operational test firings.

Research, Development, Test & Engineering (RDT & E) Costs

This category includes any preprocurement costs to be expended before additional equipments can be acquired. Hence this part of the cost analysis interfaces with one of the most difficult parts of the mobile system design problem; this is the modification or redesign of the various subsystems of the existing Minuteman missile so that they will provide the estimated performance characteristics being proposed when operating in its new mobile mode under certain new environmental conditions such as greater prelaunch shock and vibration. This is particularly true regarding the missile accuracy and reliability. In addition, certain system elements which are not currently available (such as mobile missile launchers) must be designed, tested, and these costs included.

Initial Investment Costs

The cost elements to be included are as follows:

1. Prime mission equipment (PME); since the unit cost of these items such as the modified missile subsystems and new launchers will be a function of the volume procured, a cost-volume curve should be constructed.

2. Additional facilities; many of these facilities are needed for storage of parts of the system and may therefore depend on the usage rate. An example of such facilities would include fuel tanks whose size is a direct function of the rate of fuel consumption.

3. Unit support aircraft, such as any aircraft or helicopters for transportation between the units and other points.

4. Aerospace ground equipment (AGE); all equipments necessary for missile checkout and for locating faults due to equipment malfunctions. The complexity and the numbers required are determined by the maintenance concept as part of the system design description.

5. Miscellaneous equipments; these might include the trucks and vans needed for equipment mobility.

6. Initial stocks; any initial stocks of equipments needed for the base.

7. Initial spares; the initial inventories for maintenance purposes as well as for training purposes.

8. Personnel training; includes retraining of current Minuteman crews in the use of mobile launchers.

9. Initial travel; the costs involved in transporting new personnel from their original station to the new base.

10. Initial transportation; the costs of moving all new equipment from the initial storage facility to the base, with the exception of prime mission equipments and petroleum, oil, and lubricants (POL), since transportation is included in these original prices.

Personnel Costs

Four main classes of personnel are required: operating, maintenance, administrative, and support. The number of each category and its cost will be determined as follows:

Operations Personnel. This consists of the missile launch crews needed to operate the prime mission equipments (consists of checking out the missile prior to launch and subsequently launching the available missiles). In addition, we must include the command structure, consisting of all personnel who operate the command, control, and communications portions of the system. Thus, the first step is to calculate the total number of operating personnel by classes needed at each of the launch facilities (even though these facilities may be mobile). This information can be obtained by melding information from two sources. The first is the existing Minuteman organization chart which shows currently the numbers, by class of personnel, required on a unit basis (i.e., wing, squadron, flight, or launch control facility). The second input is the system designer's prediction of the new numbers of personnel by class needed for the mobile system. Although the functions to be performed and probably the mix of officers to enlisted personnel will be somewhat the same for the mobile system as for the current Minuteman system, a comparison of the new system duties of the various personnel with the current duties performed must be made to see if additional (or fewer) personnel are required for the new system. For example, continuous movement is required in the new system which will result in more difficult duty than in the present, fixed system; truck drivers must continually operate their equipment; even those personnel who are passengers on the vehicles will operate under more strain than if they were on duty at a fixed base. This means that the duty hours may have to be shortened as compared with current duty. In any case, the number of personnel required must take these factors into account.

This system must operate continuously (i.e., 730 hours per month). Other systems having only intermittent operation could be calculated by considering just how many hours per month are standard. For example, in an air defense command interceptor squadron, a certain number of flying hours and ground alert hours per month are required. In industry, where systems are operating for eight or more hours per day, the extent of operation determines the number of personnel required.

Maintenance Personnel. The amount of equipment usage per month or other period of time, as well as the reliability of this equipment, determines the amount of maintenance which will be required. This can be scheduled, preventive maintenance, as well as unscheduled maintenance required for equipment malfunctions. Such predictions can be made by extrapolating from the current Minuteman maintenance needs and organization content. By further comparing the operational job to be done with the system design, one can predict the peacetime mean time between failures of the equipment and in this case it would be safe to assume that there will be more maintenance for the existing missiles when they are used in a mobile mode than if they were sitting in fixed silos. The amount of additional maintenance needed can be predicted by examining other mobile systems and attempting to find a relationship between the reliability of the system when fixed as compared with its degradation when it moves around. Obviously, not all of the maintenance (in fact probably little) will be performed by the mobile squadron. Thus, the analyst must determine how many maintenance personnel are required at the mobile unit as compared with the numbers necessary at fixed maintenance depots. This, again, should have been calculated by the system designer. This figure not only affects the duty hours of the personnel, and consequently the number of personnel required, but also the number of logistics personnel and equipments needed. For a more accurate analysis, the total system can be broken down into subsystems. In addition, the ratio of officers to enlisted men to civilians required with the new system can be considered.

Administrative Personnel. The functions to be performed consist of all aspects of administration. The number and mix of assigned personnel is a function of the operations and maintenance personnel needed by the system. These are obtained by using standard functional relationships which exist by the table of organization, and can thus be readily calculated.

Support Personnel. The support functions to be performed include the feeding, housing, and training of system personnel, providing their transportation and supplies, and protecting all weapon systems and bases. Here again, the number and the mix of support personnel are a direct function of the so-called "direct personnel" required by the weapon system, where direct personnel are those who operate, maintain, and administer the system.

Operation and Maintenance Costs

The cost categories are as follows:

1. Facilities replacement and maintenance; a cost estimating relationship which is generally used for this factor is 5% of the base value plus $450 for each military person on the base.

2. PME replacement; this consists of replacing the attrition caused by operational testing of the missiles.

3. PME maintenance.

4. PME POL; this is the POL required to keep the prime mission equipments in operating condition.

5. POL and maintenance of the unit supply aircraft.

6. Maintenance and replacement of aerospace ground equipment (AGE).

7. Personnel pay and allowances; this is obtained by knowing the total number of personnel and their classifications.

8. Personnel replacement training; this takes care of the predicted turnover rate of system personnel.

9. Annual travel; this is the transportation required for replacement personnel.

10. Annual transportation; this involves the cost of all of the logistics transportation required to keep the system operative.

11. Annual services; this consists of the anticipated costs of materials, supplies and contractual services for all support functions connected with

the organization (other than those already described). In general, this is predicted on a cost per military man basis. The Air Force uses $400 per military man.

DEVELOPING COST-ESTIMATING RELATIONSHIPS: AN EXAMPLE

In order to predict the cost of future programs it is the function of the cost analyst to gather and structure data relating to the costs of various parts of past programs. These functional relationships between cost and system characteristics are called cost-estimating relationships (CER). This can be looked upon as a statistical correlation problem, in that the analyst is attempting to correlate previously accrued costs of some system characteristic with the way the cost will change as this characteristic is altered. Obviously, the more factors included in the cost-estimating relationship, the more accurately the future cost can be predicted.

We shall now describe an example of the development of some cost-estimating relationships using the cost of prime mission equipment (PME) as the primary parameter. The CER to be described is derived from experience in command and control systems. As such, it represents one form of estimating which may be better than merely intuitive estimation if other cost-estimating relationships are not available.

RDT&E Costs

1. System Engineering. This includes all governmental system engineering and technical direction as well as that provided by the prime contractor. Estimate to be used: 15% PME costs.

2. Systems Management. Costs are provided by the contractor and do not include in-house, military, or governmental management costs. Estimate to be used: 7 to 10% PME costs.

3. Subsystem Design and Engineering. Estimate to be used: 10% PME costs.

4. System Test and Evaluation. This includes category 1 and 2 tests performed in the plant as well as the over-all system operating tests in the field. Thus, it includes the use of field engineers, technicians, and the logistic support for field tests. Estimate to be used: 30 to 50% PME costs.

Initial Acquisition (Investment)

1. Prime Mission Equipment. Estimate: 100% PME costs.

2. Costs Associated with PME:

(a) Aerospace ground equipment, including all test equipment required (estimate: 5% PME costs); (b) Technical data, including all engineering drawings, handbooks, and other documentation needed to describe system operation and maintenance (estimate: 7 to 10% PME costs); (c) Installation and checkout of PME (estimate: 10% PME costs); (d) Transportation of PME from plant to field sites in the United States (estimate: 1.5% PME costs); and (e) Initial spares, including sufficient spares for a 90-day supply in the field, as well as filling up the logistics pipeline (estimate: 20% PME costs).

3. Facilities: These are purchased as part of the military construction program and include the following: (a) Operations building including the computer facility (estimate: $40 per square foot of floor space); (b) Support buildings, including other buildings such as warehouses, barracks, etc. (estimate: $25 per square foot); (c) Real estate, using an average planning factor (estimate: $5,000 per acre); (d) Power, including, back-up power equipment (estimate: depends on power consumption); (e) Roads and fencing (estimate: depends on length of road and perimeter of area to be fenced); and (f) Sewage (estimate: calculation).

4. Computer Programming. This includes the following elements: (a) Preparation of operations specifications; (b) Construction of operations program; (c) Construction of support programs (e.g., data reduction); (d) Installation and checkout of program; (e) Training of personnel; and (f) Over-all documentation (estimate: this is one of the most difficult elements to estimate the cost. Past history indicates it is equal to or greater than 50% PME costs.)

5. Training. This includes an overhead training facility including a minimum amount of equipment necessary for training personnel. The estimate depends on the amount of equipment required and the size of the facility.

6. Operator and maintenance personnel. The estimate is $5,000 per month for a six month training program.

7. Annual operating costs. These include (a) Maintenance facilities (estimate: 5% of facilities costs); (b) Equipment maintenance such as spares replacement (estimate: 10% PME costs per year); (c) Computer program maintenance, including updating the program and any field support required (estimate: this is obviously a function of the amount of updating required and hence cannot be estimated parametrically); and (d) Personnel costs (estimate: $12,000 per officer and $5,000 per enlisted man, with one officer for each five enlisted men). The shift factor that may be used for 24-hour per day, seven day per week operation is as follows: The total number of personnel required will be equal to five times the number per shift. This includes all indirect costs such as guards, cooks, etc.

COST-SENSITIVITY ANALYSES

Perhaps the most important aspect of cost analysis is the structuring of data which show the cost implication of varying a certain technological performance or other characteristic in a system design. Two types of cost-sensitivity analyses can be performed: one for the purpose of intrasystem tradeoffs (primarily for use by the technologist/systems planner); the other, showing structured cost information (primarily as an aid to the decision-maker).

Cost-Sensitivity Analyses for the Systems Planner

To aid the systems planner in providing a given level of effectiveness at lowest cost, various transfer functions are needed, each relating some technological or performance characteristic to cost. Examples of such functions are the relationship of bridge material used (which affects bridge life and maintenance cost) to investment cost, and bridge capacity to investment cost. Obviously, such information which shows the economic impact of alternative courses of action (different engineering designs) is required if the systems planner is going to properly consider a range of design alternatives. The final choice of bridge design would require such transfer functions to calculate total cost. Obviously, the particular shape of the correct transfer function to be used will affect the final result. Similarly, in the ICBM planning problem one of the important transfer functions is how silo investment cost varies with the degree of hardening provided (or design overpressure).

Thus the systems planning process fundamentally consists of the formulation of these performance versus cost transfer functions for all feasible alternatives under consideration, and the combination of these functions to obtain system effectiveness and total cost. Further examples of the use of such functions for tradeoff analyses in systems planning form the basis of the subsequent chapters in Parts V and VI.

Structuring the Information for the Decision-Maker

The cost information previously described can be structured in several ways which may be meaningful to the decision-maker. As indicated earlier, the key characteristic which relates the level of effectiveness obtainable to the cost required is that of Figure 5-7, since constant cost systems were configured. To construct such constant cost systems, the analyst needs to determine the cost as a function of number of units procured. This information can be structured as shown in Figure 9-5. In addition to the RDT&E costs involved, Figure 9-5 illustrates two different manufacturing situations and

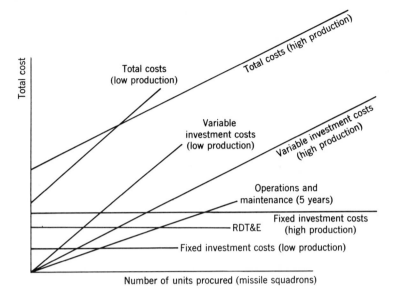

Figure 9-5. Static analysis of system costs.

the resulting fixed and variable manufacturing costs involved. The first is a low production example, and the second a high production example. The latter case involves the use of automated (but more expensive) tooling (a higher fixed cost of manufacturing) which will permit lower costs of unit production. The variable manufacturing costs (such as direct labor and materials) are assumed to be a linear function of the number of units procured, although many times a nonlinear function is used to take into account job "learning." Thus the total cost for the two cases can be computed as a function of number of units procured. Notice that the "break-even" volume, where total costs are equal, can be determined by indicating the two regions where each of these two production techniques is preferred.

The decision-maker is also concerned with system costs as a function of time as mentioned previously, and shown in Figure 9-3. These three main elements are the so-called "cost streams" incurred by the system. While these cost streams can be summated by a present worth calculation, the decision-maker may still wish to review time expenditures to look for abnormal peaks which occur and cause undue strain.

V

SUBSYSTEMS PLANNING

In the previous parts, we have examined the problems involved in analyzing systems which perform an operational mission, and we developed methods for measuring the effectiveness, costs and risks of proposed systems for performing the mission. We then investigated methods for predicting the total costs for procuring and operating these systems over the total system life. This part of the book is concerned with the application of systems analysis to the problems of subsystems planning. By a subsystem we mean that system which performs some functional part of the operational mission rather than the total mission. Hence Part V deals with systems planning at the functional level. Functional systems may be better understood by referring to the system design model of an entire operational system and its subsystem elements as shown previously in Figure 4-4. This figure shows that any system consists of an operating system (the device which actually does the operational job) and the support systems (consisting of the many support elements which keep the operating system in working order and control its operation). Thus, in the previous example, the strategic ICBM system performed the over-all strategic offensive mission. Its operating system consists of all elements connected with the missile and its operation, such as the missile itself, the missile launcher, and the missile operators. Each of these is then a functional subsystem of the operating system. One of the primary functions of the support system is to maintain the entire system in an acceptable state of operation or readiness for operation. This involves having trained personnel and equipment available for scheduled and unscheduled maintenance. These personnel also require the spare parts necessary to replace those that have malfunctioned in operation. However, many times these parts must be ordered and transported from a supply depot and, occasionally, a higher than normal demand for some part may saturate the spare parts inventory at the location of the operating system, thereby forcing the system to be inoperative for some time. Thus, proper systems planning requires a balancing of tradeoffs among the maintenance system, supply system, and the reliability of the operating system.

Thus the objectives of Part V are the following:

1. To show how to plan or design a system, taking into account the subsystem considerations. These include how to consider the various functional parts of the system, as shown in Figure 4-4.

2. To show how to obtain the proper balance of these subsystem elements so that the mission objectives can be achieved at lowest total system cost.

3. To show how to properly plan or design a functional subsystem in which other parts of the system cannot be changed or varied as they can in total systems design.

10

System Selection: Helmet Case

This first subsystem planning case was chosen primarily to illustrate one of the most important principles in systems analysis: how to select between two functional subsystems, each of which provides different levels of performance and each having a different cost. The traditional way of coping with such a problem is to combine all of the key system performance characteristics into one value called "system worth." This is not a simple task, but even when this can be done, the evaluation is confronted with a second problem as illustrated in Figure 10-1. This figure is intended to show that even when evaluators and decision-makers can agree that one system (S_2) is superior to another (S_1), but the costs are different, it may still be difficult to select the preferred system on some rational basis (i.e., to determine if the increase in benefits is worth the increase in cost which can be justified to a higher level decision-maker). We shall show that the only way to make such a selection is to analyze the higher mission level systems problem.

The particular system chosen to demonstrate the principles involved is the simplest system available; it is a one element device—a helmet for a

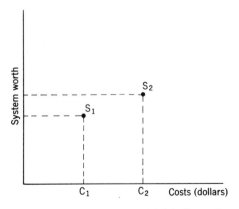

Figure 10-1. The selection problem: Which is the preferred system?

combat soldier. The purpose of the chapter is to show that even in this simple case, unless systems analysis methods are utilized, it is not possible to make a selection in any way but intuitively.

PROBLEM AS GIVEN

You are a systems analyst with an Army development laboratory whose mission is to develop new Army support equipment. You are approached by a systems planner who asks for your assistance in providing a cost-effectiveness analysis of a new lightweight titanium helmet recently developed by his group in response to an Army directive requesting that all development groups find ways of reducing the weight of the equipment carried by combat soldiers. In addition to the weight reduction problem it is reported that some soldiers have not used their issued steel helmets in combat in Southeast Asia during the high temperature months, complaining about the heat. Several have suffered heat prostration, blaming the steel helmet. For these reasons, a titanium helmet development program was initiated. Two titanium helmet designs were developed, each of which gives some degree of increased system performance. The first titanium helmet, weighing two pounds, three ounces, provides the same strength and protection as the current steel helmet, yet is one pound lighter in weight. However, the cost of this helmet is $30 compared to $5 for the steel helmet. The second titanium helmet weighs three pounds, three ounces, the same weight as the steel helmet, but provides increased protection against all current ballistic hazards; its cost is $35.

It was decided that the lighter weight helmet would be submitted for approval, since the reduced weight would not only aid in the heat problem but would also increase the combat efficiency of the soldier.

A cost-effectiveness analysis, showing the value of the proposed equipment change, is required to accompany the proposal to higher headquarters. The systems planner has endeavored to determine the value of the improved helmet by using a point scoring method considering four primary factors of (a) weight, (b) combat efficiency, (c) cost, and (d) increased usage by the soldiers, and has asked you to review his evaluation approach, providing whatever assistance you can.

PROBLEM AS UNDERSTOOD

As the analyst strives to obtain an understanding of the problem, he engages the systems planner in further discussions which reveal great insight into the systems planner's intuitive views. One of the first topics discussed is the evaluation procedure which the helmet designer is planning to use.

Point Scoring Approach to Evaluation

The evaluation method which the systems planner intends to use involves the use of a point scoring procedure as illustrated in Figure 10-2. This method is a fairly standard type of evaluation procedure in which the key evaluation factors to be considered are listed as one dimension of the matrix, and the alternatives being considered as the other dimension. Each of the key factors considered is given some relative weight which determines a maximum number of points that any alternative may receive for that factor. Thus, in the problem under consideration, four main factors are considered important and their weighted importance is: helmet weight (20%), combat efficiency of the helmeted soldier (30%), the amount of usage the helmet would provide (10%), and helmet cost (40%). These numerical values chosen are either based on operational data (e.g., the tests which measure the combat efficiency of a soldier as a function of the helmet employed) or on other available analytical data. If neither is available, the values can be based on the intuitive judgment of the evaluator. Since the total points to be awarded is arbitrarily chosen to be 1000, the maximum score for each factor is the proportionate amount, as shown in Figure 10-2. Each alternative is then evaluated with respect to each factor in order to determine how many of the maximum points allocated will go to each alternative. A total score for each alternative is then obtained by arithmetically summing each of its factor scores.

This method of evaluation has several difficulties. First, the effectiveness of a helmet was never defined in operational terms and, while the key factors contributing to the worth of a system may be identified (i.e., omitting cost), the use of weighting factors as the method of combining factors is always

	Weight 20% (200 points)	Combat Efficiency 30% (300 points)	Soldier Usage 10% (100 points)	Cost 40% (400 points)	Total Score (1000 points)
Steel helmet					
Lightweight titanium					
Standard weight titanium					

Figure 10-2. Point scoring evaluation table.

subject to challenge by some evaluators or decision-makers. Thus, the system analyst's first problem is to determine how these characteristics of weight, combat efficiency, soldier usage (and perhaps others) can be combined in a more rational fashion to give an over-all effectiveness measure.

The second difficulty inherent in the point scoring method is even more serious. While this method combines cost with the important technical or performance characteristics, there is no operational justification for the use of such a "figure of merit" approach to evaluation (i.e., choosing that system whose total score is highest is subject to attack by higher-level evaluators). Actually, the only acceptable method of choice when selecting from among alternatives (in both general and specific forms, as indicated below) is to select a system alternative (helmet design) from among those proposed, which will either result in the performance of the operational function at the lowest total cost to the organization (Army) (i.e., pivoting on equal effectiveness); or will result in the highest performance of the operational function at a given total cost to the organization (government) (i.e., pivoting on equal cost, taking into account the risks and uncertainties of the system design).

When the evaluation criteria are explicitly stated, it becomes apparent that there is some question as to whether or not the proposed point scoring evaluation procedure actually satisfies the stated selection objective. Thus, some other approach is needed.

Measuring the Effect of Weight Reduction

Intuitively it is felt that as the helmet weight is reduced, two benefits will be obtained. First, the number of soldiers wearing the helmet should increase, and hence the number of casualties should be reduced. (Some means of quantifying this relationship will be needed.) Secondly, the systems planner feels that the lighter helmet will enable the soldier to have a greater "combat efficiency." But how is combat efficiency measured, and how does this efficiency vary as a function of helmet weight? To answer this, many tests have been run by Army laboratories, measuring such things as the energy which a soldier expends in walking, and running, and other exertions resulting from combat, as a function of the load which he carries. Such data, involving treadmill-type experiments, is currently available. In addition, other aspects of the energy versus equipment weight experiments indicate that one pound carried on the head is equivalent to five pounds carried at the waist. Interestingly, none of these experiments and measures were used for the measure of combat efficiency. The measure which the helmet designer used was the ability of the soldier to sight his rifle on a target. Tests had indicated that as the helmet weight increases, the soldier's ability to maintain his rifle on target decreases. Thus the measure of combat effectiveness which the planner was using as part of his point system of evaluation was the rifleman's accuracy as a function of helmet weight. The omission of the other aspects of combat effectiveness mentioned above was questioned by the analyst (and is reconsidered later in this chapter).

Weight Reduction versus Heat Reduction

Certain information would be helpful for considering other alternatives with respect to the heat discomfiture problem. For example, why are the soldiers not wearing their helmets? Is it because of heat discomfiture, weight discomfiture, or both? Are there other more important causes of this heat prostration, such as discomfort from the rest of the soldier's clothing? What is the relationship of helmet weight (or total equipment weight) to soldier heat discomfiture?

Further discussions with the systems planner indicate that this helmet development project exploring the use of lightweight titanium is really in response to an over-all Army directive which is seeking a general weight reduction in all combat soldier equipments. This project was not primarily being addressed to heat prostration, but rather to the over-all weight reduction directive. The analyst also learns that there is another development group that is working on the problem of a new helmet suspension system (i.e., the means by which the metal helmet is attached to the soldier's head, the current method consisting of a plastic helmet liner and a chin strap arrangement). The system planner feels that a better suspension system may be a good source of reducing the heat prostration problem. Similarly, another group is working on a longer term project, examining other types of helmets such as the possibility of a solid nylon helmet or a seven-ply nylon helmet, especially designed to resist ballistic fragments. Hence, the immediate "problem to be solved" is to evaluate the two titanium helmets versus the current steel helmet from a cost-effectiveness viewpoint in a manner which will be meaningful to higher-level evaluators.

ANALYTICAL APPROACH TO EVALUATION

Based on the information developed thus far, an intuitive, qualitative approach to the evaluation task could be formulated as follows: a lighter weight helmet would, presumably, provide greater protection to a larger number of soldiers, resulting in fewer casualties. In addition, a lighter weight helmet would result in a more efficient rifleman since he would be more accurate in his aiming ability. However, even when it is clearly established that the effectiveness of the lightweight titanium helmet is greater than the steel helmet, the basic question is whether the benefits are worth the additional $25 involved. Again the selection problem involved can best be illustrated by Figure 10-1.

The analyst's dilemma is that there is no way to determine if the superior helmet is worth the $25 cost merely on the basis of helmet characteristics. Since there is no way of pivoting on either equal effectiveness or cost by looking solely at a helmet, the analyst is forced to enlarge the problem to a higher level. This enlargement is accomplished by considering how the helmet would be used. A good question to ask is "What does it cost the government if it does *not* buy the improved helmet?" or "How is the soldier better off if he does have the improved helmet?" Here, cost can be expressed in units of dollars as well as casualties. The analyst tries to structure the characteristics of the helmet to its higher-level, combat environment context.

A further insight into this analysis can be obtained by tracing the flow of operational events which occur to a helmet and a soldier during precombat, combat, and postcombat activities. These activities, incidentally, can be either defensive or offensive in nature. Since the primary function of the helmet is head protection, we first shall focus on this use.

The Use of a Helmet for Soldier Protection

The structure of the operational flow model showing the events pertaining to how and to what extent a helmet is used by a soldier is shown in Figure 10-3. Initially, all soldiers are issued helmets. However, when they go into combat a certain percentage (undefined as yet) choose to wear their issued helmets and the remainder do not. This results in numbers of helmeted and unhelmeted soldiers going into combat. During combat, each class of soldier accomplishes some offensive action, resulting in damage to the enemy. In addition, each soldier's head is subjected to ballistic fragments, and has a certain vulnerability to these fragments, depending on whether he wears a particular helmet or not. This vulnerability results in casualties inflicted on both classes of soldiers. These casualties are of several types: death on the battlefield, or injuries which result in hospitalization. Of the battlefield casualties which return for hospitalization, a certain percentage of these soldiers survive and the remainder do not. Of the surviving soldiers, some are sent back to the United States for other duty; others are sent back for discharge after rehabilitation. The remainder of the rehabilitated soldiers are reassigned to duty in the combat along with replacements for the fatalities, and the same process recurs for the next combat phase.

Need for Additional Information

Obviously, a major part of the analytical task will be to gather appropriate data which will show numerical estimates of the various parameters il-

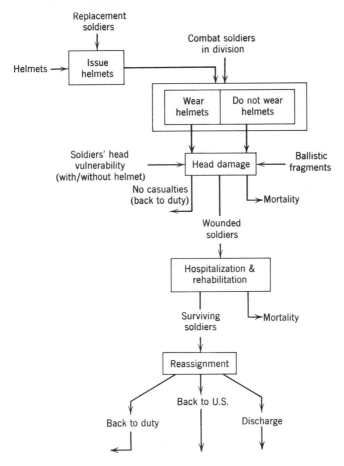

Figure 10-3. Structure of operational flow model for combat infantryman.

lustrated in Figure 10-3, and furthermore, to estimate these data as a function of which helmet type (steel or titanium) was used. These data input needs are as follows:

1. How many soldiers are involved?
2. Of this total, would the improved helmets be issued to all soldiers, only those in combat, or only those in combat in hot, tropical zones?
3. What percentage of the soldiers are currently not wearing helmets into combat?
4. What percentage of soldiers not wearing helmets are receiving head wounds?
5. How serious are these head wounds?

6. What percentage of soldiers who wear helmets receive head wounds?

7. How serious are these head wounds?

8. To what extent will reducing the helmet weight result in more soldiers wearing their helmets? That is, what is the relationship between helmet weight and helmet usability?

9. How well does the current helmet protect the head against ballistic fragments, and how much better would the thicker titanium helmet (same weight as steel helmet) protect the soldier's head?

Based on the greater insight obtained through the development of the higher-level operational flow model, it can be seen that we are interested in performing the combat task at lowest total cost, where such costs include not only the cost of helmets, but also the cost of casualties (as expressed both in terms of dollars, but also lives lost or men wounded). Thus one approach to the analysis would be to configure all systems to a given level of effectiveness and determine the total costs involved. For example, based on the analysis thus far, we could compare an Army unit (say a division) equipped with steel helmets to one equipped with lightweight titanium helmets, putting each in a combat environment and determining the total costs of each based on the casualties sustained by each.

Development of the Cost Model

What are all of the elements of cost which must be included in the analysis? These elements may be identified by constructing a higher-level system design model, as illustrated in Figure 10-4, which indicates all of the components required to accomplish the over-all objective (e.g., maintain and support a combat division). This system design model is constructed beginning with a combat soldier with helmet and rifle, then progressing to units of soldiers to the organizational level under study.

A good way of cross-checking to make certain that all of the cost elements have been included is to go back to the operational flow model and make certain that all of the system components have been included.

Based on the preceding models, the analyst can now estimate the differentiating cost in dollars and lives to maintain and support the combat unit as a function of the type of helmet used in combat. These costs may be indicated as follows:

1. Cost of helmets ($5 for the current steel helmet, $30 for the lightweight titanium helmet, and $35 for the standard weight titanium helmet).

2. Hospitalization and rehabilitation. These are incremental costs incurred by the Army medical corps as well as the Veterans Administration for ministering to soldiers with head wounds.

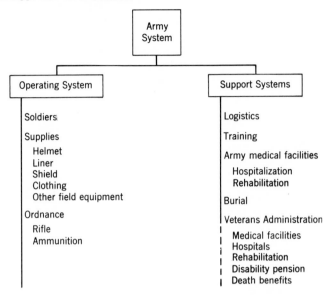

Figure 10-4.   Higher level system design model.

3. Veterans' disability and death benefits. These include disability pensions and all expenses attributable to fatalities, including survivors insurance. This is shown as indirect support. While not a part of the Army, its cost is directly related to the number of casualties.

4. Training and replacement of casualties.

To quantitatively consider the preceding advantages the analyst could hypothesize that if the units with the lighter weight helmet will have greater firepower due to its higher accuracy and lower casualties sustained during combat, a lesser number of these units is needed for a given combat task than the required number of units equipped with standard helmets. Thus the analyst can pivot on a different level of effectiveness (this time equivalent firepower needed to perform a combat task) and determine the size and cost of each unit based on all differentiating factors, offensive as well as defensive.

Evaluating Additional Improvement Obtained

We shall now consider other operational gains provided by the lightweight helmet which might be included in the analysis. As indicated previously, the rifleman's accuracy increases as his helmet weight decreases and rifle test firing data could be gathered which would indicate the extent of such improvement. This increase in accuracy could produce several operational effects, depending on the operational combat scenario chosen. There could be more enemy casualties, particularly from a first rifle round fired, due to the advantage of surprise. This could result in even fewer American casualties, because the enemy would have less opportunity to fire back. To quantify such considerations is not simple, and would require operational data from wargames. As a minor point, there should also be an additional cost saving in ammunition, with resulting logistic savings.

The worth of one pound of helmet weight could also be evaluated by using the substitution principle. This would involve examining what other equipment could be given to the soldier instead, if his total load were kept constant. In this fashion, a worth can be placed on the one pound reduction by noting its use. For example, since tests have shown that one pound of load added to the head is equivalent to five pounds added to the body, reducing the helmet weight by one pound would be equivalent to permitting five pounds of water, food, or ammunition to be added to the soldier's load. The question to be examined is how many times would these additional quantities be used in a combat scenario. For example, how many times do soldiers run out of ammunition or get close to running out and consequently have to withdraw from a battle (or suffer additional casualties)?

Another potential saving due to reduction in weight might be a saving in transportation or storage costs. This aspect was examined, but such types of savings were not achieved since the main requirement for transportation is due to the volume of the helmets rather than their weight. In other words, no additional helmets could be added to a truck even if the weight of the helmet were reduced.

Helmet Protection Versus Cost

Another important relationship to be considered is the amount of protection which the helmet offers as a function of its technical characteristics. In this case, how much extra protection will the thicker titanium helmet, having the same weight as steel, afford as compared with the lighter weight, titanium helmet (or as compared with wearing no helmet at all)? Such a transfer function of protection versus helmet thickness (and the accompanying cost) is shown in Figure 10-5.

In this case, it is interesting to note that the standard weight titanium helmet provides a 40% improvement in protection against the standardized ballistic fragment tests, as compared with the lightweight titanium helmet. This also comes at a cost of only $5 extra. The analyst would insert this information into the operational flow model of Figure 10-3, thus determining the decrease in casualties for this helmet, as compared with other possibilities. While a 40% improvement appears to be a large increase, the percentage must be converted into a decreased number of casualties before it has meaning.

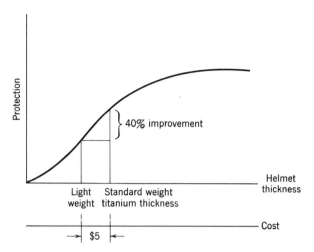

Figure 10-5.   Cost-performance transfer function.

SYSTEMS PLANNING IMPROVEMENTS

The preceding discussions focused on system improvements which could be obtained through the use of lighter weight material. Now we shall discuss other improvement possibilities, based on the over-all operational flow model of Figure 10-3.

Heat Improvement

As indicated previously, a parallel development of an improved helmet suspension system was also underway to see if this is a means of improving the heat problem. Another possibility which might be explored would be that of placing holes in the helmet to increase the air circulation inside the helmet. Here, the systems planner would have to determine the amount of protection lost as a function of the material removed from the helmet.

As indicated in the operational flow diagram, methods for inducing more soldiers to wear helmets, irrespective of their design, should be undertaken. Studies should be undertaken to determine the reasons why soldiers are not wearing these helmets. Such studies might disclose additional information useful to the designer.

STRUCTURING THE INFORMATION FOR THE DECISION-MAKER

The results of the quantitative analysis are structured in a form which would be meaningful for a decision-maker, as illustrated in Figure 10-6. Here in a comparison of helmet A and helmet B, the first measure is the total dollar cost involved to perform the same level of effectiveness, taking into account all costs which are associated with the type of helmet used. The second and third categories concentrate on the lives lost and disabilities incurred within the Army unit for each of the helmet alternatives. Figure 10-6 shows a dominant case for helmet B since helmet A is more costly not only in terms of dollars but also casualties. On the other hand, it is possible to have a situation in which helmet A would be of least cost in terms of dollars but more costly in terms of lives and disabilities. The decision-maker must now use his intuitive judgment to determine which of these alternatives is preferable to him, taking into account such things as morale. Under no circumstances should the value of a casualty be combined with dollar costs. In fact, the actual "out-of-pocket" costs for replacement training, hospitalization, and death benefits have already been included. However, the intangible factor of morale is inherent in the problem and this factor can be isolated by keeping the casualty figures isolated. The basic question which the decision-maker faces is the determination of how much he is willing to spend in dollars to reduce casualties.

At this point the reader may feel, why not spend any amount of money to decrease casualties? However, other alternatives which also reduce casualties must be considered. These include body protectors such as ballistic fragment resistant vests, and more and better offensive weapons such as automatic rifles. The point is that the marginal effectiveness of each alternative (as measured by casualties reduced per dollar) can be calculated, and decisions made on this basis.

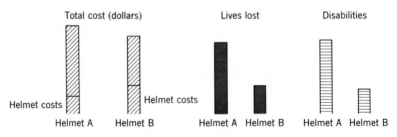

Figure 10-6.   Evaluation results.

COPING WITH UNCERTAINTIES

Since the ballistic vulnerability of each helmet can be readily tested at an ordnance test laboratory, the key uncertainties connected with this problem are:

1. Predicting the percentage of soldiers who will use each type of helmet.
2. The combat environment assumed, from which the calculations of the number of troops required and casualties were derived.

The usage factor can be explored through human engineering tests, perhaps. However, some parametric analysis can be performed to determine the range of uncertainty over which the proposed alternative is superior. In addition, scenarios of various combat missions can be explored to determine how scenario-dependent the recommended alternative is. Of course, risks can be minimized by procuring a limited number of improved helmets, gathering experience and then deciding if additional procurements would have sufficient value, using the models previously developed.

SUMMARY OF KEY PRINCIPLES

This chapter has described a method for evaluating two system alternatives where the benefits or effectiveness of one was superior to the other, but the cost of the first was also greater. The first key principle involved equating a level of system effectiveness required of each (helmet) (or a level of cost permitted for each), and thus choosing on the basis of lowest cost (or highest effectiveness). Since it was not possible to use either of these approaches for the particular system involved, the analysis had to be carried out at a higher system level by examining the combat task to be performed by an army unit.

Another way of viewing the problem is to ask the question, "If I spend money for some improvement in the system, how much do I save elsewhere in the system?" Unless the net savings are positive (in achieving a given level of effectiveness), spending for a system improvement cannot be justified.

A second principle demonstrated was the use of both dollars and human casualties as separate units of cost. No attempt was made to combine them directly (except that casualties caused a direct dollar outlay in terms of pensions, hospitalization, replacement training, etc.). Instead both elements of cost were presented to the decision-maker for his intuitive consideration and ultimate choice.

11

Satisfying an Uncertain Demand: Spare Parts Management Planning Case Determining the Capacity of a Bridge

Many systems problems take the form of configuring a system which will meet a future demand function. Examples of such systems include a supply system which must meet future equipment failures, a distribution system consisting of merchandise which is stored and then moves from the factory warehouse, to the distributor, to the retailer, at proper times to meet future customer demands, and a manufacturing system which contains stored inventories of raw stocks as well as in-process subassemblies as a "buffer" to permit an efficient operation of a factory while meeting future customer demands. Each of these systems contains the common element of storage of material, which will permit a sufficiently rapid response to an uneven, fluctuating customer demand. The problem common to all of these applications, which this chapter addresses, is how to properly design a system to meet a future demand in an acceptable fashion. Two specific applications are analyzed to illustrate the systems design principles involved. The first is the design of an inventory/supply system where the key question to be addressed is, "What size quantities of spare parts should be stored at different locations to efficiently meet the future fluctuating demands for these parts?" The second problem is a continuation of the bridge problem of Chapter 9, and deals with determining the proper capacity of a system (the bridge) to satisfactorily meet an uncertain, longer term future demand.

There are two other objectives to this chapter. The first is to demonstrate methods of quantitatively predicting a future demand based on past data

268

and any other information available. The second is to show how to design a "balanced subsystem" without having to perform a mission level analysis. Such an approach is particularly beneficial to the subsystem planner since fewer resources are required. The third objective is to illustrate how to build flexibility into a systems design so that the system objective can be met at lowest cost in spite of the uncertainty in the future demand.

SPARE PARTS MANAGEMENT PLANNING CASE

The Problem As Given (PAG) *

You are a systems analyst for a firm which consults with Air Force headquarters and you have been asked to conduct a study to determine the proper spare parts stocking policy for a new Air Force interceptor aircraft which is to be phased into the inventory in several steps to achieve its full operational capability. One hundred and eighty of these aircraft have been operating during the past ten months with eighteen aircraft at each of ten different bases across the United States. Approximately the same number of flying hours per month have been flown at each base. Data on the failure rates of the hundreds of parts involved with these 180 aircraft have been accumulated.

The maintenance and logistics procedure which has been utilized is as follows: An initial stock of spare parts is stored in the maintenance supply room of the ten air defense bases. These spare parts may either be components or entire assemblies. If any of the aircraft fails preflight or other inspections, the defective part is located and removed, grounding the aircraft until the part is replaced. If the replacement part is found in stock, the replacement can be made during the same day so that the aircraft availability rate is not reduced. However, if the replacement part is "out of stock," the aircraft remains grounded until the next time that new parts arrive by a supply aircraft, which is always on the same day of each month. Many aircraft parts such as transistors or small modules are inexpensive and, when they malfunction, can be readily isolated, using first echelon maintenance techniques. These parts are thrown away when found to be defective. Other, more expensive parts cannot be repaired at the airbase; hence they are sent back to a higher echelon maintenance depot for reconditioning and subsequent reuse. Thus the monthly supply aircraft performs several functions. First, it picks up all defective, repairable parts which have failed during the

* This problem and its solution are an elaboration of a problem originally presented by Chauncey F. Bell of the RAND Corporation to the Defense Weapons Systems Management Course in October 1964 at Dayton, Ohio.

past month, for return to the depot and subsequent repair. In addition the aircraft picks up the requisition for the number of nonrepairable parts which have failed during the past month. Finally, the aircraft delivers a set of reconditioned and new parts equal in number to those which failed during the previous month, as ordered 30 days earlier.

Having aircraft grounded because of lack of replacement parts has never occurred during the past ten months, since a sufficiently large number of each spare part type was stored at each base. This was deliberately done since there was no maintenance experience with these new aircraft on which to base a proper stocking policy. However, over the next two years a large buildup of aircraft will be taking place. Hence the very conservative (and proportionately costly) spare parts stocking policy used during the past ten months can be maintained no longer. Thus, the main purpose of this analysis is to determine the preferred number of spare parts of each type which should be stocked.

Your first task in this study is to conduct a statistical analysis to determine what the demand for each part has been in the past ten months. From each of the ten bases you are able to obtain data regarding the monthly demand for each part during each of the previous ten months of operation. Your first reaction is to attempt to analyze the data from each base separately since you notice that the average demand for each part at each base is not identical. However, since the number of flying hours at each base were approximately the same and since the maintenance crews were equivalent, it is decided to combine all 100 samples of the monthly demands for each part to provide better statistical confidence in analyzing the demand samples. While there may be some errors introduced in combining the data from all ten bases to aid in predicting the *average demand* among bases, you conclude that the stocking level decision is not an irrevocable one, so that even with the stocking levels predicted from the averaging of ten bases, a reapportionment could be made at a later date to give a larger amount of stock to those bases that might actually have a higher demand than other bases.

A proposal has been made that the inventory policy should consist of stocking each base with the average value of the monthly demand for each part. While there are many hundreds of parts involved, you decide to analyze the past demands associated with five different parts (A, B, C, D, and E, as shown in Table 11-1) to determine if the proposed stocking policy is a wise one, or if there are other stocking policies which should be considered. Parts A and B were chosen for the analysis since they are inexpensive and are designed as throw-away units. Parts C, D, and E are repairable. Because of the high cost of Part D, and its average demand being less than one, there is some question as to whether one unit of D should be stocked at each of the bases or not. There is a similar question about stocking one unit of A, even if it is of low cost.

Table 11-1. *Frequency of demands per month* *

Item	Demands †															Avg.	Unit cost
	0	1	2	3	4	5	6	7	8	9	10	11	12	13	14		
A	74	23	3	0	0	0	0	0	0	0	0	0	0	0	0	0.29	$120
B	0	1	4	8	12	15	16	14	12	8	5	3	1	1	0	6.3	$300
C	4	13	21	22	14	16	6	3	1	0	0	0	0	0	0	3.22	$3000
D	33	38	20	6	3	0	0	0	0	0	0	0	0	0	0	1.08	$30000
E	54	34	11	1	0	0	0	0	0	0	0	0	0	0	0	0.59	$6000

* 100 base months' experience.
† Number of base months for which a given number of parts failed.

THE PROBLEM AS UNDERSTOOD (PAU)

A better understanding of the problem can be achieved by both construct-ing the system design model and examining the system operation in further detail. This time we shall begin with an analysis of system operation.

Describing the System Operation

A better understanding of the system operation can be obtained by refer-ring to Figure 11-1 which illustrates the relationship between the number of units of a particular part which is stored in the supply shop and the failure and availability of operational aircraft. For simplicity assume that only fail-ures involving this one particular part occur for the 18 aircraft during the past month. Also assume that there are three parts initially stocked in the supply bin at the beginning of the month. Thus, as can be seen by referring to Figure 11-1, there are 18 aircraft initially available at the beginning of the month. At a particular time the first aircraft fails and there are, thus, only 17 aircraft operationally available until the aircraft is repaired and placed back into service. The amount of time to repair (TTR), when the aircraft is unavailable, is determined by the time taken to remove the defec-tive part, obtain a replacement part from the supply bin, and replace the part in the aircraft. When the part is replaced, the number of operational air-craft goes back up to 18, but there are now only two parts left in the supply bin. Later in the month when a second aircraft fails and the same procedure is followed, the supply bin is now reduced to one part available. Later, when a third aircraft fails, the supply stock for this part is reduced to zero. Notice that during all of these times the number of aircraft is reduced to 17 for a short period of time.

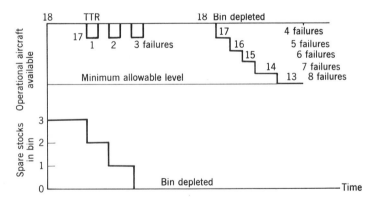

Figure 11-1.   Part depletion and number of aircraft available during the month.

When and if a fourth malfunction of this part occurs later in the month, there are no spare parts available to replace the defective part; hence the number of available aircraft remains at 17 until further aircraft malfunctions cause the number of available aircraft to be reduced to 16, then 15, and 14, etc., until finally the monthly supply aircraft arrives with the same number of replacement parts returned to the depot for repair the previous month. In this particular example, assume that there were eight failures, but only three parts available for that month. Further, assume that four replacement parts arrive at the end of the month. Thus four out of five "red-lined" aircraft could be put back into operation, providing a total of 17 aircraft (and no spare parts) available. The logistics supply aircraft would have picked up eight defective parts and the next month would return the eight reconditioned parts under this particular logistics procedure.

System Objectives

Again the analyst is confronted with a hierarchy of system objectives. The supply officer is concerned with filling the demand for parts, and his success in doing so could be measured by the fill rate or the amount of back orders of unfilled demand as measured by the number of days that a demand for a part is unfilled. However, as indicated in Figure 11-1, the primary objective of the entire logistics system is to keep the operational aircraft in proper repair and hence available for operational use. Thus an appropriate system effectiveness measure is Availability or Readiness Rate, or In-commission Rate, and this could be applied to the problem as follows: given a number of flying hours per month assigned to each base, how many aircraft are available for duty at any time? Obviously, the larger the number of spare parts which are initially stocked, the higher the number of available aircraft. To attempt to understand the relationship of having a spare part available (and consequently the possibility of having an aircraft available) to the higher level objective, the analyst must examine the mission of the aircraft system.

In a wartime situation, if the interceptor aircraft were not available more bombers would penetrate the defenses; hence there would be greater destruction of our forces, bases, or population. Thus, the worth (not the cost) of an additional available aircraft could be calculated as the expected value of destruction which would take place if this aircraft were not available for combat against an invading bomber.

A similar analysis could be made of a fighter-bomber in Vietnam whose mission is interdiction against supply channels coming from the North. Here the worth of an unavailable aircraft could be calculated by considering the additional damage which enemy forces could inflict against our forces after the enemy received the additional supplies which penetrated into South Vietnam because of the lack of interdiction sorties.

In a peacetime situation, however, there would be no direct loss during the period that the aircraft was not available, since destruction would take place only in wartime. However, the purpose of peacetime forces is to achieve a state of deterrence, a function of the degree of peacetime readiness. Hence the primary measure that can be placed on a peacetime system is an heuristic one (i.e., requiring that the system equal or exceed some minimum availability level at practically all times). For example, the commander could specify that 80% of the aircraft must be available for ground or flight alert 95% of all days. If the base commander finds that his aircraft availability goes consistently below this level, he knows his efficiency rating may be reduced.

However, as indicated in Figure 11-1, it is possible to have an abnormally high demand at certain times under any logistics system. For example, there is some (small) chance that all 18 aircraft could be down in one day and, once a replacement part is not at the base, the aircraft is down for the rest of the month. Thus, if this minimum 80% level (14 aircraft) is to be maintained most of the time, it is necessary that the logistics system include some means for obtaining additional spare parts rapidly if abnormal demands for spare parts occur and the part cannot be repaired at the base. Several ways exist for obtaining this rapid delivery when needed. One way is to use the transport aircraft at the base to fly to an adjacent base and borrow one or more needed parts. However, not only is this also a transportation cost incurred, but it reduces the number of spare parts at this base and may result in reducing its availability rate later in the month.

A better method may be to fly to the depot with one or more of the defective parts and obtain replacements. We shall assume this procedure is followed. We shall further assume that this special delivery procedure can be accomplished fast enough to obtain and install the new part in less than 12 hours. Thus, if we define an operational aircraft day as one in which the aircraft has been available for 12 hours or more, we can wait for special delivery of a part until the level diminishes to 13 aircraft and, based on the above assumption, guarantee a *100%* probability of meeting the minimum availability level.* This seems to be a reasonable assumption since the depot could have an additional part stock on hand, or have already reconditioned the part from the previous month's input, particularly if alerted by a call from the air base before the special delivery aircraft leaves. Finally, the defective part itself could be repaired when it arrives at the depot since repair personnel could be taken from their normal duties on such special calls. Of course, cannibalization of an aircraft which is waiting for some other part is also possible, but initially this will be considered as an emergency situation

---

* Later in the chapter we shall indicate how to analyze the problem without making these assumptions.

only and not part of the regular procedure. Later we shall analyze such an approach.

The structure of the operational flow model which describes these activities is shown in Figure 11-2. This structure will be quantified so that the analyst can determine the time response (and cost required) to accomplish the objective using alternative logistic policies.

The following accountable factors can be listed as being important to this problem:

1. Demand or failure rate of each part.
2. Number of spare parts stocked of each type at each base supply shop.
3. Total cost of malfunction, including (a) base maintenance cost, (b) logistics cost of ordering a part, (c) normal monthly transportation costs, (d) special delivery transportation cost, (e) depot repair cost, and (f) special delivery depot repair cost.

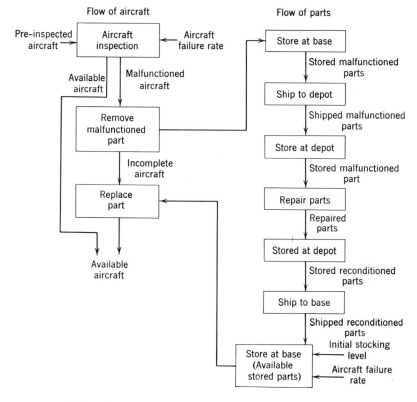

Figure 11-2.   Aircraft maintenance operational flow model.

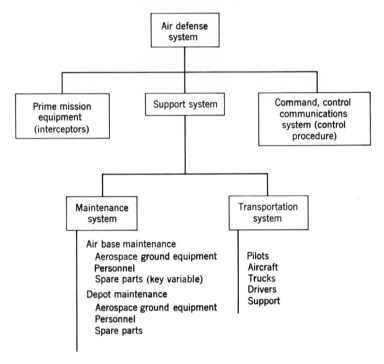

Figure 11-3.   System design model.

4. Time aircraft is unavailable, entailing (a) time to locate and remove a defective part, (b) time to get a replacement part, and (c) time to reinstall a reconditioned (or new) part.

5. Specific logistics policy.

Constructing the System Design Model

A structure of the various elements of the system is shown in Figure 11-3. These elements consist of the prime mission equipments (i.e., the interceptors), the maintenance system at both the air base and the depot, including the spare parts at each base and depot, the transportation system consisting of the depot supply aircraft which replenishes each base monthly, and the command, control, and communications system which implements the logistics policy through controlling the supply aircraft and handling replacements for the spare parts.

PROBLEM TO BE SOLVED AND APPROACH TO BE TAKEN

The PTBS is defined and structured as follows: given a number of aircraft at each base, and an average of monthly flying hours to be implemented, find the lowest cost method of achieving a given minimum level of aircraft availability. Although the *total* logistics cost includes the cost of initial spare parts inventory, the cost of replacement or reconditioning of defective parts, the cost of ordering parts, the cost of normal part transportation and the cost of special delivery transportation, we are most interested in the results obtained by varying the key differentiating costs of spending funds on the initial spare parts stocking level as compared with the cost of special delivery trips (considering all other elements of the system as fixed). More precisely, we are interested in determining that mix of spare parts and special delivery service which will provide at least the minimum acceptable availability level at lowest total cost. Furthermore, for a given logistics policy, the only variables which need be considered are spare parts investment cost and special delivery costs, since all other costs are constant and depend on the average part failure rate.

The analysis has been structured as follows: We shall first determine the relationship between aircraft availability and spare parts stocking level. We shall then determine the relationship between special delivery of parts and availability. Finally, these two relationships shall be combined.

UNDERSTANDING THE DEMAND FUNCTION DATA

We shall now examine the past demand for spare parts to gain an understanding of the meaning of these data, how to structure them, and how it may be possible to use the data to predict a future demand.

The data samples for part C are plotted in Figure 11-4. In attempting to use the actual data samples to predict a future demand, the analyst runs into the following problems:

1. The past frequency of demand function plotted appears to be a relatively smooth one with the exception of the discontinuity in the region of demands for four to six parts. The analyst should examine the past operations to try to determine if there is any logical reason for such a discontinuity, for if it is due to the limited number of data samples available, it would be better to construct a smooth distribution (envelope) straddling the points involved.

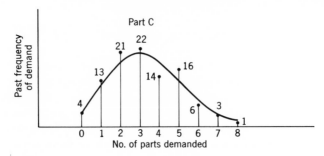

Figure 11-4.   Sample frequency demand function for part C.

2. Notice that there was never a demand for nine or more parts. Is there any reason to believe that there will never be a demand for 9, 10, or 11 parts? Obviously, it is possible that these demands will occur even though the chance of their occurring may be small. In fact with the logistics policy of calling for special delivery transportation when there is an availability of less than 80%, theoretically there can be a demand for 18 parts per day, or 540 per month. Obviously the probability of this occurring is very small.

There is the need for some probability distribution function which attempts to match the data samples obtained and can be used as a prediction of future demands. The problem is what function shall we use?

To choose an appropriate probability distribution function, the analyst attempts to enlarge his knowledge of failure rates beyond the 100 monthly samples per part that he now has. He does this by examining the demands for spare parts of similar types of aircraft and considers many possible probability distribution functions which might be used to approximate the demand. For example, of the many distribution functions from which the analyst can choose, there are the normal, log-normal, chi-square, gamma, etc. However, in many practical applications involving failure statistics, the Poisson distribution,* which contains only one parameter (average failure rate), is the simplest and most convenient to use. Mathematically the distribution is expressed by the following equation:

$$P_{\bar{T}}(n) = e^{-\lambda T} \frac{(\lambda T)^n}{n!}.$$

This function gives an expression for the probability $P_T(n)$ of n failures occurring within a time period T (T = one month). The function involves only one parameter, λ, which is usually interpreted as the average failure rate in the time interval T.

* W. Feller (1957), Chapter 17.

More complicated models, such as the Weibull distribution * with two or more parameters, are sometimes used to improve the model accuracy but at the cost of expending more effort in parameter estimation. Thus, it is reasonable to use the Poisson distribution to indicate future malfunctions as a function of time rather than using the actual data. However, note that the analysis makes two major assumptions: (a) that the future demands will follow a Poisson distribution, and (b) that the future monthly demand rate of any part will equal the past demand rate. Later in the chapter we shall discuss how to cope with the uncertainties of these two assumptions.

Since the Poisson distribution is based on only one parameter (i.e., the average failure rate which occurs during the period under examination, one month), the analyst has merely to go to a set of Poisson tables to obtain the predicted probability of failure (the frequency function) for each part, based on its average failure rate. This is shown in Table 11-2 and is plotted for

Table 11-2. *Poisson frequency functions for different parts* †

X	Part A $\bar{A}=0.29$	Part E $\bar{E}=0.6$	Part D $\bar{D}=1.1$	Part C $\bar{C}=3.2$	Part B $\bar{B}=6.3$	X
0	0.7482636	0.548812	0.332871	0.040762	0.001836	0
1	0.2169964	0.329287	0.366158	0.130439	0.011569	1
2	0.0314645	0.098786	0.201387	0.208702	0.036441	2
3	0.0030416	0.019757	0.073842	0.222616	0.076527	3
4	0.0002205	0.002964	0.020307	0.178093	0.120530	4
5	0.0000128	0.000356	0.004467	0.113979	0.151868	5
6	0.0000006	0.000036	0.000819	0.060789	0.159461	6
7		0.000003	0.000129	0.027789	0.143515	7
8			0.000018	0.011116	0.113018	8
9			0.000002	0.003952	0.079113	9
10				0.001265	0.049841	10
11				0.000368	0.028545	11
12				0.000098	0.014986	12
13				0.000024	0.007263	13
14				0.000006	0.003268	14
15				0.000001	0.001373	15
16					0.000540	16
17					0.000200	17
18					0.000070	18
19					0.000023	19
20					0.000007	20
21					0.000002	21
22					0.000001	22

* W. Weibull (1951).
† Poisson's Exponential Binomial Limit, E. C. Molina, D. Van Nostrand Co. Inc., N.Y., 1942.

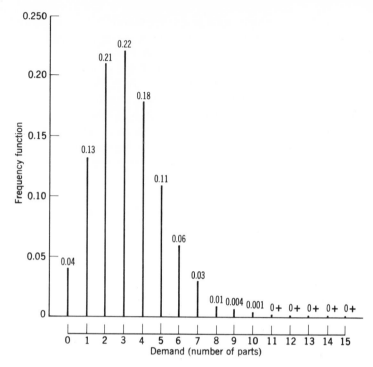

Figure 11-5. Frequency function of the demand for part C (from Table 11-2).

part C in Figure 11-5. Table 11-3 shows the cumulative distribution for each part (i.e., the probability that the number of demands will be of value X or more), in which the distribution for part C is plotted in Figure 11-6.

Information Obtainable from These Tables

We shall now examine how the analyst can use this information. Table 11-2 and Figure 11-5 show the probability that any particular monthly demand will occur. Table 11-3 and Figure 11-6 show the cumulative distribution (i.e., the probability that the demand will be for this demand level or more). Thus, for example, there is a probability of 0.13 that there will be a demand of exactly one part of C. Also there is a probability of 0.96 that there will be a demand of one or more parts of C.

It is possible to use Tables 11-2 and 11-3 to calculate the expected number of demands for a part to be met per base per month as a function of the number of parts stocked. For example, if one part, C, were stocked, this part would be used in attempting to satisfy a demand for one or more. Obviously, it will not meet *all* of those demands which exceed one. However, it will be used to *partially* satisfy all of these demands. Thus, the possibility of one stocked part being used can be obtained from Table 11-2, and is equal to the sum of all probabilities of demand for 1, 2, 3, etc. This can be calculated readily by noting that the demand for which part one will *not* be used is the demand for zero parts, which is 0.04. Thus, the probability of using this part is $1 - 0.04$ or 0.96. Note that this can also be obtained from Table 11-3 which has already made the accumulation.

Similarly, if two part C's are stocked, the *additional usage* of the second part can be found from Table 11-2 and is equal to the sum of the demands for 2 parts, 3 parts, 4 parts, etc. Thus, the demand for the second part is equal to unity minus the demands for zero and one part, or $1 - (0.04 + 0.13)$ $=0.83$. This can also be obtained from Table 11-3 under the cumulative probability demand for two parts. The total expected demands filled when stocking two parts is the sum of $0.96 + 0.83$ or 1.79. Thus, Table 11-3 can be used to determine the *incremental* expected demands filled as additional parts are stocked. The specific values for this transfer function for all five Parts A, B, C, D, and E are tabulated in Table 11-4, and are plotted in Figure 11-7 for part C, thus relating the ability to meet demands for C as a

Table 11-3. *Poisson cumulative distribution for different parts* *

X	Part A $\overline{A}=0.29$	Part E $\overline{E}=0.6$	Part D $\overline{D}=1.1$	Part C $\overline{C}=3.2$	Part B $\overline{B}=6.3$	X
0	1.0000000	1.000000	1.00000	1.000000	1.000000	0
1	0.2517364	0.451188	0.667129	0.959238	0.998164	1
2	0.0347400	0.121901	0.300971	0.828799	0.986595	2
3	0.0032755	0.023115	0.099584	0.620096	0.950154	3
4	0.0002339	0.003358	0.025742	0.397480	0.873626	4
5	0.0000134	0.000394	0.005435	0.219387	0.753096	5
6	0.0000006	0.000039	0.000968	0.105408	0.601228	6
7		0.000003	0.000149	0.044619	0.441767	7
8			0.000020	0.016830	0.298252	8
9			0.000002	0.005714	0.185233	9
10				0.001762	0.106121	10
11				0.000497	0.056280	11
12				0.000129	0.027734	12
13				0.000031	0.012748	13
14				0.000007	0.005485	14
15				0.000001	0.002217	15
16					0.000844	16
17					0.000304	17
18					0.000104	18
19					0.000034	19
20					0.000010	20
21					0.000003	21
22					0.000001	22

* Poisson's Exponential Binomial Limit, E. C. Molina, D. Van Nostrand Co. Inc., N.Y., 1942.

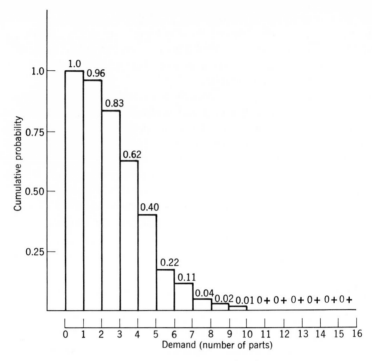

Figure 11-6. Cumulative distribution of the demand for part C equaling or exceeding a given amount (from Table 11-3).

function of the number of parts stocked or the cost of these parts. Note that this function is again one which shows diminishing return.

EVALUATING ALTERNATIVE STOCKING POLICIES

Table 11-4 may be used on a trial and error basis to determine the total expected demands filled for alternative stocking policies for Parts B, C, and D. For example, the results of the proposed stocking policy (i.e., stocking six B, three C, and one D) is shown in Table 11-5. This stocking policy requires a total investment cost per base of $40,800. Note that the total expected demands filled per month would be 8.24, the sum of the individual expected demands filled for each part. However, there is a series of other stocking policies which can be generated, each also costing $40,800. These alternative policies, together with the calculated total expected demands filled, are also shown in Table 11-5. Notice that the policy of stocking 46 parts of B, 9 parts of C and zero parts of D gives a total maximum expected demands filled of 9.50. In fact, it is possible to achieve essentially the same result by stocking 14 parts of B, 9 parts of C and zero parts of D, which would cost only $31,200, thereby saving investment costs.

Table 11-4. *Total expected demands filled*

Stock Level	Part B $\bar{B}=6.3$	Part C $\bar{C}=3.2$	Part D $\bar{D}=1.1$	Part A $\bar{A}=0.29$	Part E $\bar{E}=0.6$
1	1.00	0.96	0.67	0.25	0.45
2	1.99	1.79	0.97	0.28	0.57
3	2.94	2.41	1.07	0.29	0.59
4	3.81	2.80	1.10	0.29	0.60
5	4.56	3.02	1.10		0.60
6	5.16	3.13			
7	5.61	3.17			
8	5.90	3.19			
9	6.09	3.20			
10	6.20	3.20			
11	6.25				
12	6.28				
13	6.29				
14	6.30				
15	6.30				
16	6.30				
17	6.30				

This trial and error basis may be satisfactory for a small number of parts; however, it would be almost impossible to make such calculations for thousands of parts. Hence a simpler optimization algorithm must be developed. The concept of marginal effectiveness as described in Chapter 7 provides this means of optimization in relatively simple form.

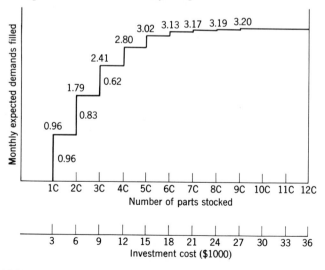

Figure 11-7. Demands filled for part C as a function of number of parts stocked.

Table 11-5. *Spare parts allocation procedure (trial and error method)*

Alternative Stock Levels			Expected Demands Filled/Month			
B ($300)	C ($3,000)	D ($30,000)	B	C	D	Total
6	3	1	5.16	2.41	0.67	8.24
106	3	0	6.30	2.41	0	8.71
96	4	0	6.30	2.80	0	9.10
86	5	0	6.30	3.02	0	9.32
76	6	0	6.30	3.13	0	9.43
66	7	0	6.30	3.17	0	9.47
56	8	0	6.30	3.19	0	9.49
46	9	0	6.30	3.20	0	9.50
36	10	0	6.30	3.20	0	9.50
26	11	0	6.30	3.20	0	9.50
16	12	0	6.30	3.20	0	9.50

Marginal Effectiveness Method

Consider now the marginal effectiveness obtained by the addition of parts A, B, C, D, or E, adding these one at a time. In this case, marginal effectiveness is measured as the additional demands filled per dollar expended and is tabulated in Table 11-6. This table is obtained as the ratio of the additional expected demands filled by each subsequent part (given by Table 11-3) to the cost of each subsequent part stocked (which is the investment cost for each part). However, the investment cost is actually equivalent to a monthly cost of one-sixtieth the investment cost, if it is assumed that the part will be used over an anticipated five-year or sixty-month period of operating life. This monthly cost was actually used in constructing Table 11-6. This will not hinder the later calculations, since monthly costs will be used throughout. The advantages of doing so will become more apparent later in the chapter.

With this table available, the analyst can establish a decision priority rule for stocking parts by taking the highest marginal effectiveness available at each decision point, as also shown in the table. Thus, the first part stocked would be the first B, giving a marginal effectiveness of 0.20. The second part B also provides 0.20 (some round-off error has been introduced). The third part chosen should be the third B and this part continues to be chosen until the sixth decision point, where the first A is chosen. These decision steps are tabulated in Table 11-7, which shows not only the decision number and the part chosen, but also the incremental effectiveness obtained (marginal effectiveness times monthly cost) as well as the subtotal effectiveness obtained with the addition of each part. In addition, the incremental costs and the subtotal of costs are tabulated. This is done to take care of the case of there

Table 11-6.   *Marginal effectiveness (additional demands filled per dollar expended per month)*

ith Part Stocked	Part B $300 ($5/mo.)	Part C $3000 ($50/mo.)	Part D $30,000 ($500/mo.)	Part A $120 ($2/mo.)	Part E $6000 ($100/mo.)
1st	0.200 (1)	0.0192 (12)	0.00134 (24)	0.126 (6)	0.0045 (19)
2nd	0.200 (2)	0.0166 (14)	0.00060 (28)	0.017 (13)	0.0012 (25)
3rd	0.191 (3)	0.0124 (15)	0.0002 (32)	0.0016 (23)	0.00023 (31)
4th	0.175 (4)	0.0080 (17)	0.000051	0.00011 (34)	0.000034 (36)
5th	0.152 (5)	0.0044 (20)	0.00001	0.000006	0.0000039
6th	0.120 (7)	0.0022 (22)		0.0000003	0.00000039
7th	0.088 (8)	0.00089 (27)			
8th	0.060 (9)	0.00034 (30)			
9th	0.038 (10)	0.000011			
10th	0.022 (11)				
11th	0.011 (16)				
12th	0.0056 (18)				
13th	0.0026 (21)				
14th	0.0011 (26)				
15th	0.00044 (29)				
16th	0.00016 (33)				
17th	0.00006 (35)				
18th	0.00002				
19th	0.000006				

being an upper threshold in cost. For example, if the decision-maker would not be permitted to spend more than $40,800, he could work up through Decision 23, spending a subtotal of $25,560. Note that Decision 24 (i.e., adding the first part D) would require $30,000 which would exceed the max-

Table 11-7. *Spare parts allocation procedure*

Decision Number	Part Chosen	Marginal Effective- ness	Incremental Effective- ness	Subtotal Effective- ness	Incremental Cost	Subtotal Cost
1	1st B	0.200	0.998	0.998	$300	$300
2	2nd B	0.200	0.987	1.985	$300	$600
3	3rd B	0.190	0.950	2.935	$300	$900
4	4th B	0.175	0.874	3.809	$300	$1200
5	5th B	0.152	0.753	4.562	$300	$1500
6	1st A	0.126	0.252	4.814	$120	$1620
7	6th B	0.120	0.601	5.415	$300	$1920
8	7th B	0.088	0.442	5.857	$300	$2220
9	8th B	0.060	0.298	6.155	$300	$2520
10	9th B	0.038	0.185	6.340	$300	$2820
11	10th B	0.022	0.106	6.446	$300	$3120
12	1st C	0.0192	0.959	7.405	$3000	$6120
13	2nd A	0.017	0.035	7.440	$120	$6240
14	2nd C	0.0166	0.829	8.269	$3000	$6540
15	3rd C	0.0124	0.620	8.889	$3000	$9540
16	11th B	0.011	0.056	8.945	$300	$9840
17	4th C	0.0080	0.397	9.342	$3000	$12,840
18	12th B	0.0056	0.028	9.370	$300	$13,140
19	1st E	0.0045	0.451	9.821	$6000	$19,140
20	5th C	0.0044	0.219	10.040	$3000	$22,140
21	13th B	0.0026	0.013	10.053	$300	$22,440
22	6th C	0.0022	0.105	10.158	$3000	$25,440
23	3rd A	0.0016	0.003	10.161	$120	$25,560
24	1st D	0.00134	0.667	10.828	$30,000	$55,560
25	2nd E	0.0012	0.122	10.950	$6000	$61,560
26	14th B	0.0011	0.005	10.955	$300	$61,860
27	7th C	0.00089	0.045	11.000	$3000	$64,860
28	2nd D	0.00060	0.301	11.301	$30,000	$94,860
29	15th B	0.00044	0.002	11.303	$300	$95,160
30	8th C	0.00034	0.017	11.320	$3000	$98,160
31	3rd E	0.00023	0.023	11.343	$6000	$104,160
32	3rd D	0.0002	0.100	11.443	$30,000	$134,160
33	16th B	0.00016	0.0008	11.4438	$300	$134,460
34	4th A	0.00011	0.0002	11.4440	$120	$134,580
35	17th B	0.00006	0.0003	11.4443	$300	$134,880
36	4th E	0.000034	0.0033	11.4476	$6000	$140,880

imum cost of $40,800. Since this decision could not be made, we would skip to Decision 25 which adds another $6,000 to the previous total of $25,560. This decision is illustrated in Figure 11-8. Thus, if $30,000 had been spent, the marginal effectiveness (the slope) would have been 0.00134. By spending only $6,000, one obtains a lower marginal effectiveness (0.0012). As shown, the slope is slightly less but this decision does enable the analyst to keep

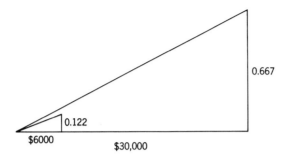

Figure 11-8.   Marginal effectiveness for decisions 24 and 25.

within the threshold limit of cost. Thus, while this threshold will prevent the optimal effectiveness to be achieved, it will still be optimal compared to other alternatives which also keep within the threshold limitation.

Note that with this type of analysis it is possible to program the decision logic described onto a computer which can cope with a large number of parts, and that the algorithm described is quite simple. The marginal effectiveness for each part added is calculated and stored in the computer; the computer is then programmed to compare the marginal effectiveness of each of the next parts which could be chosen, and chooses that which provides the highest marginal effectiveness, taking care not to exceed any cost threshold.

Other elements could have been included in the analysis. For example, the Navy uses an "essentiality factor"; this is a worth coefficient which multiplies the marginal effectiveness, thus taking into account the importance of this particular part. For example, an engine turbine blade would have a relatively high ranking compared to a pilot light, since the former could ground the aircraft, whereas the latter would not.

The information tabulated in Table 11-7 shows the optimal spare parts allocation procedure. This information may be plotted, showing expected demands filled as a function of total cost for spare parts, as shown in Figure 11-9.

Achieving a Balance Between Inventory and Special Delivery Costs

The previous discussion, culminating in Figure 11-9, has been concerned with a method for obtaining the highest level of effectiveness (expected demands filled) for a given allocation of funds used for spare parts inventory at a base only. If the analyst were asked by the decision-maker to indicate the "proper" costs for spare parts, however, he could not provide

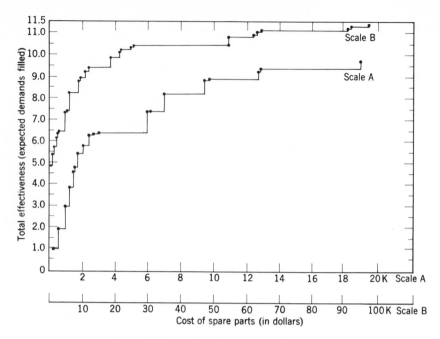

Figure 11-9. Optimal spare parts allocation.

any answer except to display the function shown in Figure 11-10. Thus, the final decision would have to be made on an intuitive basis. For example, the decision-maker could note that as inventory expenses are increased, the marginal effectiveness is reduced and he would realize he should not get too far into the saturation region. However, by expanding the system design model to take into account other ways of satisfying the demand for spare parts, the systems analyst may find a lower cost solution than the one shown. Such a solution may be obtained by comparing the marginal effectiveness of increasing the spare parts inventory with the marginal effectiveness of special delivery transportation.* Thus, the desired system is that in which the total cost for both spare parts investment and special delivery trips is minimized.

Costs of Each Special Delivery Trip

To determine the incremental, "out-of-pocket" costs associated with each special delivery trip, the analyst must consider three aspects of the problem.

---

* Each of these would increase the availability of the fixed number of aircraft. The marginal effectiveness of adding aircraft could also be considered, which would also tend to increase the total number of aircraft available.

First, he must explicitly delineate the operational procedure involved in obtaining the part on a rapid, special order basis, and the implication for the over-all system design in implementing the procedure. Second, he must consider the cost factors involved in the procedure and third, how many of these costs are truly incremental or out-of-pocket costs which would not have to be borne if the special delivery trip were not taken.

Table 11-8 contains an outline of the system procedure to be followed when making a special delivery trip. It also contains the accountable factors which would have to be considered in determining what incremental costs would be involved in this trip. Determining the incremental costs involved is a complex task for the analyst. Many, if not all, of the functions involved in implementing the special delivery trip procedure can be performed with existing personnel or existing equipments, if the number of trips per month, for example, is not high. Hence, only small incremental costs may be involved. On the other hand, if the number of special delivery trips becomes great, additional personnel and other system components will be required, raising the incremental costs.

For example, reviewing the accountable factors indicates the following:

1. Communications from air base to depot may use the existing communication links.

2. The logistics aircraft already assigned to the base may be used for this purpose.

Table 11-8. *System procedure used for special delivery*

SYSTEM PROCEDURE	ACCOUNTABLE FACTORS
1. Call depot to see if part is available or if it can be made available rapidly.	1. Communications from airbase to depot, or if part is not available, communications to other locations.
2. If so, prepare special delivery aircraft for flight.	2a. Procure special logistics aircraft. 2b. Preflight service of aircraft. 2c. Assign an available pilot to fly aircraft.
3. Fly aircraft to depot.	3. Petroleum, oil, lubricants (POL) required.
4. Depot prepares part for shipment.	4a. Depot inventory of parts or rapid reconditioning of part perhaps requiring special maintenance personnel. 4b. Special shipment handlers if required.
5. Load part on aircraft.	5. Depot loading personnel.
6. Return to base.	6. POL required.
7. Unload part.	7. Part handlers.
8. Transport to maintenance building.	8. Ground transportation.

3. Preflight servicing of the aircraft may be done with the normal base personnel, unless the number of flights becomes excessive.

4. The special aircraft pilot could be an officer who is interested in obtaining additional flight time and, hence, would involve no incremental cost.

5. Petroleum, oil, and lubricants used for the trip could be charged against training if such training is required to keep flight personnel at a given level of performance.

6. Inventory of parts at the depot would require an additional investment.

7. Rapid reconditioning of parts, if required, could necessitate the use of overtime personnel. However, since all parts will have to be reconditioned anyway, much of this could already have been done with the normal maintenance personnel since special trips will most probably occur during the latter part of the month. Also there is a tradeoff between this and the number of parts kept in depot inventory.

8. The task of handling special shipments might be done by the regular handlers.

Thus, the analyst would have to determine incremental requirements as a function of the number of special delivery trips per month.

At this point it is possible for the analyst to be in a dilemma for the following reason. The optimal stocking policy which he wishes to determine is a function of the cost of a special delivery trip. However, the cost of a special delivery trip is a function of the number of special delivery trips required each month, which in turn is a function of the number of spare parts stocked. One way which the analyst may use to eliminate this "round robin" is to assume that only a small number of special delivery trips per month would be required; hence he can use most of the components within the existing system for the few special delivery trips required. Such an assumption might lead to the conclusion that a special delivery trip has an incremental cost of $500, as an example. The analyst can then continue the analysis, using this estimated cost, and thus determine the optional stocking policy based on this figure. Continuing on, the analyst can determine the average number of special delivery trips per month required, and then recheck to see what the expected cost of the special delivery trip would be, based on this new information. If it turns out that the cost would be radically different from the assumed special delivery cost (i.e., $500 per trip) the analyst can do the analysis again, based on the new figure or an extrapolation thereof. Thus, using a trial and error method as indicated, an optimal operating point can be obtained.

To continue with this exercise, we shall now show how this information can be used to determine the proper stocking levels. Recall that when a special delivery trip is made, it is because a part is absolutely needed to bring the total number of available aircraft back up to the minimum level (i.e., from 13 to 14). Thus, one special delivery trip results in an absolute gain in incremental effectiveness of one demand satisfied. Notice that this is absolutely satisfied and not an expected value. Moreover, since we are assuming the cost of the trip was $500, the analyst can compute the marginal effectiveness of the trip, which is 0.002 demands met per month per dollar. By comparing this marginal effectiveness with the decision table of Table 11-7, the analyst can note that it is more efficient to use special delivery than to stock beyond Decision 22. (Notice that Decision 23 yields a marginal effectiveness of 0.0016, which is less than 0.002.) Moreover, the logistics policy would probably state that when a special delivery trip is required, all (or certainly more than one) of the other repairable parts will also be returned to the depot on the same trip, along with an order for all or some replacements for the nonrepairable parts which have failed.* Such a procedure will reduce the chance of needing an additional special delivery trip during the remainder of the month since the other unavailable aircraft could now be repaired. With this type of policy, the marginal effectiveness of a special delivery trip will be greater than 0.002, since there is some probability that one or more additional aircraft would have failed before the end of the month. Determining the exact value of the expected demands met, and hence the value of the marginal effectiveness of a special delivery trip, is difficult to do analytically. However, it is quite simple to solve the problem by means of a simulation exercise, discussed in the following.

MONTE CARLO SIMULATION

Monte Carlo simulation offers a method of simulating all of the events which occur during actual operations over a long period of time. In the simulation described here, the events are restricted to one base and the depot. The base will be provided with 18 aircraft and an initial stock level of spare parts. The same probabilistic demand for each part previously made explicit will again be used. Each base will be replenished each month with a supply of parts which equaled the number of parts requisitioned during the previous month. If the base runs out of parts and the number of available aircraft is reduced to 13, two courses of action are available: cannibalization of a

---

* The number of additional parts which should be requested on the same special delivery trip can be determined by the same simulation procedure described here, and will certainly be a function of any additional handling costs involved as well as the probability of using the part during the remainder of the month.

grounded aircraft to obtain the part or, if this is not permitted, a special delivery trip to the depot for spare parts.

To simplify the simulation somewhat, it is assumed that the depot has a sufficiently large supply of parts to meet any special delivery demands by any one base. Since a record will be kept of what these demands actually are, this information will indicate the amount of inventory necessary to meet this demand.

The procedure which would be employed in performing the simulation is as follows:

1. Choose alternative stocking policies of interest. For example, the analyst might be interested in knowing what the anticipated results would be of following the policy of stocking up to Decision 22 of Table 11-7, compared with stopping at Decision 19 or Decision 16. As indicated before, it would not be efficient to go beyond Decision 22; therefore, the simulation exercise will be constructed to show the anticipated results obtained in following each of several alternative stocking policies hoping that the optimum policy can be bracketed somewhere between the policies selected for analysis.

2. Set up accounting files which will keep records of (a) the number of available aircraft for each day of the simulation, (b) the base inventory each day for each part, and (c) number of special delivery trips taken over time.

3. Set up a data base which describes the demand for each part on a probabilistic basis. This data base should consist of the cumulative distribution function for each part.

Operation of Simulation

Initial Setup. The first step is to set in the initial conditions of the number of available aircraft as well as the initial stocking level for each part at the start of operations. Using a Monte Carlo random draw, determine the demands for each part for the first interval of time, based on the cumulative distribution function describing the demand for each part. From here the game can be run either one of two ways. The first (and most accurate) way is to divide the total time of operation (five years) into small enough time intervals so that upon using the Poisson distribution of the demand data, there is only a very small probability of demanding two or more of any given part during the time interval. As an example, a time interval of one-half day (or one hour) could be used. Based on this time interval, the new cumulative probability of demand (using the new average rate over the new time interval) can be shown, as in Figure 11-10, indicating that the probability of two or more demands is exceedingly small. In this type of operation, the drawing of a random number will determine if there is a demand for this given part during this time interval.

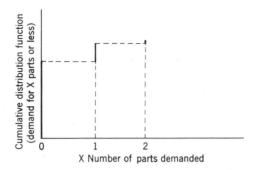

Figure 11-10. Probabilistic demand for X number of parts during the time interval.

Determine if Special Delivery Trip is Required. At the end of each interval, determine if a special delivery trip to the depot is required. This is determined by comparing the demand which has occurred for each part with the total number of each part on hand at the beginning of the interval. Then see if the number of available aircraft is reduced below the minimum allowable level of 14. If so, a special delivery trip is required. Cannibalization of needed parts from grounded aircraft can also be included in the analysis, if desired.

Using the logistic policy agreed upon for returning to the depot and replacing parts, replenish the parts from the depot to the base and continue the process to see if a second special delivery trip is required later in the month. If so, repeat the special delivery trip. In addition, normal monthly replacement of parts is made in accordance with the replacement policy and the appropriate amount is added to each supply stock.

For each policy, the simulation should be run for enough trials so that the analyst obtains statistically significant results. As indicated in Chapter 7, the number of trials which constitutes statistically significant results can be obtained by plotting a variable such as the total special delivery costs (which is expected to vary randomly). In this case, the analyst could plot the average of the special delivery costs, as well as the probability distribution shown in Figure 7-5. Enough trials of each policy should be made until the standard deviation of each policy cost is small compared to the differences among the averages of the costs.

Data Recorded. The following data should be recorded:

1. The number of special delivery trips required over the total operating life of the simulation.

2. Total cost of special delivery trips. The analyst may use special delivery costs which are either fixed or are a function of the number or type of parts reconditioned if this is applicable.

3. Available aircraft by day. These data may be desired if the analyst wishes to calculate the *average* availability or as a probability distribution. This measure may be significant in trading off average availability versus total dollars spent even though the minimum availability for each policy will always be at the level of 14 aircraft per base out of a total of 18.

4. Specific monthly demands and the order in which the parts fail. These data should be recorded, as they can be used for sensitivity analysis in comparing alternative stocking policies. By always using the same monthly demands as alternative stocking policies are examined, this degree of randomness can be removed from the analysis so that any differences in results will be attributed directly to the logistics policy as opposed to any randomness of demand. On the other hand, if a random generator is used which will always produce the same order of numbers for the Monte Carlo draws, the same part demand will always occur; hence the specific monthly demands for each part need not be recorded.

By rerunning the simulation for different type stocking policies, the results may be compared on the basis of *total* costs expended (for both spare part stocks and special delivery). These costs can further be compared with the average availability if desired. If this is not desired, the final choice among alternative stocking policies will be made on the basis of minimum total cost to obtain at least the minimum level of availability.

A second simulation method which could be employed uses the time interval of one month. Thus, the cumulative probability of any given demand used for the Monte Carlo draw will be that of Table 11-3. In this case the random draw using the above data would determine the exact number of demands for each part of the entire month time interval. We now compare the total demand for each part with the available supply for each part. If in any monthly interval the number of failed aircraft exceed the minimum acceptable limit, one or more special trips will be needed. The exact number will be determined by examining in greater detail when each part failed. This can be done by dividing the month into small intervals of time (say 100), assigning each interval a number from zero to 99. Then by assuming that it was equally likely that each part could fail in any interval and by choosing a random number for each of the failed parts, we may determine in which interval each failed. In this way we can discover when the first special delivery trip was made, which parts were taken back for replacement, and if an additional special delivery trip was needed during the remainder of that month. While the first simulation method using small time intervals is more exact as long as the time interval is small enough so that only one part fails in each interval, the second method (using monthly intervals) is simpler to implement, and requires less computer running time, particularly if the stocking policy is such that only a small number of special delivery trips is needed.

Considering Time To Repair

The preceding analysis made several simplifying assumptions regarding the total time required to repair an aircraft. First it implicitly assumed that this time was short compared to the time between failures and, second, that the total time to repair (including a special delivery trip if needed) would always be less than twelve hours. We shall now reconsider these assumptions so that we show how a more exact analysis can be performed.

First, for each part type we must construct a probability distribution for the total time to repair (given that the spare part is available at the base) considering all activities involved to final check-out inspection when the aircraft is placed back into service. This distribution may be constructed using whatever data are available to determine the mean time to repair and some assumed function. Similarly, a probability distribution is constructed for the special delivery time required to obtain a part from the depot if the part is not available at the base.

During the Monte Carlo simulation previously described, a random draw is taken each time a particular repair or special delivery trip is to be made. This determines the time when the aircraft is placed back into service. The results of such an analysis will be aircraft availability, expressed as a probability distribution. However, since now the availability may be below the previous minimum of fourteen, the analyst and decision-maker can determine the probability that a minimum availability will be equaled or exceeded. If this analysis shows that this probability is not high enough, the systems planner could change the logistics policy (e.g., call for a special delivery trip earlier, such as at the 14 or 15 level), or more money expended for spare parts, and rerun the analysis to see the effect of such change on the availability function.

COPING WITH UNCERTAINTIES

It should be emphasized that the entire analysis which has been performed has been based on a frequency of demand for each part based only on 100 past monthly samples (ten from each of the ten bases). These limited data are the main source of uncertainty in this analysis.

Before indicating ways of treating this uncertainty it should be noted that even if the true statistical demand were known perfectly, there is always the chance that some base will have "bad luck" one month. Thus if the minimum level of operational effectiveness is to be maintained, various "backup" capabilities must be provided to cope with this bad luck if it should occur. Examples of this backup capability consist of the following:

1. An excess number of aircraft; i.e., 18 per base, whereby only a minimum of 14 is required.

2. Special delivery trip capability to fly from the base to the depot or to other bases, to obtain spare parts if the normal replenishment supplies were insufficient for this month.

Other means could be developed for further backup, if required.

The analyst can also conduct a sensitivity analysis to ascertain quantitatively the extent of the uncertainties, and how they affect the predicted results. This is discussed in the following.

Uncertainties Due to Limited Data

The basic uncertainty problem can be described as follows: Given that only 100 monthly samples of data have been collected, what information can be extracted regarding future demands from these samples? Again the analogy which is useful in understanding this problem is the following: Consider an urn which contains an infinite number of data samples of the *true* demand distribution from which the 100 data samples have been drawn at random. The basic question which the analyst faces is, "What can be said about the true distribution based on the 100 samples drawn?"

The approach which we took in structuring the 100 data samples was to make two assumptions: (a) the true data has a Poisson distribution, and (b) the parameter of the distribution is the average demand rate of the 100 samples obtained (the most likely value). Such an assumption would provide the probability distributions of the demand for a given part as illustrated by the function of Figure 11-5. However, since only 100 data samples were available, it is possible that the Poisson parameter may differ from the most likely value assumed.

It is possible to determine the confidence limits associated with the Poisson parameter (i.e., the mean value of the 100 samples) analogous to the way it was done in Chapter 8 for the binomial distribution. Thus, it can be shown that for a 90% confidence coefficient, the true Poisson parameter lies somewhere between the upper and lower limits given by the following equation:

$$\bar{m} - (1.96) \sqrt{\frac{\bar{m}}{n}} < m < \bar{m} + (1.96) \sqrt{\frac{\bar{m}}{n}} \, ,$$

where m = the true Poisson parameter (i.e., the true mean value),
$\bar{m}$ = the observed mean based on the 100 samples.
n = the number of data samples.

Thus, if the observed mean value of part B, for example, is 6.3 and 100 samples are available, the analyst could state with 90% confidence that the true mean value lies between the values of 5.81 and 6.79. Thus we can say with 90% confidence that the true Poisson distribution lies somewhere between the two Poisson distributions whose frequency functions are determined by the values of these parameters.

It is possible to perform the same type of confidence limit analysis without the assumption that the distribution is Poisson. In this case the upper and lower distributions can be obtained as before, but now wider upper and lower limits are obtained.

Uncertainty in Life of Part

Another source of uncertainty is the assumption that malfunctions occur at a constant frequency rate and that the future demand for parts will approximate the past demand. Actually, failure rates of many components may change with time. As we discussed in Chapter 6, there are essentially three phases which occur over the operating life of equipment. The first portion is the phenomenon known as "infant mortality" wherein components exhibit a higher than average rate of failure. During the "normal" life of the component, the failure rates are fairly constant. Finally, the "wear-out" phase is reached in which the frequency of failure starts to rise rapidly.

In general, for those equipments such as aircraft systems where the cost of failure is high, maintenance policies are constructed so that these components are operated in the lowest failure rate region only. This is accomplished by running initial "break-in" tests of the components at the factory before they are inserted into the actual system. This eliminates most of the initial failures. Wear-out time is then determined in one of two ways. The length of service time is recorded and when an arbitrary service time is reached (determined by past tests), all of these components are replaced as preventive maintenance.

The analysis performed in this problem assumed that the future rate of demand would be constant. Some probability distributions take an increasing or decreasing future demand into account, and this could be done if we knew how the demand would change with time. There are two reasons why this course of action was not taken. First, we did not know how the demand would change with time; secondly, it is more difficult to use these other distributions.

Engineering changes are another cause of uncertainty. In this case various parts may become obsolete and hence may not remain in operation for the five-year period as first assumed. Hence, the actual monthly cost of the stocking inventory may be larger than the assumed or predicted monthly cost. Thus the marginal effectiveness would decrease, and the decisions made in Tables 11-6 and 11-7 would be incorrect.

Some analyses take this factor into account by increasing the cost of the part to include an "obsolescence charge," generally some percentage of cost which depends on the risk of obsolescence. Another slightly different approach is to predict total expected cost of the part over the entire five-year period, taking into account the initial cost, probability of obsolescence or need for remodification, and reprocurement or remodification cost. One sixtieth of the expected total cost would provide the expected monthly cost which could be used as previously described.

Conclusions Drawn from this Analysis of Uncertainties

It should be emphasized that the analyst is always faced with uncertainty regarding the future demand for a stocked part, no matter at what level he stocks these parts (whether for a base or depot). The analyst's real problem is whether it is better to stock on the high side or the low side. If the inventory stocking level decision could be delayed, the analyst could obtain more data on which to base a decision. However, he is forced to make a decision recommendation now; thus his best approach is to explore the various risks involved in stocking on the high side or the low side.

If he stocks on the high side he will have reduced the number (and cost) of special delivery trips. However, he will have over-spent on the stocking investment cost. This overstocking is not very critical for those parts which are nonrepairable since when he perceives that the demand is actually less than his original calculations, he can begin to reduce his stocking level of nonreturnable parts by using these for continuous replacements as required. However, this cannot be done in the case of repairable parts, unless a part is finally scrapped after so many repairs.

The second risk is that associated with engineering changes. If these occur and certain parts become totally obsolete, the inventory of these parts may be a total loss to the organization, since the parts may have to be scrapped.

Note that if the stocking level is set on the low side, the risks previously mentioned are reduced. However, the number of special delivery trips and, hence, this total cost increase. The question is then, what happens to total cost? While the analyst previously determined the optimum stocking level based on a cost of special delivery, his next task is then to determine the broadness of this optimum (i.e., the minimization of the total cost in spending funds for both spare parts and for special delivery trips). This optimum is shown in Figure 11-11, which illustrates the various ways of achieving the minimum availability level (14 aircraft per base) as required. Economists call the function shown an "isoeffectiveness" curve which is the locus of all points representing all of the different methods of achieving the same effectiveness or availability rate (shown as decision points D14 through D25 in Table 11-7). The curve is also known as an "indifference" curve since there

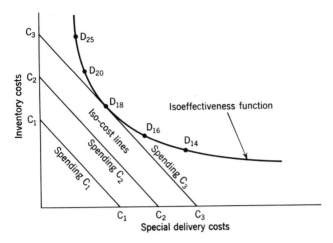

Figure 11-11. Trade-off comparison for obtaining a given level of availability.

is no difference in the output obtained even though the funds spent for stocking spare parts and special delivery trips may vary. This function is obtained by taking the cost results of the various simulations run previously.

The optimum operating point is obtained by constructing a series of "iso-cost" lines of constant expenditures for both inventory and special delivery trips, as shown in Figure 11-11. Thus the line which is just tangent to the isoeffectiveness curve provides the proper point, where total costs are minimum for the given level of effectiveness. This figure also shows how broad this optimum is; namely, how the total cost varies if other operating points are chosen. Thus if it turns out that this optimum is fairly broad, the analyst's problem becomes less critical since he can then intentionally stock on the low side without accruing large increases in total costs.

Coping with Uncertainties

One conclusion that can be drawn from the uncertainty analysis is that ordinarily the analyst would like to stock on the low side during the early phase of operations (e.g., during the first year) and then increase the stocking level if this is needed. This strategy provides several advantages; first, it reduces the risk of obsolescence caused by engineering changes which are greatest during the first year, and second, it delays the decision until more data are available. The amount the stocking level may be reduced is determined by how rapidly total costs rise as the optimum stocking level is reduced. (These additional costs may be attributed to "insurance" during the first year.) Lower stocking policies can be allowed by expanding the logistics system to include the factory as an inventory buffer, particularly for repairable parts. Thus, a suggested policy may be as follows: During the first year stock on the low side at both the base and depot, but maintain a rapid special delivery capability from the factory to the depot and/or to the bases to fill unexpected demands. Thus the factory could operate with some overtime as needed to increase the availability of parts if a high demand is requested. This factory capability could be used not only as an inventory store, but also to repair parts. This definitely reduces some of the risks since the factory is producing these for aircraft assembly anyway; hence it is a capability which can readily be brought into the system.

ADDING MORE AIRCRAFT AT EACH BASE

Recall that over time more aircraft will be added to each base. As this is done the demand rate per base will increase in proportion, if the flying time per aircraft remains the same. Thus the new average demand rate, the Poisson parameter, can be computed and new Poisson probability distributions obtained. Thus the previous analysis would be reworked using the new distributions. This increase in aircraft also reduces the risk of overstocking on parts; some initial overstocking could now be permitted since this would be reduced as the new aircraft came into the inventory. In this case the possible losses due to overstocking would be: (a) interest lost from premature overstocking, and (b) losses due to obsolescence caused by engineering changes. A further study of those parts which are more likely to be changed should be undertaken to attempt to minimize these losses.

SUMMARY OF KEY PRINCIPLES

The main purpose of this problem has been to focus on ways of determining the value (worth versus cost) of various amounts of inventory located at one or more stations as a means of satisfying a future demand at a station when this demand is not known deterministically. Another application of a probabilistic analysis was demonstrated, showing how the characteristics of available data can be structured on a probabilistic basis when the demand function is not known deterministically. It was shown how demands serve as a buffer storage, thus permitting the deliveries of parts to the stations to be made in such quantities and times as determined by other time and cost constraints, such as optimum lot size and transportation arrangements, rather than solely on the exact demand characteristic. This problem is identical to that faced by other applications such as a manufacturing facility and a distribution system; hence it has wide application.

The most important principle demonstrated in this problem involved the tradeoffs which must be performed among system factors such as inventory size, logistics characteristics, and operational effectiveness, in properly balancing resources.

This problem illustrated another application of the important concept of marginal return or marginal effectiveness, showing how optimal inventory decisions involving many thousands of parts may be made using this one principle. Also discussed was the fact that the best operational effectiveness measure of defense units which are operating in a peacetime environment is combat availability, while meeting its requirements of performing a given amount of operational hours per month.

Another application of Monte Carlo simulation was demonstrated in which many more variables were included in the model than could be easily handled analytically.

DESIGNING A BRIDGE FOR AN UNCERTAIN TRAFFIC DEMAND

While the preceding case example indicated how past data could be structured to provide an estimate of the future demand, there are many situations in which there will be an even greater uncertainty to the future than is represented by the random fluctuations of the past demand. Such uncertainty may be caused by a different system environment even in the early phases of system operation, or because the system is being designed for extremely long life (say 30 years), and it is difficult to accurately predict the demand so far in the future.

However, decisions such as the choice of the system capacity must be made using whatever data are available at the time of the design of the system. The following problem, an extension of the bridge problem of Chapter 9, discusses a method for doing so, including the use and evaluation of system expansion techniques as a means of gaining flexibility.

Problem As Given

A certain turnpike authority is currently building a part of the state's super highway system which will replace a nearby road, as shown in Figure 11-12. The existing road currently ends on each side of a river and the auto passengers must now take a ferry to cross this river. The ferry currently crosses every 30 minutes and each crossing takes 12 minutes; thus it is too slow to be used as part of the new transportation system. Hence a toll bridge is to be constructed, linking the new roads on each side of the river. Obviously the choice of the width (or traffic capacity) of the bridge will be dependent upon the traffic expected over its future life. Data of the traffic over the

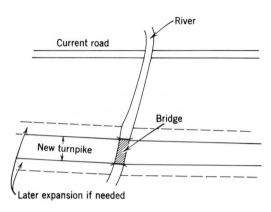

Figure 11-12.   The operational situation.

old road have been collected over the past few years and are shown in Figure 11-13. The data indicate an increasing trend in traffic volume; there is every reason to believe this rate of increase will continue and perhaps increase because of all the new construction. Hence if too small a bridge capacity is chosen, a second bridge will soon have to be constructed. On the other hand, if too large a bridge capacity is chosen, extra expenditures will have been incurred which may never be recovered from the toll revenue expected to be collected.

It should be noted that the turnpike road designers have the same problem of selecting a proper highway capacity (i.e., the number of lanes), but have one additional variable which aids them. To cope with the uncertainty of future highway growth, an additional width of land was acquired so that additional lanes of highway could be added to each side of the highway at some future time if the traffic growth continued. Unfortunately the bridge designers do not have the same degree of freedom.

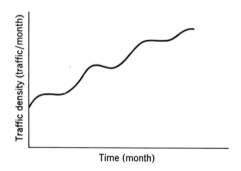

Figure 11-13.   Past traffic density.

Problem As Understood

Traffic information available on which to make an estimate of the future traffic demand consists of the monthly averages, showing a seasonal tendency as well as hourly traffic data during major holidays when peak traffic is expected. Based on such data involving these seasonal and peak demands, the traffic engineers constructed relationships which determine the bridge (or road) capacity needed to satisfy the given peak demands, given that the average annual traffic demand is known. We shall assume the existence of such functional relationships. Hence a major problem the analyst faces is to find a way of predicting the average annual traffic to be encountered over the future life of the bridge. Nearly all the traffic engineers agreed that as soon as the new road and bridge were opened, there would be an immediate rise in traffic of some amount, as some motorists changed their usual adjacent route in favor of going through the new rapid highway system. The problem was, how much of an increase would occur? In addition, some traffic growth over time was also expected, but how would the future growth compare with the past growth? Each of the various traffic engineers was asked to examine the past numerical data available, as well as traffic trends in this part of the country, and construct his best estimate of the annual traffic for each of the next 30 years. When the analyst plotted the various individual estimates of future traffic as a function of time that each of the traffic engineers submitted, it resulted in a spread of data points for each year so that the resulting "demand function" appeared a "best estimate" of future traffic demand (based on the mean or median value obtained for each point in time) in the midst of a band of uncertainty as illustrated in Figure 11-14.

Notice that there was general agreement that some maximum traffic demand would eventually occur, taking into account other planned roads in the region. The problem concerned what maximum value would be reached.

It was evident that this uncertainty in estimating the future traffic was going to cause a problem in setting the design capacity of the bridge and in being able to defend such a choice. To attempt to resolve the problem, a meeting of the project team was called, and the following design alternatives were considered.

Proposed System Alternatives

The two key questions which face the planners are:

1. How large a capacity should the bridge have? The larger the capacity, the smaller the chance of exceeding the bridge capacity and needing a second bridge, but the larger the bridge's cost will be.

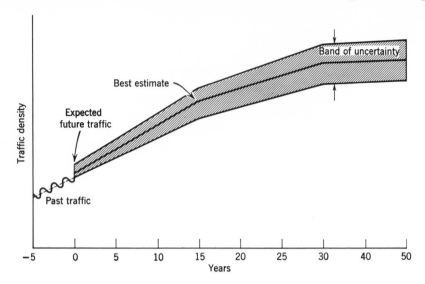

Figure 11-14. Uncertainty in estimating future traffic demand.

2. For what useful life should the bridge be designed? The longer the life, the more expensive the construction price.

Thus, the three main characteristics to be considered in the problem are: (a) the relationships among system performance, (b) bridge capacity and life, and (c) cost. Since we previously considered the relationship between life and cost in Chapter 9, we shall not include such considerations here. Rather, we shall assume that each bridge has an indefinite life as reflected by an initial cost and subsequent yearly maintenance costs. With this in mind, the different feasible strategies or system alternatives which were proposed might be classified as follows:

1. "Play it safe" strategy. One strategy considered was to attempt to satisfy the future demand by building one large bridge initially. It was proposed that the capacity of such a Bridge A be at least as large as the maximum capacity T_1 shown in Figure 11-15. However, the investment cost of such an alternative would be expensive and the bridge would be much larger than required for most of the years.

2. "Two bridge" strategy. This strategy was formulated to meet the maximum demand shown in two steps. Bridge B could be designed for some lower capacity as shown in Figure 11-15. When this capacity was reached, a second bridge would be built to handle the additional load. This strategy offers the advantage over Strategy 1 of "buying time." We could begin operations with a smaller Bridge B, more closely meeting the initial demand of

which we are more certain. Since the actual capacity of Bridge C would be chosen at a later date, its capacity would be based on the additional traffic information available at that time, giving greater flexibility.

3. "Take a chance" strategy. This strategy would involve choosing a capacity for Bridge D large enough that there would be a reasonably good chance that the capacity would not be exceeded. If this level were exceeded, a second bridge would be built later to handle the overflow. This strategy may offer an advantage over Strategy 1 since it may be possible to meet the uncertain demand with one lower capacity bridge. If not, we can still follow Strategy 2 in building the second bridge, based on additional data.

4. "Flexibility" strategy. This strategy would combine the advantages of Strategies 1 and 2 by designing an initially small bridge capable of being expanded to T_1 (or greater) by adding a lower floor at some later date, if required. This bridge would be similar in design to the George Washington Bridge in New York City, with piers and supports to accommodate the addition of the second, lower level floor. The extra cost for the additional material required as compared with Strategy 2 would have to be tabulated. However, it would probably not be high. The key principle here is that many times flexibility can be initially designed into the equipment, if previous thought is given to the need for flexibility regarding future jobs which the

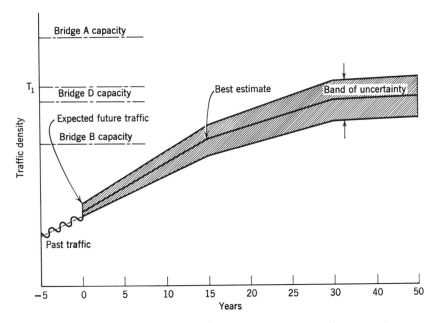

Figure 11-15. Uncertainty in estimating future traffic demand.

system will be required to handle. Note that system flexibility does cost extra. However, this cost may not be as great as adding a complete second unit.

Creating the System Demand Function

The system planners are now faced with several problems in evaluating each of the strategies described. First, what bridge capacity should be specified for each alternative? Second, how should the alternatives be compared? Again we are faced with the same problem of source selection as represented by Figure 4-10b; the greater the performance (capacity), the greater the chance of satisfying the future demand, but the greater the cost. How should a proper choice be made?

The key to this problem is to find a way of determining the chance that a given bridge capacity will be needed, since buying excess capacity which is never used results in a wasted overdesign. While there may be uncertainty in estimating the future demand function, we may be able to quantify this uncertainty (i.e., express the demand function in probabilistic form) and use this model to predict the expected cost of each alternative. Thus the next step undertaken by the systems analyst is to establish a group consensus of what the future traffic is expected to be. To do this the analyst asks each of the traffic engineers to provide three estimates of traffic for each of the 30 years. The first number was to be thought of as that value of traffic for which each engineer felt there would be the same odds that the actual traffic would exceed this value as the traffic would be less than this estimate. Thus, as shown in Figure 11-16, the estimator is saying that he believes there is a probability of 0.5 that the actual traffic for that year will exceed this estimate.

To indicate quantitatively the estimator's degree of uncertainty in making this estimate, two other estimates are to be provided for each year, as illustrated in Figure 11-16. These were to establish practical values of the minimum and maximum annual traffic to be considered for each year. Here the term "practical maximum traffic," for example, would be that value for which there was only a small chance of the actual traffic exceeding this value. Thus in deriving a cumulative distribution of the traffic from these three estimates, as shown in Figure 11-16, we shall assume that there is a zero probability that the actual traffic will be equal to or exceed this value. While in practice there is always some chance, however small, that the traffic will be greater than this value, this simplification aids the analysis, and, from a practical point of view, can be justified by the assumption that the engineer does not wish to consider any traffic greater than this, since there is only a very small chance that it would occur. In a similar fashion the "practical minimum traffic" is estimated and there is a probability of 1.0 that the actual traffic will exceed this value. Other values of the cumulative

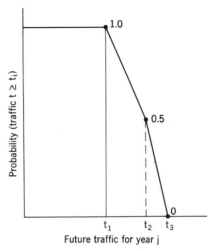

Figure 11-16. Traffic estimate at one point in time.

probability function for a given year might also be derived. These numerical estimates are all based on both a statistical analysis of the data samples of past traffic, as well as a subjective estimate of future growth. For ease in subsequent calculations, it will be assumed that the probability distribution is linear between the various estimates, as shown. The next step is to obtain group consensus on the system demand function using the various probability estimates obtained. Group conferences is one of several methods used to obtain such a consensus. Here, each estimator explains to the group the logic used to arrive at his estimate. After all discussions are completed, each estimator is asked to provide a new estimate which factors in the new information obtained. The "Delphi Method" * is an attempt to obtain group consensus by reducing some of the deficiencies of the group conference approach through revised estimating after additional group information is circulated within the group anonymously in written form. In the final analysis, a decision-maker will have to meld together the set of separate estimates into one consolidated estimate on the same basis. There are many different approaches for doing so. For example, the 0.5 probability level could be taken as the median of all estimates for the year under concern. While some sort of a "weighted" mean could also be used for this or the other values, in the final analysis it is up to the decision-maker to decide what values will be used for this probability distribution. This final system demand function, in probabilistic form as shown in Figure 11-17 will be used as the standard for designing and evaluating all alternative bridge designs proposed.

* N. C. Dalkey (1967), N. C. Dalkey and O. Helmer (1963), B. Brown and O. Helmer (1964), R. Campbell (1966).

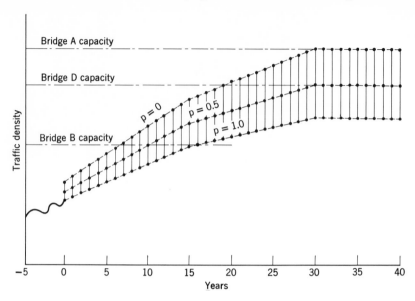

Figure 11-17.   Uncertainty in estimating future traffic demand over time.

Calculating Expected Cost of Each Alternative

We shall now use this system demand function to calculate the cost which we would expect to pay for each of the strategies described earlier. The total cost of each strategy can be obtained by using the present worth of each of the yearly costs of the total cost stream. Thus, for Strategy 1, which consists of a new maximum capacity Bridge A, only one bridge is needed since the demand function never exceeds its capacity. Its cost stream consists of the initial investment cost (C_i) and the yearly maintenance costs, as shown in Figure 11-18. (The salvage cost will be ignored since it is assumed that the life of all bridges will be indefinitely longer than the 30-year period covering the analysis.) Design costs (C_d) could also be inserted as part of the cost stream preceding investment costs, as shown.

Strategy 2 consists of two bridges—Bridge B to be constructed initially and Bridge C to be placed in operation when the capacity of Bridge B is exceeded. The present worth of the Bridge B cost stream may be obtained as already described. While we know the cost stream of Bridge C, we do not know when it begins, because of the uncertainty of the future traffic. However, since we have quantified this uncertainty by expressing the traffic in probabilistic form, we can use this probabilistic information to compute the expected cost of each alternative as follows:

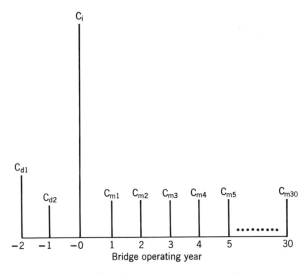

Figure 11-18. Cost stream of expenditures.

$$C_t = p_1C_1 + p_2C_2 + \ldots p_n C_n,$$

where C_t = present worth of total cost,
C_i = present worth of cost expenditure in the ith year,
p_i = probability of spending this cost.

Thus, for each year this expected cost consists of the product of the cost which would be required if the event occurred and the probability of the event occurring. But the probability of the event occurring (building the new bridge this year) is the joint probability of needing the bridge this year and the probability that it was not built in a previous year. That is,

$$P_b = P_n \cdot \overline{P}_{bp},$$

where P_b = the probability of building the bridge this year,
P_n = the probability of needing the bridge this year,
P_{bp} = the probability that the bridge was built previously,
$\overline{P}_{bp}$ = the probability that the bridge was not built previously,
$= 1 - P_{bp}$.

Thus we need to calculate the probability of requiring Bridge C for each particular year of the analysis. This may be obtained from Figure 11-17 which shows the design alternatives superimposed on the traffic demand function. Figure 11-19 indicates the probabilities of needing Bridge C (i.e., the traffic exceeding Bridge B) as a function of operational year. Thus the

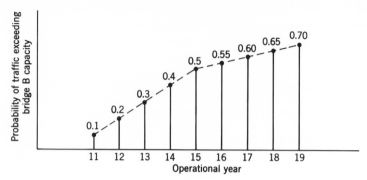

Figure 11-19. Probability of traffic exceeding bridge capacity for each operational year.

probability is 0.1 during year 11, increasing to 0.5 at year 15 and to 1.0 at year 30. Let us assume that this probability function increases linearly between these three available points as shown in Figure 11-19.

Figure 11-19 provides the information to generate the expected costs per year, i.e., the expected cost stream, for Bridge C. For example, each year an expected investment cost could be inserted by taking the product of the investment cost and the probability of building a bridge that year, where the calculations for the probabilities of having to construct Bridge C during the first five years are illustrated in Table 11-9. Similarly, the expected maintenance cost for any year is obtained as the product of the maintenance cost and the probability that the bridge has already been constructed by that year (P_{bp}).

The expected cost streams of the other strategies are obtained in the same manner. However, for Strategy 3, the optimal size of the first bridge must be chosen. This is done by considering not only how the probability of needing a second bridge decreases as a function of bridge capacity, but also how the cost of considering these factors varies with capacity. A trial and error solution can provide the proper bridge size which will minimize the total expected cost.

Thus, once the cost stream for each alternate strategy is obtained, the

Table 11-9. *Probability of building bridge in each year*

Year	P_n	P_{bp}	$\overline{P}_{bp}$	P_b
11	0.1	0	1.0	0.10
12	0.2	0.1	0.9	0.18
13	0.3	0.28	0.72	0.22
14	0.4	0.50	0.50	0.20
15	0.5	0.70	0.30	0.15

present worth of the expected cost can be calculated. Therefore the selection criteria can be that strategy whose expected present worth is lowest. However, such an approach does introduce a possible fallacy which would not be present in the real situation. In the preceding analysis, an arbitrary choice of 30 years of service was taken as the level of equal effectiveness around which each strategy was designed. This cutoff time was taken since it may be difficult to estimate what may happen to the need for the traffic system if we look much further into the future. Because of this restriction, the analysis indicates that if an additional bridge were needed in year 29, for example, we would still include the expected investment cost in the analysis, even if only one year of service would be obtained. In real life, if the bridge were needed in year 29, it would be built only if its need could be justified for some reasonable length of time, say 15 years, into the future. For example, if highway plans called for the construction of another road in the immediate future which would drain off some of the first road's traffic, the decision would probably be to live with the bridge overload until the new road could be constructed.

Hence, a better analysis might be as follows. If, in the analysis of different strategies, a new bridge is required, add maintenance costs for the system for a minimum amount of operational time, say 15 years, even if the 30-year total operational time is exceeded. This now presents a problem of comparing strategies on some equal basis. Since it may be difficult to use a time equal to the least multiple of the different operational times, another approach would be to calculate the total cost per year of total operational time, and select the strategy which requires the lowest total expected cost per year of service. Such an approach is appropriate under those situations where an indefinite long service life is expected.

SUMMARY OF KEY PRINCIPLES

As shown in the preceding example, one of the most important steps in planning for a system designed to meet an external demand function is to construct a quantitative system demand function using past data and other information available. Any uncertainties in the demand function should be expressed in probabilistic form. In this fashion various system alternatives may be evaluated by pivoting on a level of constant effectiveness; that is, all systems are designed to meet all demands, even the maximum, and the selection method can be to choose that system which meets all demands at lowest total cost.

The use of a flexibility such as expansion capability, as a system design concept, was explored in this chapter. Generally, such flexibility is an important design tool in coping with uncertainties.

12

Information Systems Planning: The Strategic Force Management System Case

Automated information systems, consisting primarily of sensors, communication, data processing and display equipments, are rapidly increasing in their use as parts of industrial as well as defense systems, primarily for the purpose of decision-making and control of the operating elements of a system. Examples of these include management information systems, production control systems, inventory control systems, and distribution systems, all used in a business. These are called command and control, force management and military information systems in defense systems.

Automated information systems offer the main advantage of gathering and using a given quantity of data more rapidly than a manual system, or conversely, of handling larger quantities of data in the same time as a manual system. This generally results in better control over the entire process which may result from better decision-making. Sometimes automating the information system will achieve net cost savings by reducing the number of personnel required to operate the system, such as in automating a payroll operation. However, many times the problem of comparing the worth of information (e.g., better decisions made) to the cost of the information system is not so obvious. This is the problem to which this chapter is addressed. Just as we showed how to determine the proper balance between spare part stocks, and special delivery costs and the rest of the logistics system in Chapter 11, we shall now show how to determine the proper balance between the degree of sophistication of an information system and other system elements of a higher-level system, without having to perform a total mission level analysis.

In addition, for the defense-oriented reader we shall describe how to analyze a command and control system as well as how to analyze another scenario relevant to the strategic force planning problem of Part III.

PROBLEM AS GIVEN

Assume it is now 1972 and you are a systems analyst with a firm which provides technical support to the command and control division of the Air Force systems development command. Assume that you have been assigned to a strategic planning study and learn that recent intelligence estimates indicate that the Soviet Union appears to be building up the size of its ICBM capability. Since they have developed high energy propulsion systems for their spare program, we believe they have the capability of combining these higher energy rocket engines with a high megaton warhead and a more accurate guidance and control system to obtain a missile with increased lethality. In light of this intelligence information, and to counter the possible shift in the balance of power, advanced planners within the Air Force have proposed several alternatives. The proposal from the missile systems division is to procure additional missiles to maintain the same balance of power as before. However, there is some fear that this may lead to a new arms race as this balance of power "seesaws." There is also some concern about the so-called "decapitation" attack; that is, an assumed attack on the command structure which, if successful, might prevent the forces from being employed in a counterstrike. To cope with such an attack, two possible additions of command, control, and communications capability, which will improve the force management and hence the effectiveness of the existing ICBM force, have been proposed by the command and control division.

These improvements are as follows:

1. Survivable communication links will be added. They will connect a strategic air command (SAC) survivable command post to the missile launch facilities. The links will transmit the "go word" directly to the missile launch facility if the communications between the launch facility (LF) and its launch control facilities (LCF) or higher headquarters is disrupted.

2. An automated reassignment capability will be added at the survivable, central command post, using a computer to rapidly reallocate the ICBM force based on post-attack availability status of each missile in force.

The technical feasibility of each of the preceding alternatives has been demonstrated by the system designers.

It will be noted that this problem is actually part of the higher-level problem of strategic force planning as was described in Part III. However, in this case we are dealing with a different scenario than the attack on the missile force assumed in Part III.

In addition, instead of comparing alternative weapon systems against one another as in Part II, the analyst in this case must find a way to compare weapons with information systems, each of which is a part of, and contributes to, the higher-level ICBM weapons systems.

PROBLEM AS UNDERSTOOD

An understanding of the various proposed improvements to the ICBM weapon system may be gained through the construction of the system design model, shown in Figure 12-1. As indicated in this figure, the total ICBM system consists of weapons and support to the weapons. Improvements to any

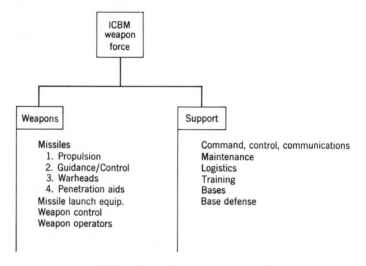

Figure 12-1. System design model.

of these elements will result in an increased effectiveness of the entire system. We shall now examine qualitatively how this would be accomplished.

Proposed Solutions

The first proposed solution for this case consists of purchasing additional missiles, each of which is identical with the missiles which are already part of the weapon system.

Need for Information

Two other proposals have been made to improve the communication dependency of the current system. These proposals can best be understood by examining Figure 12-2 which shows the various SAC elements which make up the ICBM force structure and connect SAC headquarters with the individual ICBM missiles. Here it is important to note that to protect against unauthorized use, none of the missiles can be fired unless a coded and validated command is received by the individual missile from the SAC headquarters through the various lower level SAC command elements. Thus, if any individual launch facility becomes isolated (i.e., out of communication with the higher-level command headquarters), its missile cannot be launched. Obviously we would like to reduce the chance of isolation of the launch facility.

A second problem presents itself as missiles are removed from the force either by destruction or isolation. The reader will recall from Chapter 7 that the analyst had previously used an optimizing algorithm or rule to assign the

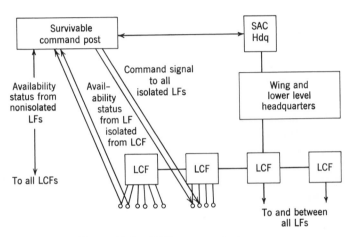

Figure 12-2.   SAC command elements.

proper number of missiles to each target, based on the worth of the target and the probability of kill of the missiles assigned. Note that if some of the assigned missiles are no longer available, the allocation of the remaining missiles to different valued targets is no longer optimal. It is possible to improve the effectiveness of the remaining missiles if a centralized reassignment capability is available.

Proposed Information System Alternatives

The first information system alternative proposed is basically an alternate ICBM command post for SAC headquarters, and provides a full capability for coping with the problems previously described. This system would consist of the following elements, shown in Figure 12-2.

1. A survivable command facility located in an aircraft or a hardened underground command post, or even a space command post. The term "survivable" is a relative one; nothing is completely survivable from a determined enemy. However, these two examples are relatively more survivable than the current above-ground facility.

2. Communication links between the command center and SAC headquarters (as well as the National Command Authority) so that the center can determine when to assume command of the ICBM force when there is a lack of communications with SAC headquarters.

3. Additional communication links from each launch control facility and launch facility to the command center. If an airborne center were used, airborne radio relays would also be needed. These links will be called "up links." Their purpose is to provide status information of all available missiles for reassignment purposes.

4. An automated missile reassignment capability to rapidly reallocate all surviving missiles.

5. Additional communication links from the command center to each LCF and LF to transmit the launch command or "go word" as well as the new missile assignments. Again, airborne radio relays may be required; these links will be called "down links."

The second information system proposed is basically a minimal command post omitting the automated reassignment capability and the up links from the LF's. Thus, its only elements are:

1. Up and down links to each LCF, thereby obtaining status of all alive and nonisolated missiles.

2. Down links to all LF's so that the "go word" and an assigned target can be transmitted directly to any LF when the communications between the LF and the LCF (or higher-level headquarters) is disrupted.

All this second, minimal proposal provides is some bonus value so that any isolated, but alive, missiles may be fired. Of course, there is no precise way of knowing if the command has been carried out.

Each of these systems is technologically feasible, but obviously the full capability system costs more than the minimal system. The problem here is to determine what benefits or increased effectiveness each system provides and then compare these with the cost of providing the added capability to determine if either of these proposals has greater value as compared with the proposal of adding more missiles.

Description of System Operation

To help answer these questions, the analyst constructs a flow diagram such as shown in Figure 12-3, demonstrating how the system would operate

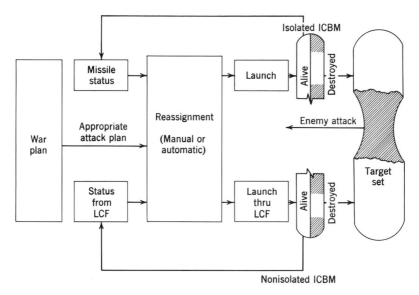

Figure 12-3. ICBM command and control functions.

under wartime conditions. First, a series of war plans or options for many possible contingencies would be formulated in peacetime and stored for possible future use. Each of these options indicates for each contingency considered the particular assignment of every missile to a particular target of the target set. We shall now assume an initial enemy missile attack launched against the ICBM force. This enemy attack divides the ICBM force into four unequal parts. First, some of the ICBM's are destroyed while others remain alive. In addition, the enemy attack may disrupt communications between the ICBM launch facility and its higher-level LCF, thus resulting in a set of isolated, but alive, LF's. Following the attack the National Command Authority makes an assessment of the damage done to this country as well as an assessment of the enemy's attack plan. This results in the selection of an approved counterattack plan from among the set of available contingency options. If SAC headquarters has been destroyed, the alternate command facility now assumes command of the ICBM force. Complete missile status is then obtained by the command center, either through the LCF's or directly from the surviving LF's which are isolated from their LCF if these LF's have up links. Rapid missile reassignment of available missiles is then accomplished and launch commands are issued, either through the LCF's or directly to each LF.

If the minimal command post is used, missile status of all alive and non-isolated missiles is obtained from each of the alive and nonisolated LCF's.

Based on this status and the attack information provided by the National Command Authority, an appropriate preconceived attack plan is taken from the files and implemented. Launch commands are transmitted to each non-isolated LCF, which controls its LF's, as well as to all other LF's involved (since some may be alive but isolated).

APPROACH TO BE FOLLOWED

Problem to be Solved

To summarize, the decision options to be included in this analysis are limited to:

1. Adding more missiles.
2. Alternate command post (with reassignment capability).
3. Minimal command post (no reassignment capability).
4. Doing nothing.

This case is limited to these options since all other possibilities that could have been included in this study have been analyzed in the strategic planning case of Part III.

Operational Flow Model and Model Exercise to be Used

Since the objectives and the operational events of this case are the same as those of the strategic planning case of Part III, the same operational flow models and various methods of model exercise, ranging from a dynamic wargame to a simple expected value model, could again be employed. For any of these methods, the effectiveness of the "do nothing" option would first be evaluated using as a measure the "worth of the enemy targets destroyed," for example. In the case of a two-sided wargame exercise, the measure would also be the "worth of the American targets destroyed."

Next an equal cost force structure would be constructed for the remaining options. For example, assuming a constant expenditure of two billion dollars, the analyst would determine the number of American missiles which could be purchased for this amount, taking into account all costs of investment and five years of operations and maintenance (assuming no RDT&E required). In the case of adding the alternate command post, since the total costs would be less than the two billion dollars available, an additional number of missiles could be purchased to arrive at the same cost limit. In the case of the minimal command post, an even larger number of missiles could be purchased. By exercising each equal cost force structure, the effectiveness of each alternative could be obtained and the alternative providing the highest effectiveness would be chosen, as was shown in Figure 5-7.

Several problems are associated with this approach. First, as indicated in Part III, the most accurate way of analyzing this problem is to construct a two-sided wargame in which all of the weapons are included. However, this can be time consuming and can involve more resources than may be available to lower-level planning organizations. That is, only the Joint Chief of Staff level or OSD could justify such expenditures. Yet to ignore these other weapon units may reduce the credibility of the analysis. A second problem is that the effectiveness obtained may be highly sensitive to the scenario assumed. In this case it would be highly dependent on the number of missiles assumed to be initially destroyed or isolated. This, of course, is under the control of the enemy.

There is a way out of this dilemma and this is the use of the subsystem comparison method of evaluation. This method has wide application in systems analysis since it provides a means of performing a credible analysis without being forced to do the entire higher-level job.

Subsystem Comparison Method

This method consists of measuring the relative ability of a support subsystem as compared with its effector (i.e., weapon or operating unit), to contribute to the mission effectiveness. This is done by determining how many effectors (i.e., missiles) would be required to obtain the same incremental increase in effectiveness as would be obtained by adding the improved support subsystem. Determining the relative costs of each way of accomplishing the same output (i.e., mission effectiveness) indicates the most efficient, lower cost alternative. This provides the information needed to compare the support system with the operating system while pivoting on a level of constant effectiveness.

We shall describe this method further in the context of the specific problem being analyzed. Consider the two force structures shown in Figure 12-4. Force Structure 1 consists of the existing missile force plus the alternate command post in support. Undoubtedly, this force structure would provide some higher level of effectiveness than the missile system without the improved command and control capability. On the other hand, a Force Structure 2 could also be configured to provide this same higher level of effectiveness by adding more missiles instead of the improved information system. The same type of analysis could be performed for comparing the minimal command post with its additional missile equivalent. The results of these analyses could be structured for the higher-level decision-maker by providing him with the information shown in Figure 12-5, as follows:

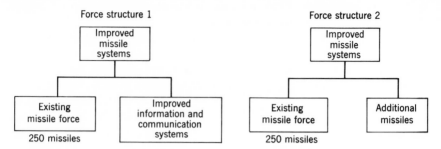

Figure 12-4. Comparison of alternatives.

1. The level of effectiveness obtained by doing nothing (costing C_0 over a five-year period of time).

2. The additional system effectiveness obtained by adding either the minimal command post or additional missiles, and the total cost associated with each alternative.

3. The additional effectiveness obtained by adding either the complete alternate command post or additional missiles, and the total cost associated with each alternative.

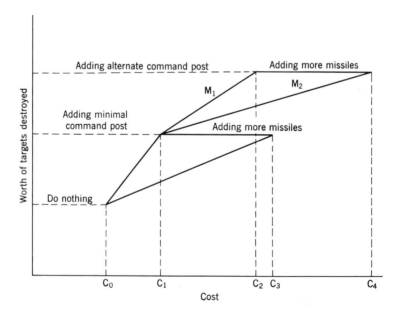

Figure 12-5. Marginal effectiveness of the various system improvements.

The decision-maker can then decide (intuitively) if the increase in effectiveness is worth the increase in cost.

By pivoting on an equal level of effectiveness in the analysis, the total worth of each information system can be established; that is, it is equal to the additional missiles which have to be purchased to obtain the same level of effectiveness. There is, however, a cost associated with each alternative. Hence the *net worth* of adding the information system can be determined as the equivalent additional missiles minus the cost of the information system (as measured in missile units). Suppose, in this example, that the information system can do the same job as adding 20 missiles; however, its cost is the same as two missiles. Thus its net worth would be the same as obtaining 18 missiles free.

The previous method measures the worth of a support system (i.e., the information system) in units of the equivalent additional effectors (missiles) it provides. It should be mentioned that there is another way of measuring worth. This is in terms of "missiles saved." Consider a problem related to the case originally presented in which the enemy threat is *not* increasing so that enough American missiles are available to accomplish the deterrence function. Now by adding an improved information and control system we can have an increase in system effectiveness (in fact, to a level which is more than necessary for deterrence). Hence some missiles can be removed or "saved" to perform some other mission, such as counterforce for damage limiting purposes. Thus the number of "missiles saved" by the addition of the support system can be a measure. This is a good method to use where resources saved, such as people, can be used for another purpose. However, the first approach is more appropriate for the present problem.

Another way of expressing the preceding information to the decision-maker is in terms of the marginal effectiveness of each alternative (shown as the slopes in Figure 12-5). If, as before, $C_4 - C_0$ is the cost of adding 20 missiles to the existing force, and the alternate command post achieves this same level of effectiveness at a cost of $C_2 - C_0$ equivalent to two missiles, the slope M_1 is 10 times the slope of M_2. Hence the marginal effectiveness of alternate command post information can be said to be equivalent to 10 times that of a missile. Obviously, this is a simplification and is only true if one considers buying missiles in a block of 20, rather than as single units.

Unfortunately, the equivalent net worth and marginal effectiveness is a number which is a function of other variables. As indicated earlier, the net worth of the information system will be a function of degree of missile survivability and isolation. Hence a parametric analysis will be needed to express the net worth as a function of these parameters.

ANALYSIS OF ALTERNATIVES

The various decision options will now be analyzed. First, the "do nothing" option will be considered as a "base case" reference to establish the level of effectiveness achieved without any system improvements being added. Each of the two information system additions will then be considered in turn to determine the increase in effectiveness which each provides. However, since each system provides an additional operational function which can be added incrementally, the analysis will be performed on an incremental basis. The first increment of improvement to be considered is the minimal command post which adds down links to launch any isolated missile. Hence this capability will next be analyzed to determine its increased effectiveness with respect to the "do nothing" base case. Finally, the alternate command post adding a reassignment capability, thus providing an optimum allocation of missiles, will be analyzed to determine the increase in effectiveness with respect to the minimal command post. In both cases, the net worth of each improved alternative will be determined.

Analysis of the "Do Nothing" Option

For illustration we shall use the following scenario whose operational flow model structure is illustrated in Figure 12-6.

1. A missile force of 250 missiles is available for assignment to a series of targets having a given hardness and worth, as shown in Figure 7-13.
2. The missile kill probability is 0.9 for a soft target and 0.4 for a hard target, assuming the missile is available for launch.
3. The 250 missiles are initially assigned to the target set, based on each missile having a total probability of a missile being available and launched, and destroying its target of p_{k_p}.
4. These missiles are then subjected to an enemy first strike, out of which is derived a probability D that a missile will be destroyed, and hence a probability $S_1 = 1 - D$, that the missile will survive.*
5. Of those missiles which survive the attack, there is a probability I that a missile will be isolated from the higher-level commander and cannot be launched. Hence there is a probability $S_2 = 1 - I$ that the surviving missile is capable of receiving the launch command. Of the total missiles originally

---

* A second, related parameter which could have been chosen is the *fraction* of missiles destroyed (or surviving), and a related analysis can be used for this approach, as described later.

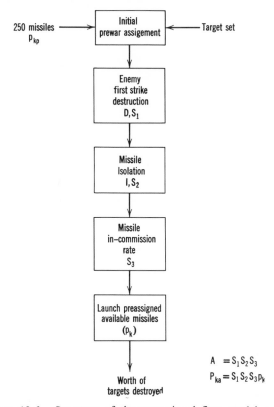

Figure 12-6. Structure of the operational flow model.

available, there is a probability S_1S_2 that each missile will both survive and receive the launch command * (assuming independence of the two functions).

6. Of these missiles, there is a probability S_3 that a missile is in commission and is thus capable of being launched. Hence the total probability that a missile is available for launch is $A = S_1S_2S_3$ * (assuming independence of the three functions).

7. An available missile which is launched has the probability of destroying the target of p_k, consisting of the joint probability of (a) in-flight missile reliability, (b) missile survivability through the enemy ABM defenses, and (c) warhead destruction of target.

8. The worth of the expected targets destroyed will be the effectiveness measure.

* Again, the *fraction* of missiles could have been chosen as the parameter.

Prewar Missile Assignment Technique

The same missile allocation technique described in Part III can again be used for this problem. This optimal assignment technique is based on two main factors: the worth of targets, and P_{ka}, the *total* probability of a missile destroying the target. For all illustrative purposes, we shall assume the same target set of Part III as shown in Figure 7-24; in this case p_k will be the missile probability kill, given that the missile *is completely available for launch;* that is, the missile has survived the first strike, is nonisolated, and is in commission. The value p_k will equal 0.9 for a soft target, and 0.4 for a hard target, given that the missile is available.

The next step is to assign the 250 missiles to the target set in some fashion. Recall that, if a blind firing operational procedure is to be followed, all available missiles will be launched immediately after the enemy's first strike, and there will be no reassignment of missiles based on knowing which missiles are actually available. Hence to properly assign the missiles, a prewar prediction must be made of the percentage of missiles which will be available for launch following an enemy first strike.

In fact, an optimal prewar assignment can only be made if a perfect prewar prediction of total probability of kill of the missile (including postwar availability) can be made. However, no one can forecast the actual damage which will occur during a first strike enemy attack, since this damage is a function not only of many enemy weapon characteristics but also of enemy intentions, such as the number of weapons initially launched. How can the analyst responsible for proper prewar missile assignment make his assignments under such uncertainties?

One approach is to ignore the problem and assign missiles on the basis of an assumption of 100% availability, using the logic of Tables 7-18 and 7-19a. A second approach is to assume some lower value of availability and by constructing tables similar to Figures 7-18 and 7-19a, using the assumed reduced kill probabilities to make the appropriate assignments. Each of these possibilities produces some reduction in effectiveness, which is a function of the amount of error in estimating the total kill probabilities, especially the availability. For purposes of illustration we shall assume that prewar assignments are made on an assumption that $A = 1$, and later discuss the improvements which could be made if some way of assigning on the basis of the true availability were at hand.

A third possibility is to rapidly reassign all of the surviving nonisolated missiles using a missile assignment computer and the optimization procedure illustrated in Figure 7-18. (This is the alternate command center approach proposed.)

Calculation of Effectiveness (Using Prewar Assignments)

If a postwar missile availability of 1.0 were assumed for purposes of prewar missile assignments, the optimum missile allocation algorithm illustrated previously in Figures 7-18 and 7-19a could be used. Figure 7-19a indicates that the 250 missiles were allocated by assigning missiles down through Decision 9, obtaining the total force assignments shown in Table 12-1. The total effectiveness which would be achieved if all the assumptions made were correct, would have an expected value of 7144. The true effectiveness achieved will differ as the predicted number of missiles surviving varies. One way of showing how sensitive the results are to enemy uncertainty (which causes the actual missile availability, and hence kill probability, to be uncertain) is to conduct a parametric analysis with respect to missile availability. For illustration, consider how one of the data points of the parametric analysis would be obtained. Assume the missile availability of each of the 250 missiles following the enemy attack is 0.5; hence the probability that each missile will be both available following the enemy attack and destroy its target is 0.45 and 0.2 respectively, depending on the hardness of the target. Thus to find the expected results, a logic table based on these new kill probabilities must be constructed as shown in Table 12-2. This table shows the expected incremental effectiveness provided by each succeeding missile assigned to a target. By summing the marginal effectiveness of each missile, as previously assigned in Table 12-1, a total effectiveness of only 3860 is obtained for an actual availability of 0.5 as shown in Table 12-3.

The expected results which would be obtained if the availability were to take on other values could be obtained by performing the same type of calculations, using these values. Performing such calculations for the total span of availability values would provide the desired parametric results which could be displayed for the decision-maker, as shown in Figure 12-7a.

Table 12-1.   *Force Allocation (Based on Availability = 1.0)*

Target Class	Target Worth	Number Targets	Missile Assignment Per Target
1	100	10 Soft	1
2	60	30 Soft	1
3	40	20 Soft	1
4	100	20 Hard	3
5	60	40 Hard	2
6	40	50 Hard	1

Table 12-2. *Force Allocation Procedure (Availability = 0.5)*

No. Targets	1st	2nd	3rd	4th	5th	6th
10 Soft	$100(.45)=45$ ①	$55(.45)=24.8$ ③	$30.2(.45)=13.6$ ⑧	$16.6(.45)=7.5$ ⑱		
30 Soft	$60(.45)=27$ ②	$33(.45)=14.8$ ⑦	$18.2(.45)=8.1$ ⑮			
20 Soft	$40(.45)=18$ ⑤	$22(.45)=9.9$ ⑫	$12.1(.45)=5.4$ ㉒			
20 Hard	$100(.2)=20$ ④	$80(.2)=16$ ⑥	$64(.2)=12.8$ ⑨	$51.2(.2)=10.2$ ⑪	$41.0(.2)=8.2$ ⑭	$32.8(.2)=6.5$ ⑲
40 Hard	$60(.2)=12$ ⑩	$48(.2)=9.6$ ⑬	$38.4(.2)=7.7$ ⑰	$30.7(.2)=6.1$ ㉑	$24.6(.2)=4.9$	
50 Hard	$40(.2)=8$ ⑯	$32(.2)=6.4$ ⑳	$25.6(.2)=5.1$	$20.5(.2)=4.1$		

Table 12-3. *Total Effectiveness Achieved for A = 0.5 (Pre-war Assignment Based on A = 1.0)*

Target Class	Number Targets	Missiles Assigned Per Target	Effectiveness per Target Class	
1	10 Soft	1	10 (45)	= 450
2	30 Soft	1	30 (27)	= 810
3	20 Soft	1	20 (18)	= 360
4	20 Hard	3	20 (20 + 16 + 12.8)	= 976
5	40 Hard	2	40 (12 + 9.6)	= 864
6	50 Hard	1	50 (8)	= 400
			Total effectiveness	= 3860

As indicated previously, if a missile reassignment could be made based on knowing the exact probability of a missile being available following the enemy attack, a gain in effectiveness could be achieved. The amount of this gain will be calculated later in this chapter when the reassignment capability is discussed.

An alternative analysis could have used as its key parameter the *fraction* of missiles actually available for launch. In this case, the expected worth of each target destroyed is found as the product of the probability that n as-

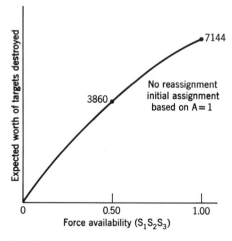

Figure 12-7a. Missile effectiveness based on no reassignment and prewar assignment assuming A = 1.

signed missiles will survive to be fired and the expected worth of the destroyed target given n missiles were fired on the target. Using the previous missile assignments of Table 12-1 and the incremental expected worth of Figure 7-18, we can determine the expected worth of targets destroyed as a function of the number of n surviving, assigned missiles as shown in Table 12-4. Given that 125 randomly selected missiles survive out of the 250 missiles originally assigned, the probabilities that exactly n missiles will survive is also shown in Table 12-4. These probabilities are obtained from the hypergeometric distribution.* From this table it can be determined that the total effectiveness is 3861, a value very close to that of the other method of calculation, since most targets had only one missile assigned to each of them.

ANALYSIS OF MINIMAL COMMAND POST

We shall now consider the problems of evaluating the effectiveness of the same force structure (i.e., 250 missiles) in which a minimal command post capability has been added. The minimal alternate command post provides "down links" to communicate the "go word" to all isolated alive missiles (but does not provide a reassignment capability for optimal employment of these missiles). Since the only benefit provided by the minimal command post is to reduce the probability of isolation to zero, these benefits are obtained only when missile communications are destroyed to some degree (including the extreme case of the command center being completely destroyed as in the "decapitation" attack scenario).

In Figure 12-7a, the expected worth of targets destroyed was graphed only as a function of availability ($A = S_1 S_2 S_3$). In order to see the effect of isolation more clearly, we would like to graph the expected worth of targets destroyed as a function of two parameters: a parameter pertaining to isolation, S_2, and a parameter for "survivability" and "in-commission," $S_1 S_3$. This can be readily done as shown in Figure 12-7b by using the function of Figure 12-7a and the following relation:

$$S_1 S_3 = \frac{S_1 S_2 S_3}{S_2} = \frac{A}{S_2}.$$

Thus if $S_2 = 1$ (i.e., $I = 0$), the functions of Figure 12-7a and in Figure 12-7b are identical, since the abscissa scales are the same ($S_1 S_3 = A$). However, if $S_2 = .5$ (i.e., $I = .5$), for a given value of the function, the abscissa of Figure 12-7b must be twice the corresponding value of the abscissa of Figure 12-7a,

* See W. Feller (1957).

Table 12-4. *Worth Calculations Using the Hypergeometric Distribution*

Target Class	Target Worth	No. of Targets	Missile Assignment per Target	Number of Targets	Worth if n Assigned Missiles Survive			Probability of n Missiles Surviving			Expected Worth
					n=1	n=2	n=3	n=1	n=2	n=3	
1	100	10	1	10 Soft	90	—	—	0.5	—	—	450
2	60	30	1	30 Soft	54	—	—	0.5	—	—	810
3	40	20	1	20 Soft	36	—	—	0.5	—	—	360
4	100	20	3	20 Hard	40	64	78.4	0.3765	0.3765	0.1235	977
5	60	40	2	40 Hard	24	38.4	—	0.5020	0.2490	—	864
6	40	50	1	50 Hard	16	—	—	0.5	—	—	400
											3861 Total

since $S_1S_3 = \dfrac{A}{.5} = 2A$. Similarly, if $S_2 = 0.25$ ($I = 0.75$), the abscissa of Figure 12-7b is four times the original scale, and if $S_2 = 0$ ($I = 1$), the function equals zero. This translation permits a plot of system effectiveness as a function of both this abscissa and the other parameter, isolation (where $I = 1 - S_2$).

Thus the results expected for the "doing nothing" case are presented in parametric form for any combination of the three main factors of survivability from the first strike, degree of isolation, and in-commission rate, each expressed as a probability. (As indicated before, in-commission rate produces the same effect as survivability and hence the joint probability can be used.) These results will be used as a "base case" for comparing the increased effectiveness obtained by adding the first increment of improvement, the minimal command post capability.

Determining the Worth of the Minimal Command Center

The primary function of the minimal command center is to establish one way communications with all of the isolated missiles, reducing their probability of isolation to zero. Since the increase in effectiveness obtained is highly dependent on I, the probability of missile isolation, a parametric analysis must again be performed. We shall now describe how this analysis would be performed using the example of a force structure consisting of 250 missiles. The worth of adding the minimal command center to the force

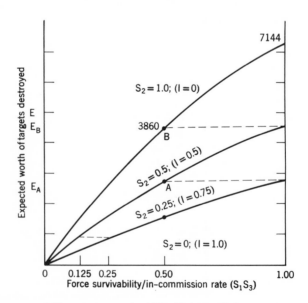

Figure 12-7b.   Force survivability / in-commission rate (S_1S_3).

structure may be obtained from Figure 12-7b which shows the effectiveness obtained for different values of missile isolation and force survivability/in-commission rate. The additional effectiveness achieved by the minimal command center is that increment of effectiveness obtained by moving vertically from a given operating point (describing the assumed components of availability) to the zero isolation function. For example, in Figure 12-7b if we assume a force structure operating condition of 0.50 force survivability/in commission rate, and 0.50 missile isolation (Point A), the addition of down links reduces the isolation to zero (Point B). Thus the incremental effectiveness gained is the amount $E_B - E_A$.

We shall now quantify the worth of this system improvement. As previously indicated, a measure of this worth may be obtained by calculating the number of additional missiles required, if the minimal command center were not added, to obtain the same higher level of effectiveness, E_B, found earlier. The scope of the problem is illustrated in Figure 12-8. Since there are several ways that the calculations can be performed, two of these methods are discussed so that the reader will have a better appreciation of the complexity involved and some of the analytical simplifications which can be made.

The first step in the analysis is to assume that X missiles are to be added to the basic force; hence a prewar assignment of the $250 + X$ missiles is made, based on an assumed availability of $A = 1.0$. We must now find that value of X such that when the $250 + X$ missiles, operating with a total $P_k = p_k A$ (where A is the actual post strike missile availability), will achieve the same effectiveness level of E_B. As will be seen, the solution must be performed on a trial and error basis. Since there will be one correct value of X for each value of the main parameter A, a series of assignments can be made for $A = 1.0$ and for different values of X without much wasted effort if the calculations are performed manually.

Consider now the calculations involved for an assumed actual availability A. One method of calculation would be to assume that all $250 + X$ missiles of Force Structure 2 are now under attack and, therefore, the availability is

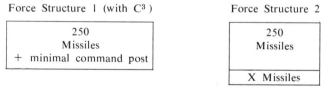

Force Structure 1 (with C³)   Force Structure 2

| 250 Missiles + minimal command post |

| 250 Missiles |
| X Missiles |

Prewar assignment based on $A = 1.0$ Prewar Assignment based on $A = 1.0$
Actual $A = 0.5$ Actual $A < 0.5$
Effectiveness $= E_B$ Effectiveness $= E_B$

Figure 12-8. Equal effectiveness force structures.

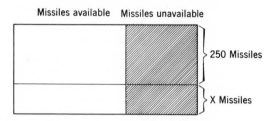

Figure 12-9a.   Available missiles in Force Structure 2 (arms race assumption).

the same as the force of the initially assumed conditions for Force Structure 1, as shown in Figure 12-9a. This assumption would be correct for the case of the enemy enlarging his force in response to ours, as in an arms race. However, the assumption made in this analysis is that we are responding to a fixed enemy increase.

Thus we shall assume that the enemy attack is of a constant level (i.e., only 250 missiles are subject to the original availability of A) with the remaining X missiles of Force Structure 2 subject to an in-commission rate of A' (no attack) as shown in Figure 12-9b.

Using this simplifying assumption, a trial and error solution to the value of X may be obtained. This is accomplished in the following fashion. First choose an arbitrary value of X and derive the optimal prewar assignment for the entire $250 + X$ missiles, assuming $A = 1.0$. Next calculate the value of effectiveness of the two missile subsets based on the number of missiles in each subset (250 and X), and the availability of each (A and A'). This calculation is not a simple one because the preattack missile assignments interrelate the two subsets. To simplify the calculation we shall construct an equivalent force structure of $250 + X$ missiles, each of which has an equivalent post-attack availability of A_{eq}, and whose effectiveness is approxi-

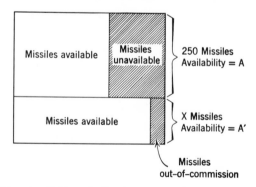

Figure 12-9b.   Available missiles in Force Structure 2 (no arms race).

mately equal to the original force structure containing the two subsets. By analyzing the two stage joint probability process, it can be shown that:

$$A_{eq} = \left(\frac{250}{250+X}\right) A + \left(\frac{X}{250+X}\right) A' = \frac{250\,A + XA'}{250+X} \,.$$

After calculating the value of A_{eq} (based on the trial solution of X), we construct a new decision logic table for this value of A_{eq} (similar to Table 12-2 for $A = 0.5$). Next, calculate the value of effectiveness for the assumed values of X and A_{eq}, and see if this effectiveness is equal to the desired value of E_R. Different values of X are then taken and the procedure repeated until this value is obtained. Obviously using a computer for such trial and error calculations reduces the effort considerably.

In this same fashion, calculations for other parametric values of S_1, S_2 and S_3 are also performed. The total results can then be structured for the decision-maker, as shown in Figure 12-10, illustrating the equivalent additional missiles which the minimal command center represents, as a function of the postwar isolation (I_2) and survivability/in-commission rate ($S_1 S_3$), which is approximately the same as the per cent of force surviving and per cent in isolation.

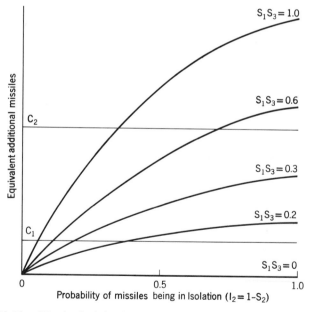

Figure 12-10. Worth of minimal command center as a function of type of attack.

Tradeoff of System Effectiveness with Cost

We shall now consider the question: "Is the addition of the minimal command center worth the cost involved?" This may be determined by finding the net worth of the proposal, where net worth is the total worth minus the system cost. This net worth could be plotted in parametric form and illustrated as shown in Figure 12-10. Thus if C_1 were the total cost of the center, the parametric analysis indicates that there is a positive worth over a large set of conditions (scenarios). On the other hand, if C_2 were the cost, the net worth is positive over a much smaller range of scenarios. The structured information aids the decision-maker in arriving at a decision based on his intuitive opinion of how likely he believes this type of attack may be.

Gain in Effectiveness if True Availability Were Known

Before leaving the analysis of the worth of this system, it is of interest to explore the additional worth that could be obtained if the missile assignments were based on the *actual* post-attack availability instead of the assumed availability of unity ($A = 1.0$), since there is a high chance that the actual availability will be less than unity after an enemy first strike. Such consideration is of interest to the analyst for two reasons. First an analyst should always be interested in determining how much degradation in effectiveness is obtained by introducing sources of error, such as simplified assumptions (e.g., A will equal 1.0). The only way to determine the degree of degradation in effectiveness is to calculate what maximum level of effectiveness is obtainable if perfect information were available (i.e., if by some means the analyst could exactly estimate the postattack missile availability). Second, if the degradation in effectiveness is not small, consideration should be given to developing a procedure for rapidly obtaining this information (i.e., an estimate of postattack availability) and using it for assignment purposes, if the total cost of such an approach is small enough compared to the gains it provides. This could involve constructing a series of war plan options, each based on a different value of availability, then using that option closest to the post-attack estimate of the actual availability.

To analyze the gain in effectiveness obtained by using such a procedure, we shall now reconsider the example in which the actual availability turned out to be 0.5. If this information had been known ahead of time, preattack assignments could have been made on this basis rather than the previous assumption of $A = 1.0$. Thus, by using Table 12-2 (based on $A = 0.5$) for the optimal assignment logic, the resulting assignment of 250 missiles (down to Decision 12) would have been as shown in Tables 12-5 and 12-6. The assignments of Table 12-6 can be compared with those of Table 12-1 (based on $A = 1.0$) to indicate the change. Note that this assignment policy for

Table 12-5. *Force allocation procedure* (*Availability* = 0.5)

Decision Number	Marginal Worth	Missiles Assigned	Subtotal Missiles Assigned	Incremental Effectiveness	Subtotal Effectiveness
1	45	10	10	450	450
2	27	30	40	810	1260
3	24.8	10	50	248	1508
4	20	20	70	400	1908
5	18	20	90	360	2268
6	16	20	110	320	2588
7	14.8	30	140	444	3032
8	13.6	10	150	136	3168
9	12.8	20	170	256	3424
10	12	40	210	480	3904
11	10.2	20	230	204	4108
12	9.9	20	250	198	4306
13	9.6	40	290	384	4690
14	8.2	20	310	164	4854
15	8.1	30	340	243	5097
16	8.0	50	390	400	5497
17	7.7	40	430	308	5805
18	7.5	10	440	75	5880
19	6.5	20	460	130	6010
20	6.4	50	510	320	6330

A = 0.5 results in Target Class 6 having no missiles assigned to it at all, while Target Class 4 now has four missiles per target. This is because the expected value of the first missile for Target Class 6 is so small in contrast to the other alternative. Thus, the total effectiveness expected would be equal to 4306, as tabulated in Table 12-7. Note that this is an increase of 11.6% as compared to the previous effectiveness of 3860, which used a predicted availability of 1.0.

By using this same analytical approach, calculations could be made by using other values of availability to demonstrate the maximum effectiveness obtainable through "blind firing" if, by some means, the true availability

Table 12-6. *Force allocation* (*Based on Availability* = 0.5)

Target Class	Target Worth	Number Targets	Missile Assignment per Target (A = 0.5)	Previous Missile Assignment per Target (A = 1.0)
1	100	10 Soft	3	1
2	60	30 Soft	2	1
3	40	20 Soft	2	1
4	100	20 Hard	4	3
5	60	40 Hard	1	2
6	40	50 Hard	0	1

Table 12-7. *Optimal Force Allocation*

Target Class	Number Targets	Total Effectiveness Achieved for A = 0.5 (Prewar Assignment Based on A = 0.5)		
		Missiles Assigned Per Target	Effectiveness per Target Class	
1	10 Soft	3	10 (45 + 24.8 + 13.6)	= 834
2	30 Soft	2	30 (27 + 14.8)	= 1254
3	20 Soft	2	20 (18 + 9.9)	= 558
4	20 Hard	4	20 (20 + 16 + 12.8 + 10.2)	= 1180
5	40 Hard	1	40 (12)	= 480
6	50 Hard	0		0
			Total effectiveness	4306

could be used for preassignment purposes. The resulting increased effectiveness obtained could be displayed as a function of the actual availability, as shown in Figure 12-11. This increased effectiveness could be obtained by constructing the series of assignment options, each based on a given poststrike availability, obtaining an estimate of the actual availability by communicating with the LCF's and pulling out the appropriate option based on this information.

While the above analysis indicates the best gain in effectiveness which can be obtained by having the correct availability estimate, the analyst should conduct sensitivity analyses to see how this gain in effectiveness varies as the accuracy of the estimate varies. This determines how much effort (and expense) is warranted in getting accurate data.

Finally, the value of this additional amount of information may be deter-

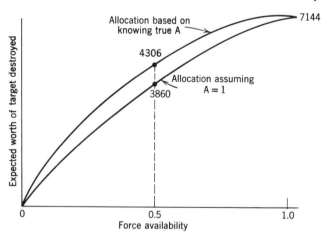

Figure 12-11. Worth of optimal initial assignment policy.

mined as measured in terms of equivalent additional missiles, using the same procedure already described.

ANALYSIS OF ALTERNATE COMMAND POST CAPABILITY

We shall now consider the effectiveness of the same force structure in which the alternate command post capability has been added. This capability provides down links so that no surviving missile is isolated, and in addition, provides up links at each missile (thus assuring status information at the command post for each surviving missile) and a rapid post-strike reassignment capability so that all surviving missiles can be reassigned to the target set in an optimal fashion.

The operational flow diagram for this capability is shown in Figure 12-12.

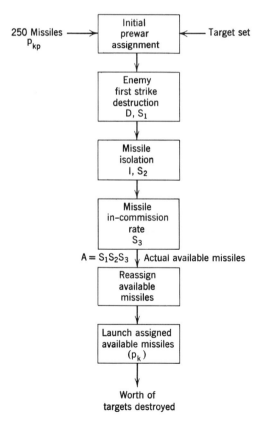

Figure 12-12. Operational flow diagram for alternate command post capability.

The same approach previously described is used in evaluating the effectiveness obtained by adding the reassignment capability to the minimal command post and comparing this resulting effectiveness against the effectiveness obtained for the minimal command post as the new base case.

Again, for illustration of one datum point of the parametric analysis, consider an actual postattack availability in which there is a probability of 0.5 of a missile surviving (and being in communication). With a reassignment capability, all available missiles would report their status and await reassignment. These can now be reassigned in an optimal fashion using Figures 7-18 and 7-19a (based on A = 1.0). Such assignments will be optimal because the availability of these missiles is actually equal to one. Figure 7-19a indicates that the expected number of 125 missiles would provide a total effectiveness of 5120. This compares with the effectiveness calculations previously made of 4306 and 3860 for the minimal command post, where preattack assignments were based on an assumption of A = 0.5 (the best that could be done for blind firing; see Table 12-7), and 1.0 (see Table 12-3) respectively. The resulting effectiveness could be illustrated as a function of missile availability as shown in Figure 12-13.

Determining System Worth

The worth of the system can be quantified by again determining the X additional missiles which would be required to obtain the same effectiveness of 5120 if the reassignment capability were not available. This is calcu-

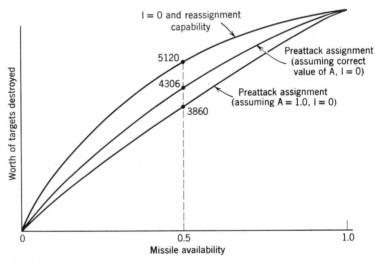

Figure 12-13. Resulting effectiveness for different levels of force management capability.

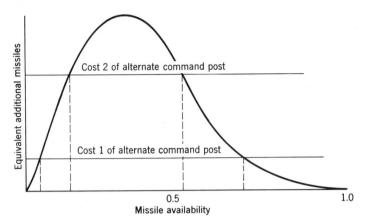

Figure 12-14. System Worth as a function of missile availability.

lated in the same manner as for the minimal command post (i.e., using a trial and error solution for X), finding A_{eq} and E until $E = E_B$.

These parametric results can be structured for the decision-maker as shown in Figure 12-14. Inserting the cost of the system (Cost 1 or Cost 2) again indicates the range over which the worth is positive.

CONCLUDING REMARKS

In this chapter we have demonstrated a method for determining the worth of an information system alternative. The worth measure used was the equivalent number of operating units (missiles in this example) which would have to be added to the total system to achieve the same increase in mission or operational effectiveness that the improved information system would achieve. This method of pivoting on a level of constant effectiveness is a particularly good approach when it is not possible to relate the measure of effectiveness to cost. However, if profit (or cost of operation) is used as the effectiveness measure, the evaluator merely has to determine if the net profit has increased following the addition and cost of the information system alternative. Some examples of this are as follows:

1. The use of radio controlled taxi cabs permits a cab to cruise in a heavily populated section of the city or stand at a high demand location such as an airport, but allows the cab to respond to a definite demand when called.

2. An automated inventory control system can be used to regulate the flow of inventories at a company warehouse, distributors, and retailers rapidly enough to meet a fluctuating customer demand (or to meet it more

rapidly). Here the analysis could consist of determining the impact on net profit by adding an improved information system. It should also include a tradeoff between the size of inventory required and type of information system employed, to meet a given amount of customer demand (as described in Chapter 11) or to produce a higher demand through faster service, and the higher level measure of company profits.

3. A reconnaissance system which keeps track of all enemy mobile targets (such as ICBM's) will prevent our firing on false targets such as empty silos or vacant airbases, thus more efficiently using our weapons.

4. By decreasing the time required to transmit a request for close air support from a forward air controller to the aircraft assignment officer, there is a higher probability that the target will be at the location when the aircraft arrives to deliver the ordnance, thus not wasting the cost of a sortie.

5. The rapidity and accuracy of providing vectoring information to direct an aircraft or guided missile against an incoming aircraft target increases the total probability of kill against the target, thus reducing the expected cost of destroying the target.

While each of the above examples can be considered an application of the use of an information system to properly control a process, the use of information systems with more accurate and timely feedback control of manufacturing processes is becoming more prevalent. Here the cost of such a control system must be compared with the higher quality level (less the number of product rejects), or higher volume which can be achieved, leading to high net profits.

In all cases, the cost of the information system must either be related to the measure of effectiveness (if it is measured in monetary units, such as net profit) or it should be related to the cost of some other system element showing how many dollars can be saved while obtaining some required level of effectiveness. This, of course, is the real test of value in proposing any new system design.

13

Intrasystem Tradeoffs: Guidance and Control System Selection Case

The subsystem planning cases of Chapters 10, 11, and 12 involved subsystems whose performance characteristics are relatively independent of the other subsystems involved. The system analyzed in this chapter involves system performance characteristics which are highly interrelated, yet are susceptible to analysis. This generally is the situation in subsystems planning.

The first objective of this chapter is to summarize and reinforce the key principles of tradeoff analysis in the context of a more difficult case. The second objective is to show how systems analysis may be used to properly set requirements or the specifications of a system by taking into account the various intrasystem tradeoffs involved. Here we shall use the same case example to focus upon two separate but related contexts which occur in systems design. First, there is the problem which a subsystem designer has in optimizing his design, and his reliance on being provided a higher-level systems evaluation model by the higher-level system designer. Second, a mission-oriented cost-effectiveness analysis, as performed during the systems planning or concept formulation phase, is a necessary first step in obtaining the higher-level systems evaluation model used in setting system and subsystem specifications.

PROBLEM AS GIVEN

You are a systems analyst with Control Systems, Incorporated, and have been assigned to a guidance and control systems project group which is participating in a detailed systems planning effort for a short range Army missile system, handling targets up to 15 miles range. The Ajax Missile Corporation has over-all system responsibility for the study program, with your

341

company having been given responsibility for the missile guidance and control system package of the total system.

The total missile systems project has been organized along functional lines, consisting of various subsystem specialities. In addition to your guidance and control subsystem group, there is a warhead group, an airframe group, a propulsion group, a logistics support group, etc. Integrating all of these inputs is a missile systems group whose task is to evaluate the analyses performed by each of the subsystem groups in order to justify the subsystem recommendations they are proposing, as well as to conduct intrasystem tradeoffs as required to arrive at the preferred missile system specifications. This group will have the final say in defining the system specifications, based on the inputs of performance and cost which each subsystem group provides.

In the initial orientation meeting with the guidance and control system team, the missile system project manager indicates the following:

"I scheduled this meeting to discuss your contribution as it relates to the total missile system planning effort. As you know, we have just received a contract for the missile contract definition phase, so that the output of this study is to be a detailed feasibility study of the missile system design which will include the set of systems specifications and our plan for implementation, and a fixed price cost proposal for the engineering development phase, as well as a total cost estimate for the entire missile system. We are faced with two types of competition. First, our study group is in competition with two other companies who are also participating in this contract definition phase. In addition, our system will be competing with the Army's field artillery shells. Since we plan to have a higher accuracy in our missile, we hope that one missile will replace a several round salvo of field artillery shells. Thus the emphasis is on a low cost system. Hence we must make every effort to configure a system which will be superior from a cost-effectiveness point of view.

"The missile system we shall configure will utilize an inertially guided missile. The operational concept which we are to use is as follows: A forward observer, located with the forward troops, detects fixed targets to be destroyed. He determines his present position by a LORAN navigational aid, and determines the target's position with respect to him by means of a laser optical range and angle finder. He inserts both sets of data into a coordinate translating computer whose output is the position of the target with respect to some referenced origin. The observer then calls for a missile to be fired to this calculated target position. The preliminary system planning phase has already been performed by the Army; their report, including the operational concept of the missile system, is available to us.

"The missile itself consists of four main subsystems: the propulsion subsystem, airframe subsystem, warhead subsystem, and an inertial guidance and control subsystem. The latter, your design responsibility, will consist of the light-weight, low cost accelerometers you have developed, and electronic integrating circuits which can accurately control the missile flight to a previously inserted terminal ground position. The flight control unit will consist of electronic amplifiers and actuators for positioning the missile fins.

"The design strategy we shall employ in this competition is to use as many off-the-shelf, proven components as we can in the system design. This should not only aid us in convincing the Army that we offer tested, reliable hardware, but should also reduce our costs since we hope to eliminate a great deal of development and testing costs. For example, we hope to be able to use the airframe and propulsion subsystem of an existing missile which the company now has in production.

"I am told that all of your components of the guidance and control subsystem, except the electronics package, are essentially available off-the-shelf. Thus your main problem will be to package the electronics within the small weight and space allocations of the airframe we are planning to use. If the size and weight of the warhead and the other subsystems are excessive, we will have to redesign the propulsion system and possibly the airframe, inducing further expense. If this program is successful, the Army expects to need several hundred thousand missiles.

"Your company's electronic circuit designers indicate that they are considering two different types of feasible designs. One approach is a repackaging of standard type micro-miniaturized electronic units now in production. A second design approach would consist of modular, replaceable, integrated circuit units which could be thrown away when found defective. Thus the maintenance concept would be minimal for the throw-away unit design but would have to include more elaborate fault location and repair equipment for the standard electronic package.

"While your company has done a great deal of development work and pilot production and reliability testing of the integrated circuit units, you will have to build additional production facilities if this packaging concept is employed. However, further development work will not be required— only repackaging. Since large-scale production of these missiles is planned for this contract, it is felt that once a large-scale production facility is established, the cost of the integrated circuit units will be lower in cost than the conventional electronic units.

"Perhaps the most important problem you are facing is that of missile control accuracy. One design possibility is that of a 'bang-bang' or discontinuous control system which you are now producing for another missile contract and which uses rather inexpensive actuators. A second design is a proportional control system whose actuators would be more expensive and whose electronics would require two additional amplifier stages. Flight control simulation tests indicate that the proportional control system has twice the accuracy of the "bang-bang" control system.

"I am asking each subsystem design team to begin configuring alternative subsystem designs based on some of the constraints we have discussed with you, and your preference among these alternatives. Meanwhile, we in the systems project office shall begin the development of a systems evaluation model which can be used as the over-all framework for dealing with the various intrasystem tradeoffs and for the cost-effectiveness analysis of the various systems which we shall have to evaluate."

CONFIGURING GUIDANCE AND CONTROL ALTERNATIVES

We shall begin the discussion by considering the first problem posed; that is, how a guidance and control system analyst can properly evaluate alternative guidance and control system designs and aid in selecting the preferred alternative.

Understanding the Guidance and Control System

The analyst may obtain greater understanding of the guidance and control system problem through constructing the system design model which indicates the various parts of the system and their relationships, and a system operational flow model which indicates how these parts contribute to the accomplishment of a given job. We shall begin with the systems design model.

As shown in Figure 13-1, the guidance and control system is composed of three main elements: (a) inertial components, (b) guidance and control electronic components, and (c) actuators.

As indicated in the case, the system designers are hopeful that they can use the newly developed, low cost inertial sensors. Thus these elements are essentially fixed in the missile design since all other inertial sensors available are too costly for this application. The fin actuators, on the other hand, may be of two different types—either proportional actuators (as in an aircraft) or discontinuous ("bang-bang") actuators which are either centered or moved to the limit in either direction. The guidance and control electronics consist of four different design types. They may provide either a proportional or bang-bang signal output, the proportional type requiring extra electronic amplifiers. Similarly, the electronic packaging may be of two primary types, either standard microminiature electronics, or the integrated circuits built with a modular, throw-away maintenance concept. It should be noted, however, that there could be hybrid circuit alternatives available in which circuits are constructed using a combination of both types of electronic packaging.

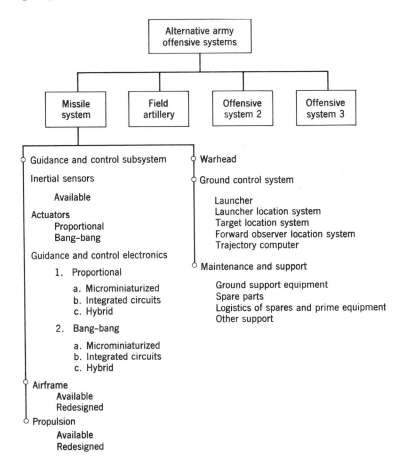

Figure 13-1. System design model.

It is beyond the scope of this book to discuss a particular guidance and control system concept in detail. We shall indicate, however, that the accelerometers keep track of the missile accelerations in the various directions. A double integration of these signals is performed by the electronic circuits, thus producing deviations from some reference flight path. These deviation signals are amplified and drive the actuators in the proper direction so as to reduce the deviations to zero. The other elements of the missile, shown in the systems design model of Figure 13-1, consist of an airframe (which may be either the one currently available from a previous missile design, or a redesigned airframe), a propulsion system (which may be the one currently available or a redesigned one), and a warhead. The missile system group indicates that the rest of the total missile system consists of two other system elements: the ground control system and the maintenance and support system, as shown in Figure 13-1. The ground control system begins with the information system needed to direct the missile, consisting of the forward observer, the laser position finder, and the LORAN position finder. Target position is then forwarded to the missile center and, together with the missile launch position, is fed into a missile trajectory computer. This computer calculates the missile ballistic coefficients needed to be inserted into the missile, providing information such as the proper missile propulsion burn time to yield the proper flight range. The ground control system also consists of a launcher and a launch crew.

The remaining portion of the system is that required for maintenance and support and consists of the ground maintenance support equipment, the spare parts, the logistics transportation required for transporting the spare parts, and any other support units needed to keep the missile system operating in the field. This higher-level system design model showing the other elements which interface with the guidance and control system can be obtained from the missile system designer.

One last portion of the higher-level system design model which might be added consists of those other systems which compete with the missile system. These include field artillery as well as other offensive systems such as tactical Air Force, close air support delivery systems.

DEVELOPING THE SYSTEMS EVALUATION MODEL

While the guidance and control system designers are synthesizing various feasible system design alternatives, their systems analyst begins the development of an evaluation model or structure which can be used to predict the expected over-all system effectiveness and cost of each guidance and control system alternative. To do this, the systems analyst must have a good understanding of those key factors of performance and environment which he, the systems designers, and the decision-maker (the missile system manager) feel are pertinent to the guidance and control problem. A list of such major factors can be compiled as shown in Figure 13-2. This list classifies the accountable factors into system performance, physical characteristics, and miscellaneous factors.

It should be noted that the WSEIAC effectiveness characteristics of Figure 4-6 described previously can be used as a checklist for the performance characteristics. However, note that some of the WSEIAC characteristics apply to more than the guidance and control subsystem. For example, the prime equipment which fits into the missile will have the same mobility characteristics as the missile itself, regardless of the packaging used. This is

SYSTEM PERFORMANCE
    Accuracy
    Reliability
    Maintainability
    Mobility
    Vulnerability
PHYSICAL CHARACTERISTICS
    Size
    Weight
    Modularity
MISCELLANEOUS
    Confidence in System Performance Factors
    Producibility
    Cost
        RDT&E
        Investment
        Operations and Maintenance

Figure 13-2. Guidance and control system accountable factors.

also true for the vulnerability of the prime equipment. All such environment specifications, such as shock, vibration, and temperature, will have to be the same for all design alternatives; and hence they may be considered fixed and nondifferentiating among system alternatives. On the other hand, these characteristics of mobility and vulnerability may vary for the spare parts needed for each design concept. For example, a system which is not as reliable as another system will need greater numbers of spare parts and therefore its mobility may be reduced.

The guidance and control system analyst's next step would be to establish measures for each of these accountable factors and evaluate each of the alternative designs with respect to each factor. For example, the accuracy for the various systems could be measured in terms of its "circular probable error" (CEP) in feet. The size and weight of each unit could be measured in appropriate units. The maintainability could be predicted in terms of mean time to repair, and the airborne reliability measured in terms of the probability of satisfactorily surviving a given mission flight time.

In a similar fashion, the costs for each of the design alternatives could be calculated, where the costs for RDT&E, investment, and operation and maintenance would have to be estimated for the newer type system electronics and the inertially guided components (which are presumed constant for all alternatives).

Relationship Between System Synthesis and System Evaluation

We shall now discuss what is probably the core problem in systems planning. Each of the alternative guidance and control designs will provide some quantifiable level for each of the key characteristics listed in Figure 13-2; this is provided at some level of total guidance and control system cost, which can be calculated by the methods discussed in Chapter 9. However, how does the system designer select the preferred system? Several problems face the designer in attempting to evaluate system alternatives:

1. Purely on the basis of the total set of noncost oriented system characteristics for each system, is there a best system? Generally, no one system alternative dominates all others on the basis of having the highest level for each of the performance characteristics, so it is not obvious which system is best. For example, the proportional guidance system may be more accurate, but is probably less reliable and takes up more space. Hence, there is a need to combine system performance characteristics in some way in order to arrive at a measure of system effectiveness.

2. In general, the higher the levels of performance obtainable, the greater the system cost will be. For example, for any given technological approach used, greater reliability can be obtained (through redundancy, for example), but at a higher cost. The proportional system, for example, results in greater missile accuracy than the bang-bang system, but the cost would also be greater. Thus, the systems designer·must determine if the higher levels of system performance are worth the additional cost. This is the evaluation problem and is the heart of successful systems planning.

We shall now describe methods for dealing with the two preceding problems.

Point System Approach to Evaluation

One method used many times for combining the different performance characteristics and cost is the "point system evaluation approach," discussed in Chapter 10. Since the main rationale behind such an evaluation procedure is intuitive judgment, it is not the most defensible approach to system evaluation and selection. It was decided instead to review the situation with the missile systems manager.

RELATING SUBSYSTEM EVALUATION TO TOTAL SYSTEM EVALUATION

The subsystem designer's problem can be summarized as follows: He can design an alternative system whose set of performance characteristics can vary over a wide range. Even if these systems could be ranked (or measured more accurately) in terms of performance preference, the designer would still have a selection problem when he considered the relationship of total effectiveness to cost, since more effective systems generally cost more. Unfortunately, a selection from among the subsystem alternatives cannot be made without approaching this problem from the next higher level (i.e., missile system) point of view. We shall now discuss how to select the proper subsystem alternative which will satisfy the higher-level decision-maker. In attempting to do so, the following dialogue might take place between the guidance and control planner and the missile systems planner.

G&C Systems Planner: "Here is a series of technologically feasible guidance and control system alternatives which I can provide. Which one would you like?"

Missile Systems Planner: "The answer to that depends not only on the performance, but also on the cost of each alternative."

G&C Systems Planner: "We have a total cost estimate for RDT&E, investment and O&M in terms of the number of units you have requested. Since we wish to provide you with a subsystem which will have the greatest worth to you with respect to its cost, can you indicate the evaluation model you are going to use in setting the total system specification?"

Experience of the author indicates that the project leader (the missile systems manager) may answer the question in any of several ways, including "I would like your best recommendation of your subsystem specifications."

With the preceding discussion as background, we shall now examine the fundamental tasks involved in the systems design approach from the missile system point of view, as differentiated from components or subsystem design.

Missile Systems Analysis

If there is such a thing as the "systems approach" to planning or design, it can be defined as accomplishing the following tasks:

1. Look at the characteristics of the job to be performed.
2. Devise alternative system combinations for performing this job.
3. Evaluate the total resources required for each of the means for doing this job.
4. Choose that alternative which requires the lowest *total* cost to the user.

While the missile systems designer will consider all elements necessary to perform the mission of destroying targets, he does not need to consider the tradeoff of alternative means for destroying targets (e.g., missiles versus Air Force close support aircraft). A higher-level systems analyst should perform this analysis. Therefore, in general, one does systems design by enlarging the system only up to the mission level. This will be defined as the "problem to be solved."

As will be seen, much of the information required can be found in the cost-effectiveness analysis originally performed as part of the concept formulation package (i.e., the final report showing the analysis performed during the phase). This report was obtained from the Army and describes operational aspects and evaluation criteria as they pertain to the systems function. We shall now discuss each aspect in detail.

Operational Aspects

It can be assumed that this missile system would be used in support of troops in a field army area. The geometry of this situation, pictorialized in Figure 13-3, indicates that there is, in general, some "forward edge of the battle area" (FEBA) which is an imaginary line separating the friendly forces from the enemy forces. Incidentally, this line may not be completely identifiable at any given time, as in a fluid situation. On the enemy side of this line there exists a set of enemy targets whose characteristics include the following factors:

1. Number of targets.
2. Range from the FEBA.
3. Separation from each other.
4. Worth of target.
5. Vulnerable area.
6. Movement or mobility.

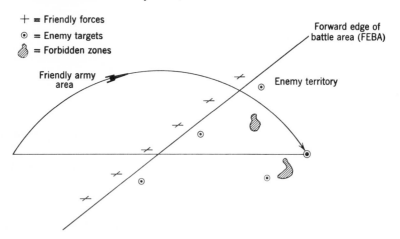

Figure 13-3. Geometry of operational situation.

In addition there are certain areas from which we wish to keep all of our missiles. These include the areas in which our own troops are located, as well as certain enemy areas (e.g., civilian population areas).

It is planned to operate the missile system in the following manner: A forward observer or reconnaissance aircraft locates a target generally through visual observation. The position of the target is then determined in one of several ways, such as correlating its position with visual landmarks of the terrain as noted on a map, or making indirect calculations. For example, if the observer knows his own position (e.g., through some navigational system such as LORAN), and then can take a sighting of the target from his position (e.g., through the use of a laser range finder and sighting device), the two positions may be added vectorially and the target position calculated and transmitted to the missile site. This target position with respect to the missile launch site is then inserted into the inertially guided missile. After the missile is fired, target destruction may be observed by the forward observer. If the observer notes that the target survives, he may call for a second missile to be fired. However, if the target is highly mobile, it may then disappear after the first round has been fired and thus is not available for the second firing.

After the missile is fired there exists the possibility that the enemy may observe the missile launch position (e.g., by observing missile launch) and may thus initiate an enemy attack on the launch site. Thus, the missile system needs to be sufficiently invulnerable to survive such an attack.

Evaluation Criteria

Based on the above operational scenario, the systems analyst can establish the following statement of the problem to serve as the criterion for choosing among missile system alternatives. All subsystems must also satisfy this higher-level evaluation criterion:

1. There exists a target set whose targets are located in some relationship to the FEBA.

2. We would like to destroy as many of these targets as possible, taking into account the relative worth of these targets. A reasonable measure for this process would be the "expected worth of the targets destroyed" (as described in Chapter 7.)

3. Since some of these targets are at close range to the FEBA, there is some danger of hitting our own troops if we fire at these targets. In addition, there are other targets which we would not want to hit accidentally.

4. Thus, we wish to choose that missile system which will accomplish a given level of destruction at lowest total cost, where cost is equal not only to

the total monetary costs of the system, but could also include the implications of destroying the wrong targets due to missile inaccuracies.

There are two ways that the analyst can handle the problem of the cost from hitting the wrong targets. One way is to compile the total measure of wrong targets destroyed, such as lives lost by our own forces, or by enemy civilians. However, the implications for world opinion as a result of accidental losses is difficult to assess. Hence, it is possible to put a realistic operational constraint on each missile system being evaluated to eliminate such considerations. This constraint would be the policy of not launching missiles at any target which is near a forbidden area. This means the system "score" (i.e., expected worth of targets destroyed) will be less for low accuracy missile systems since its effective target set has now been reduced from the maximum level. If such a constraint were not used, we would have to keep track of accidental damage.

Constructing the Operational Flow Model

We shall now explore how the missile systems project leader would evaluate alternative missile system designs, as provided by the missile system synthesizers. These alternative designs consist of different combinations of feasible subsystem alternatives as proposed by the subsystem project leaders.

Since we have already developed a means for evaluating an offensive system, we can use the appropriate elements of the Blue versus Red operational flow model (Figure 5-9) developed in Chapter 5 and the appropriate parametric transfer functions as developed in Chapter 6. Using this model, and the numerical values appropriate to each missile system alternative, the missile systems planner can determine the expected worth of the target destroyed by each system, based on those targets from the assumed target set at which the system can fire. The pertinent submodels of this offensive model have been shown in Figure 13-4, and are identical to the parametric transfer functions discussed in Chapter 6. Only the numerical values are different and are obtained from each particular system design.

Since there is no Red defense zone to penetrate, we can consider all aspects of airborne reliability of the entire missile as one transfer function, as has been shown in Figure 6-12, using the exponential distribution as the basis of the frequency function. Now the main parameter is the MTBF of the entire missile.

Uses of the Operational Flow Model

As described in Chapters 6 through 8, the operational flow model can be used as a means of determining the missile system effectiveness as a function of the various system performance characteristics, the environment

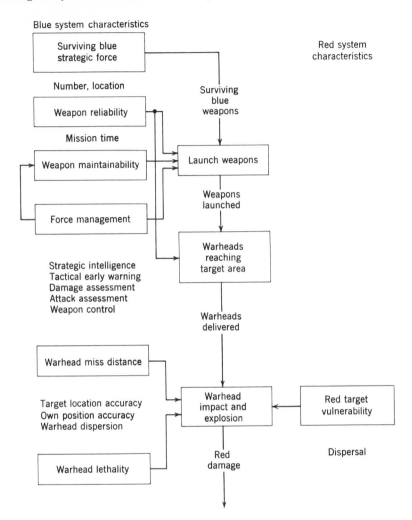

Figure 13-4. Blue versus Red operational flow model for tactical missile system.

(i.e., characteristics of enemy targets), and the number of missiles fired. The measure of system effectiveness would be the worth of targets destroyed, and could be expressed either as an expected value or as a frequency function, since we are dealing with a random process. Hence, one key use of this operational flow model by the missile system analyst would be to evaluate all missile system configurations which are generated. Perhaps the easiest way is to pivot on constant effectiveness as follows:

1. The operational flow model is used to determine the single-shot kill probability of a missile against any target.

2. Missile allocations to the target set are made to obtain the same given level of effectiveness (which may be treated parametrically) using a logic of maximizing the marginal effectiveness of each suceeding missile, as described in Chapter 7.

3. The total cost is calculated for each missile system alternative.

4. That system requiring the lowest total cost is selected.

The same operational flow model can also be used for two other system planning purposes: for subsystem evaluation, and for conducting various tradeoff analyses needed for properly determining system specifications.

Evaluating Alternative Subsystem Designs

Another related use of the operational flow model is as a means of evaluating alternative subsystem designs. For example, the guidance and control subsystem designer could use this as the evaluation model for doing a preliminary filtering of some of the guidance and control alternatives during the early stages of the systems design effort. This method of suboptimization is not too dissimilar from that employed in the helmet evaluation or the system evaluation process, and can be accomplished as follows:

1. Assume values of the performance characteristics of all other missile subsystems as well as the costs of all other subsystem components. In general, this can be a first approximation estimate provided by the missile system group.

2. Insert all of the performance characteristics of the guidance and control subsystem configuration being evaluated and, as a result of this, determine the missile p_k as well as the number of missiles required to achieve a given level of effectiveness (e.g., worth of targets destroyed). Again, multiple rounds would be fired to achieve the effectiveness level.

3. Total system cost would be used as the evaluation measure, with lowest total system cost used as the criterion for selection of the subsystem.

Several observations can be made regarding this type of suboptimization approach. First, this method enables a subsystem designer to compare two or more subsystems to see if the extra cost of one is warranted. Second, a major system planning effort, such as the missile system effort described, involves each subsystem designer trying to optimize his design, yet the optimization process is dependent on the design parameters of the other subsystems. Hence, one way of accomplishing the total process is through an iterative approach. This involves using the operational flow model as the evaluation framework, and having the missile system designers construct a "first cut" of all system and subsystem specifications, and provide these to all subsystem planners. They could then use this information to generate a refined "second cut" of their subsystem specifications which could later be distributed to the other subsystem designers for further refinement.

TRADEOFF CONSIDERATIONS

The discussion up to now has assumed a context in which the missile system designer can choose several alternative designs by inserting their performance characteristics and costs into an operational flow model containing all other performance and environmental characteristics and the costs of all other subsystem elements. This leads to a "cut and try" method of system planning. We shall now discuss a more systematic method for attempting to find the best balance among subsystem elements. The method can theoretically arrive at a true optimum system design; however, because of the many interrelationships involved in a complex system, a purely mathematical optimization approach many times cannot be implemented and some "cut and try" is needed. The concepts involved are of interest because they aid the system designer to more rapidly converge to a satisfactory (though perhaps not optimal) solution.

We shall now illustrate the approach described by concentrating on performance characteristics associated with the guidance and control planner's problem.

Warhead Lethality Versus Guidance and Control Accuracy Considerations

The first question to be examined concerns the considerations to be included in determining guidance and control system accuracy (omitting all other considerations, such as reliability). As seen in the operational flow model of Figure 13-4, guidance and control accuracy is related to the submodel of target destruction by a single missile firing and this model contains the three key variables as shown in Figure 13-5: (a) total system accuracy, (b) warhead lethality, and (c) target vulnerability. In this model, guidance and control accuracy is only one part of the total system accuracy. However, for purposes of illustrating how tradeoff considerations are made, we shall initially simplify the discussion and then gradually increase the number of variables being considered simultaneously. Thus, while in practice the analyst would consider as many variables as he could handle simultaneously, we shall add variables sequentially.

The first tradeoff analysis to be considered involves the division of resources between the missile warhead and the missile guidance and control

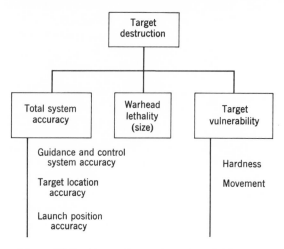

Figure 13-5.   Target destruction submodel factors.

system, assuming a fixed target of given hardness, but zero error in target position and launch position information. Errors involved in these factors will be considered later. The total amount of guidance and control accuracy needed to destroy the target is a function of the size of the warhead used, higher accuracies being required when small warheads are used; lower accuracy if larger warheads are employed. These relationships have been discussed in Chapter 6 and are now plotted in Figure 13-6. From this figure it can be seen that it is possible to obtain a given probability of kill by utilizing various possible combinations of warhead accuracy and warhead lethality. Is there a preferred combination between these two key missile system performance characteristics? There is if the planner includes cost considerations and asks what is the *lowest total cost* method of achieving this given

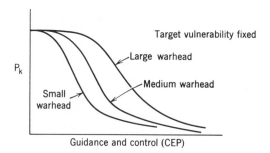

Figure 13-6.   Target destruction for varying accuracy and warhead performance characteristics.

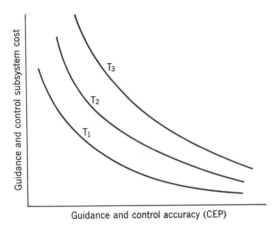

Figure 13-7. Performance-cost transfer function for different technological approaches.

level of kill probability, where total cost includes the cost of accuracy as well as the cost of warheads. Thus, the amount of accuracy required cannot be stated without an examination of the performance versus cost characteristic of its complementary element (i.e., the warhead, which interfaces with the guidance and control system). Examples of these cost-performance functions for different technological approaches are shown in Figures 13-7 and 13-8 for the guidance and control system and warhead respectively. In the illustrations shown, the average unit cost of the device is used for a given lot size, with all fixed costs such as RDT&E amortized over the lot size. Based on this data, the tradeoff between the *cost* of warhead lethality and missile accuracy can be derived, as shown in Figure 13-9. This figure plots the *most efficient* (i.e., lowest cost), technologically feasible way for achieving each of the various levels of kill probability. For example, a kill proba-

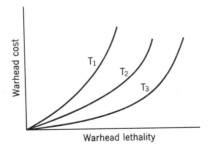

Figure 13-8. Performance-cost transfer function for different technological approaches.

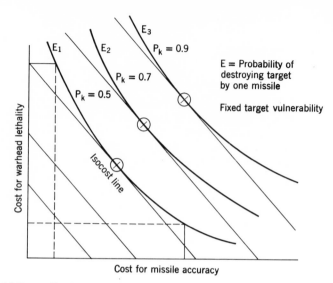

Figure 13-9. Isoeffectiveness and isocost relationships for achieving target destruction.

bility of $p_k = 0.5$ may be achieved by spending a large sum for warhead lethality (e.g., one megaton warhead) and very little for missile accuracy. On the other hand, the same kill probability may be obtained by spending a small sum for warhead lethality (e.g., small size warhead) and a high sum for missile system accuracy (e.g., one foot CPE). For a fixed target vulnerability, both combinations will achieve the same probability of destroying the target with one missile. In a similar fashion, the cost-performance transfer functions will show all other technologically feasible combinations for achieving the same kill probability. Each curve of Figure 13-9 is sometimes called the isoeffectiveness or indifference curve and each contains only the lowest cost way of obtaining a given performance (e.g., warhead size). Obviously, there may be other less efficient combinations which exist (i.e., using technologies which provide the same performance characteristics at greater cost).

This accuracy versus warhead tradeoff characteristic indicates that many times one deficiency (e.g., low missile system accuracy) can be overcome by spending more money on some other performance characteristic (e.g., larger warhead). This example illustrates the concept of complementary elements, which are those elements interfacing with one another to yield a given level of effectiveness. Thus the system design problem is to devise various combinations of performance characteristics which will yield a given level of effectiveness and to determine that combination which obtains a given level of effectiveness at lowest total cost.

This may be done by constructing a series of isocost or isodollar lines as shown in Figure 13-9. The one which is just tangent to the isoeffectiveness line indicates the most efficient solution for that level of effectiveness. In this example each of the isoeffectiveness curves has a fairly broad minimum since it is possible to operate with other combinations of missile system accuracy and warhead lethality without the total cost increasing very much. This broad minimum gives the system designer some added flexibility to consider other factors before finally choosing the proper subsystem specifications. For example, the systems designer may wish to use a smaller warhead (and greater accuracy) than the optimal to minimize the chances of accidentally hitting nearby friendly targets or forbidden areas. This is particularly true if the targets are close to the FEBA. Thus the information of Figure 13-9 illustrates that the additional cost will not be very great in achieving the gain of being able to fire on a larger number of targets.

A second tradeoff also illustrated in Figure 13-9 is the tradeoff between quantity and quality. The systems designer must still decide whether he should use a salvo of missiles each having a single shot kill probability of 0.5, or one missile having a single shot kill probability of 0.9. Since the optimal operating condition for each is indicated in Figure 13-9, additional calculations comparing the total costs (missiles and all other support) for each alternative can be made.

Tradeoffs Among System Accuracy Components

As previously discussed, the total warhead miss distance or dispersion for a fixed target is composed of three main components:

1. Guidance and control subsystem accuracy.
2. Target location accuracy.
3. Launch position accuracy.

As indicated in Chapter 6, if each of these positional uncertainties is random and independent of the others, the total warhead variance is the sum of the three individual variances, the total warhead dispersion (or uncertainty) may be found by taking the square root of the total variance:

$$\sigma_t = \sqrt{\sigma_1^2 + \sigma_2^2 + \sigma_3^2}$$

where σ_t = standard deviation of the total warhead dispersion,
σ_1 = standard deviation of the guidance and control subsystem,
σ_2 = standard deviation of the target location measuring subsystem,
σ_3 = standard deviation of the launch position measuring subsystem.

Thus, the analyst has an opportunity to balance not only performance and cost characteristics between system accuracy and warhead lethality, but also performance and cost among the separate components of system accuracy. This also enables us to explore the original statement that "the CEP of the bang-bang system was twice that of the proportional system." The real question concerns the *absolute* values of the CEP's. Are they 5 feet and 10 feet, respectively, or 500 feet and 1000 feet? Finally, how much does it cost for an improvement in accuracy compared to the corresponding costs of other accuracy improvements?

Again, the starting point for an analysis which will answer such questions is the appropriate set of cost-performance transfer functions as shown in Figure 13-10. These indicate the cost associated with each method implementing the function of obtaining a given amount of target position location and launch position location. Using these transfer functions and the operational flow model, the analyst can construct the transfer function of Figure 13-11 which shows the least total cost method of obtaining various amounts of total system accuracy. This transfer function is then used in the tradeoff analysis with warhead lethality, instead of the function in Figure 13-7 which ignored the other components of system accuracy.

Target Movement Considerations

The preceding tradeoff analyses assumed a stationary target. We shall now consider a target such as a tank which may move after missile launch. After the missile is launched, it will continue to travel in flight to the last designated target position, since new target information is no longer useful with the operational concept under study. Thus to calculate the new prob-

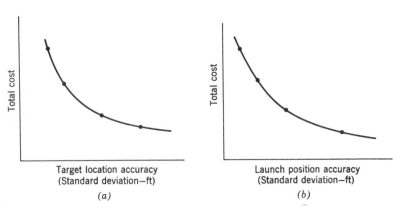

Figure 13-10. (*a*) Target location accuracy cost-performance transfer function; (*b*) launch position accuracy cost-performance transfer function.

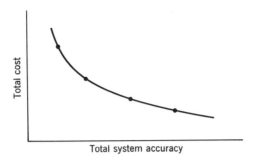

Figure 13-11.   Performance-cost transfer function.

ability of destroying the target, the analyst must combine two distances each of which is a random variable:

1. Total missile dispersion with respect to the original target position, previously calculated as a normal distribution.

2. Target travel during the total missile flight time.* This target travel is directly proportional to target speed and total missile response time. If the target travel from the last observed target position were also a normal distribution, the analyst could again add its variance to the other variances and still use Figure 13-11 to determine p_k. It is unlikely, however, that this is how the target would act. Rather, the target might travel at some operational speed in some arbitrary direction, as shown in Figure 13-12, unless the enemy had some reason to believe it was now under surveillance before

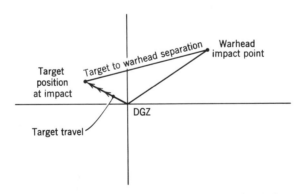

Figure 13-12.   The geometry of target movement and warhead dispersion.

---

* We are ignoring the rest of the total system response time (such as the time taken to prepare the missile for fire) by assuming an operational concept that the launch can be cancelled at any time if the forward observer sees the target move.

an attack, in which case it would probably increase its speed and head for cover.

Hence, based on assumed target movement parameters, the analyst can determine on a probabilistic basis the distance between the target and the actual warhead detonation point, and calculate the probability that this distance is less than the radius of vulnerability. Another way of dealing with target movement when a target cover such as jungle terrain is involved is to assume some probabilistic distribution in which the target stays at the location as a function of time, as in Chapter 6, for mobile targets.

How can the systems designer improve the system design so as to reduce the effects of target movement? One way is to reduce any system response time which effects this dispersion (missile flight time in this case). In this fashion, the target will have less time to move. A second way is to track a movable target and attempt to predict its future position, as a function of time, by some method of extrapolation. The ultimate approach, of course, is a totally guided missile which is continually changing its terminal position even during flight, as it receives command changes from the forward observer. Such an approach is, of course, much more complex and expensive than the simple inertially guided system proposed. A third method of coping with target movement is to increase the lethality of the warhead. This method has the accompanying disadvantage of jeopardizing nearby forbidden zones; hence it may not be an acceptable solution.

Equipment Malfunction Tradeoffs

We shall now discuss the various tradeoffs which involve equipment malfunctions or unreliability. Again, the analyst asks two key questions:

1. What will take place operationally when an equipment failure occurs?
2. What are the cost implications, as measured in dollars, time, casualties, etc., that arise from such malfunctions?

Figure 13-13 provides an overview of the answers to these questions. As indicated in the operational flow model of Figure 13-4, equipment malfunctions can result in a "ground abort" (a failure detected prior to missile launch) or an "air abort" (in which some failure occurs after missile launch) serious enough to result in missing the target. If a ground abort occurs, the ground crew must remove the defective component (or the entire missile) and fire a missile found to be in satisfactory condition. Thus there is some maintenance cost incurred, which includes not only the cost of repairing or replacing the defective component, but also all of the costs connected with the total logistics system. A secondary but important loss is that of response time which is the total time taken to fire a satisfactory missile to target. Ad-

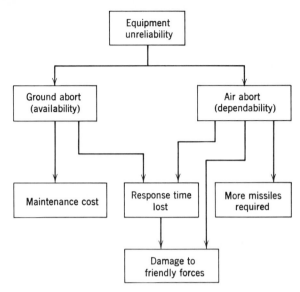

Figure 13-13. Effects of equipment unreliability.

ditional response time can result in a loss of target and/or additional damage inflicted on the friendly force by an enemy target which has not been eliminated.

Air aborts can affect the situation even more drastically. Now the entire missile may be wasted (rather than just one component as in a ground abort) as when the propulsion system explodes after lift-off; this may also result in damage or casualties to friendly forces. Again, the total system response time to get a good missile to target is lengthened, with possible resulting losses.

We shall now consider the various tradeoffs involved, starting with air aborts (covered by the "warheads reaching target area" submodel of Figure 13-4). This submodel may be expanded to include a further breakdown of all subsystems which affect total system reliability, as shown in Figure 13-14a. An even further breakdown of the guidance and control subsystem is shown in Figure 13-14b. As shown, any system may be viewed as composed of a series of subsystem elements, each of which must function satisfactorily if the total mission is to be accomplished. (This ignores any parallel redundancies which may exist.) The reliability of each subsystem may be determined by choosing a suitable probability distribution, such as the commonly used exponential distribution. Thus the reliability of a guidance and control subsystem element is calculated as:

$$R_{21} = e^{-t/\text{MTBF}_{21}},$$

where R_{21} = subsystem component reliability,

t = mission time,

MTBF_{21} = mean time between failures of the guidance and control subsystem.

In addition, the total subsystem reliability R_2 is

$$R_2 = R_{21}R_{22}R_{23},$$

and the total system reliability R_t is

$$R_t = R_1R_2R_3R_4.$$

Thus, the system tradeoff, with respect to component reliability, involves a comparative determination of the incremental effectiveness (ΔE) achieved (from ΔR) for a given incremental cost (ΔC) for each way of increasing reliability (i.e., ΔR_{21}, ΔR_{22} or ΔR_{23}), as compared with the ΔE obtained by adding additional missiles (at a cost of its ΔC). There is, of course, the factor that an unreliable missile may cause the target to disappear (and later cause damage) before the second missile can reach the target. This can be accounted for by firing a salvo of missiles (two or three) to compensate for system unreliability, as well as warhead dispersion.

Ground aborts (the launch weapons submodel) may be analyzed in the same fashion by considering the additional performance characteristics provided by the maintainability concept used. Here the key performance characteristics, shown in the transfer function of Figure 6-5a is the probability of having a weapon ready for launch as a function of time. As shown,

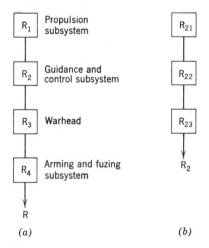

Figure 13-14. (*a*) Calculating total system reliability; (*b*) calculating guidance and control system reliability.

these characteristics are interrelated. Hence, the systems designer can examine the following alternative maintenance approaches for achieving the same objective:

1. Employ a larger number of maintenance personnel to make more frequent scheduled maintenance checks on the system.
2. Employ more skilled maintenance personnel who can perform the scheduled maintenance checks more rapidly.
3. Employ automatic checkout equipment for performing these maintenance checks more rapidly.

These must now be compared with other alternatives such as specifying components with higher reliability or having a larger number of missiles and launchers available for launch in case some fail to check out satisfactorily. Again the same type of marginal effectiveness analysis for each of these alternatives should be performed.

Further tradeoff analyses may be made between the cost of increasing the reliability of the component and the total cost of repairing a system component found to be defective before missile launch. Here we must examine all costs involved in the maintenance and the logistics functions. Here, for example, we can compare the cost of repair versus the cost of throw-away modules.

INTERRELATIONS AMONG CHARACTERISTICS

We have tried thus far to illustrate some of the key tradeoff analyses which can be performed by the analyst in trying to optimize the subsystem or system. However, it is extremely difficult to arrive at an optimum solution for any complex system because of all the component alternatives available and all the interrelationships involved (particularly among different subsystem characteristics involved). For example, if the proportional guidance and control system has greater accuracy than the bang-bang system, it may also have lower reliability. Thus while the approach described is but a guide to the system designer, it is certainly a purely "cut and try" approach. In fact, it is logically possible to program the operational flow model and the various performance-cost transfer functions onto a computer and permit much of the arithmetic work to be done by the computer rather than the systems designer. In the process of converting the system performance characteristics into physically reliable equipments, other system characteristics (e.g., size and weight) arise which also must be considered in determining the preferred system alternative. Some of these characteristics are now discussed.

Size and Weight Considerations

Here the systems planner must consider how the characteristics of size and weight affect missile system performance or cost. If the size or weight of any subsystem is increased, the maximum flight range of the missile will probably be reduced, thus lowering the number of targets which can be attacked. On the other hand, the maximum range can be held constant if the size of the propulsion system, and probably the airframe, are increased to compensate for the larger missile weight. However, this will increase the total missile cost. We shall now examine these transfer functions.

Effect on Other Subsystem Cost. The functional relationship between the total missile weight and missile cost is shown in Figure 13-15a for different maximum flight ranges. In the problem under study, however, the missile systems planners have indicated that they wish to use an existing airframe and propulsion system since these particular units have already been developed and tested, and thus these expenses would not have to be repeated.

But suppose that some new technological element were developed which would permit a high performance characteristic to be incorporated into the guidance and control subsystem package (e.g., very high reliability or very high accuracy at low cost) but which would increase the size and weight of the package over the limit allocated by the missile system designer. This element could be included in one of two ways; the first way would be to permit the subsystem size and weight to be exceeded, modifying both the airframe and propulsion system to accommodate the additional size and weight. The second way would be to increase only the airframe, without changing

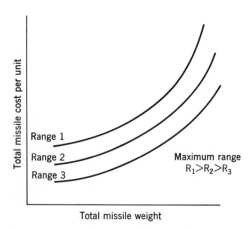

Figure 13-15a. Missile cost versus weight characteristic when designing missile from start.

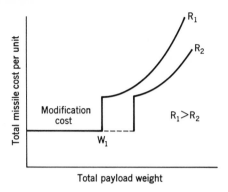

Figure 13-15b. Missile cost versus weight characteristic when modifying existing missile design.

the existing propulsion system. This would result in the maximum target range decreasing to some value R_2. The second function shows how costs increase with weight as the maximum range decreases to R_2. The additional costs for each of these alternatives is shown in Figure 13-15b. The first function shown keeps the maximum target range R_1 fixed. As the total missile weight exceeds the original design limit W_1, a fixed modification redesign cost is incurred, followed by additional manufacturing costs as the missile weight increases. The decision as to which transfer function to use is determined by considering the number and worth of targets which are no longer within the maximum range if it is decreased. Many times it makes very little difference if the range is decreased, say 10%, since the next larger size weapon may cover this range.

If the payload weight or size is decreased, it may be necessary to modify the existing airframe to take care of a new center of gravity, for example; hence this may result in some modification cost.

Effect on Logistics Cost. As the size and weight increases, so do the costs for moving and storing the total equipments involved in system operation. Thus more trucks or carrier aircraft are required and more storage facilities needed for storing spare parts. This may also result in a system availability decrease since the mobility of the system has been reduced to some extent.

Treatment of Other System Characteristics

There are a number of other system characteristics which arise in a systems design effort whose consideration may be requested by one of the decision-makers (e.g., the missile systems manager). These include flexibility, growth potential, evolutionary design, modularity, and commonality.

Since these terms often cause difficulty for the systems design, they will now be discussed to show how they may be handled as part of the systems analysis effort. One of the difficulties involved with these terms is that they mean different things to different people. Thus, it is important that the systems planner makes certain he understands what the decision-maker has in mind when he requests consideration be given these terms.

Flexibility. In general, this means the ability of the system to perform different jobs which may occur, or to perform one job under different environmental conditions.

Growth Potential / Evolutionary Design. These terms are related to that of flexibility in that there is some feeling that the job is apt to change or enlarge over time so that provision should be made to handle the larger job when it occurs at the later date, rather than let the system become obsolete at that time. As will be described in Chapter 14, many times computers are designed on a planned expansion or evolutionary design basis so that additional hardware or software elements may be added without making the initial design completely obsolete.

Modularity. This is a form of design construction which permits the system to be built as a series of interconnecting elements rather than one large piece of equipment. The benefits of this design technique may be varied. First, it permits ease of carrying or transporting the system if the system is to be transportable or mobile. Second, it may permit different systems to be constructed by merely combining the system elements in different combinations. This is the so-called "tinker toy" design approach. Thus, there is the example of the Army engineers who realize they must have the capability of constructing different sized bridges, depending on the width of the river to be crossed; therefore, they must construct the elements of their bridges so that the elements may be interconnected in different fashions to afford different width bridges of different loading capacities. In the case of a pontoon bridge, most of the elements are identical but additional pontoon elements are added in series to yield the different bridge lengths.

The same approach works for command, control, and communications equipment to be used for different levels of warfare. For a low-level system, small capability is needed; hence, only a minimal system is required. If the job escalates to a medium or high level, the capability of adding additional units in series or in parallel is desirable, rather than having an entirely new system to replace the smaller one.

Commonality. This involves the use of the same part for different jobs. For example, a large automobile manufacturer may deliberately design a door handle for use on all of his car models, irrespective of brand name.

This permits a higher volume of this particular part, thus lowering the overall unit cost of these parts. The example of a pontoon bridge design is also an application of commonality since many of the parts are the same.

Factoring In These System Planning Terms

All of the preceding terms pertain to the possibility that the system being planned may have several jobs that will have to be done during its operational life. The examples given included the variable length of a bridge, variable bridge capacity, and low or higher level of warfare. It is possible for the systems designer to deliberately plan his system design to cope with these future eventualities (scenarios) ahead of time. By designing his system in a more adaptive fashion, the systems planner can reduce the total cost of coping with many different types of jobs.

Thus, when a systems planner is evaluating different system alternatives, he must take into account the different jobs that the system may be called upon to perform during its future operational life, and make some subjective estimate of their likelihood of occurrence, as was done in the bridge problem of Chapter 11. Thus if the environment changes in some random fashion, the designer may construct the set of environments as a probabilistic function and treat this as a system demand function. Thus the designer can select that system alternative which performs the total set of demands at lowest total expected cost. However, in a competitive environment, the best way of designing a system which will cope with the different possibilities is to design the competitive matrix discussed in Chapter 8 and configure a system which is relatively insensitive to the changing environment by deliberately planning for change.

SUMMARY OF KEY PRINCIPLES

The over-all process involved in designing an optimal system must emphasize the following principles:

1. Systems planning is a process which involves not only technological experience but also economics.

2. Systems planning requires the formulation of transfer functions involving not only performance characteristics but also their cost implications. Assembling such a technological data base should be one of the primary tasks of the system designer, for without this any form of system optimization is impossible.

3. The generic logic involved in properly combining system components is illustrated by the operational flow model of Figure 13-16. As shown, vari-

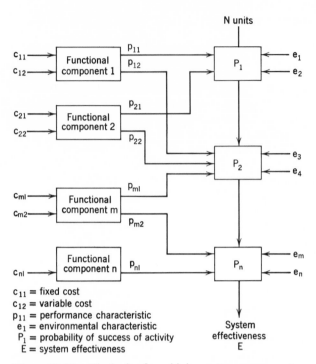

c_{11} = fixed cost
c_{12} = variable cost
p_{11} = performance characteristic
e_1 = environmental characteristic
P_1 = probability of success of activity
E = system effectiveness

Figure 13-16.   The logic of combining system components.

ous functional components are required to configure a given system. These functional components are obtained from the system design model and, in the missile system, would include some type of warhead, airframe, etc.

While the output of each functional component is one (or more) performance characteristic, $p_{i1}, p_{i2} \cdots p_{ij}$, a component cannot be obtained without expending a certain amount of cost. Each cost is a function of the type of technology and method of implementation employed, and the resulting performance characteristics associated with these. The total cost has two components: c_{1_i}, the fixed costs (such as for RDT&E) and c_{2i}, the variable costs which are a function of the number of units used. These costs have been shown as the input function to the process, since cost leads to components which lead to performance characteristics, which lead to system effectiveness. Hence, systems design may be viewed as the process of allocating a fixed level of cost among components in such a way as to achieve the highest level of effectiveness. The missile warhead, as an example, is a functional component with the two key performance characteristics of lethality and reliability. For a given level of each of these performance characteristics, the total cost may be calculated as shown in Figure 13-17 for two different values of warhead lethality. Note the following observations:

a. Technology A may require some fixed investment (for RDT&E or manufacturing tooling), but is less costly than Technology B if enough units are procured.

b. The cost versus volume curves may be nonlinear.

c. Technology D may be used for the lower lethality 1, but is unavailable at the higher lethality 2 for the time period under consideration.

d. Similar curves could be constructed for higher values of reliability.

e. The same type of transfer functions could be constructed for the guidance and control unit, whose key performance characteristics would be missile dispersion and reliability.

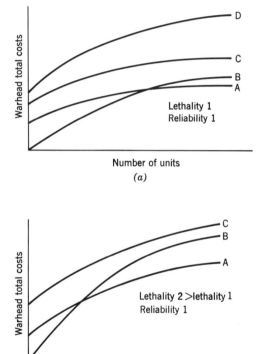

Figure 13-17.   Cost of varying lethality.

The various performance characteristics shown in Figure 13-16 are combined with the various environment characteristics, e_n, for each of the events involved in the process. The operational flow model of Figure 13-21 is the model for combining such factors. Thus the total effectiveness E of a system whose activities involve random processes is a function of the various probabilities of success and the number of units (e.g., missiles) used. That is,

$$E = f(P_1, P_2 \ldots, P_n, N).$$

Hence it is possible to achieve a given level of effectiveness in either of two ways—by increasing any (or all) of $P_1, P_2 \ldots P_n$, or by increasing N, the number of units employed. Since both of these alternatives have a specific cost associated with them, the task of systems planning is to find that combination of performance characteristics and units employed which will provide a given level of effectiveness at lowest total cost expended.

4. Obviously, designing a system with increased flexibility, modularity, commonality, or growth expansion will increase total system cost. If a number of uncertain situations are possible, however, many times these additional costs can be justified by showing that the *expected total cost* of meeting these uncertain situations is least by including such features.

CONCLUDING REMARKS

1. It is impossible to do systems or subsystems planning or design properly without evaluating at the mission level. To do this, two classes of models are generally required. The first (the operational flow model) converts the system performance and environmental characteristics, including the scenario specified, into a measure of total mission effectiveness. The second type of model relates performance characteristics and their cost implications for each of the technological elements of the subsystems. The combination of both these classes of models will yield the total mission or operational effectiveness of the system, as well as the cost implications.

2. We have indicated a method of evaluating alternative system designs and selecting the preferred one through the use of an operational flow model and a cost model, and pivoting either on constant effectiveness (and choosing on the basis of lowest cost) or pivoting on constant cost (and choosing on the basis of highest effectiveness).

3. The same procedure will work in the evaluation and selection of alternative subsystems, by inserting in the operational flow model and cost model assumed values of the performance and cost of the other subsystems.

4. It is the mission level system designers who are responsible for the entire analysis. This group, in general, has the best knowledge of the operational aspects of the job to be performed, as well as the system evaluation procedure to be used. However, they do need subsystem performance versus cost information in order to properly design the total system, considering all of the pertinent tradeoffs involved.

5. Each subsystem designer is the key individual who best understands the differences in performance obtainable from each subsystem design alternative and the cost implications associated with each. This he can provide in the form of cost-sensitivity characteristics.

6. The initial subsystem requirements given to each subsystem planner by the next higher level planner many times conflict with one another when they demand the best technical or performance characteristics, all at lowest cost. Hence these specifications should be treated merely as an initial starting point, subject to subsequent tradeoff analyses. Both the subsystem and subsystem designers should always find out what would happen to system effectiveness and cost when a specification is not met. The operational flow model and system cost model offer a method for determining this.

7. The real problem in systems design is for each subsystem designer to provide performance and cost information regarding his subsystem to the higher level system designer (i.e., missile systems project manager) thus aiding him to make the proper intrasystem tradeoffs concerning total performance characteristics versus total cost. Only by performing these various intrasystem tradeoffs can the higher level systems designer evaluate alternative subsystem proposals.

8. These tradeoffs will assist the decision-maker to properly allocate resources appropriately among the different subsystem elements so that a given level of effectiveness may be achieved at lowest *total* system cost.

VI

THE ROLE OF ANALYSIS
IN SOURCE SELECTION

The previous parts described and illustrated the use of systems analysis in the concept formulation or preliminary design phase, whose objective is to determine whether a particular system provides sufficient benefits with respect to its cost and risk to warrant proceeding into the next phase of procurement. During this phase, detailed specifications are formulated and sent to a list of qualified vendors who will propose a system designed to meet the specifications previously delineated (including delivery date) and their quoted price for doing so. An evaluation of these proposals is made and a winner chosen on some basis of optimum combination of system performance with economy.

There are a number of problems associated with the above tasks:

1. How are detailed system specifications to be set, particularly when there may be high uncertainties in the job(s) to be performed by the system?

2. How are the proposals from each of the vendors to be properly evaluated and a contract source selected when the specifications of a proposed system may not be identical to the desired specifications? Or if development or other vendor uncertainties are involved, they may give reason to believe that the resulting system may not fully meet the original specifications of performance or delivery time, or may overrun the original quoted price.

In Part VI we shall show how the same systems analysis methods and techniques described earlier can be used to assist the systems engineer in the systems procurement phase involving source selection. Two case situations are discussed. Chapter 14 involves the use of systems analysis in setting system specifications in a "request for proposal" (RFP) and in evaluating alternative vendor proposals for "off-the-shelf" electronic data processing (EDP) equipment. Chapter 15 involves the same problem for a system where development uncertainties are involved.

14

Dealing with Job Uncertainties: Electronic Data Processing Systems Selection Case

The selection of electronic data processing (EDP) systems has presented a problem to the decision-maker for many years since competition became established in the production of computing equipment. The development of computers has been characterized by a diversity of approaches, designs, and configurations.

Solutions to this problem have been varied and cover a wide spectrum all the way from essentially ignoring the technical differences to the carrying out of detailed studies utilizing sophisticated tools.

This problem is not limited to the EDP user. The manufacturers themselves must make design decisions involving a tradeoff between the cost of the equipment and its performance. For a particular user the problem is basically simpler since he can confine his evaluation to how the EDP equipment proposed by the vendor satisfies his particular requirements.

A significant amount of attention has been devoted in the computer literature to the definition of measures of system performance and effectiveness.* These definitions have become further complicated with the availability of large scale multiple-access computer systems.

The approach adopted for selection itself must represent a tradeoff between effort (time and resources) applied and credibility achieved. There are some analysts who feel that throwing a multisided die may be sufficient to select among qualified vendors. Such a process certainly reduces the expenditure of selection resources, but unfortunately it suffers in the areas of

This chapter is adapted from a technical report by J. D. Porter and B. H. Rudwick of the MITRE Corporation (March 1968), and is based on work developed by them for the Air Force under Contract No. AF 19(628)-5165.
* See (1–16) in EDP system bibliography.

repeatability, defensibility, credibility, and acceptance by the competing vendors and the selection authority who is responsible for the final decision.

A number of EDP equipment selection procedures have been described in the literature * and a far greater number have undoubtedly been used but not formally reported. None of the reported procedures have satisfactorily handled the problem of combining performance and cost. In the last analysis, all methods must make use of an explicit determination of the worth ** to the user of the variety of features proposed by the competing vendors. Such methods depend heavily upon intuitive judgment. It is the identification of these judgment areas and the degree to which they can be rendered explicit and defensible that contribute to the success of a particular selection process.

The objective of this chapter is to illustrate how systems analysis can be applied to the problem of evaluating and selecting among alternative, proposed EDP systems designed to meet a set of EDP user needs. A framework is provided to enable the EDP system evaluator to combine the selected relevant system performance measures and the related cost elements to arrive at a rational, defensible selection decision. The approach described is based on work accomplished for the EDP Equipment Office of the Air Force which is responsible for evaluation of Air Force EDP equipment. The approach has been applied to a number of EDP selections for several government agencies.

It should also be pointed out that the cost-effectiveness analysis techniques developed in this chapter specifically for application to EDP system selection are also applicable to the source selection process in other system areas.

PROBLEM AS GIVEN

You are a systems analyst with the Data Processing Administration, a newly formed federal agency, responsible for procuring data processing systems for all federal agencies on a centralized basis, and have been asked to attend a conference with Walt Johnson, the director of EDP Equipment Evaluation. When the meeting begins, Mr. Johnson says to you, "As you know, the mission of this agency is to work with the various federal agencies who require new EDP systems, help each determine the specifications of the system suited to their needs, issue the Request For Proposal (RFP) to eligible vendors, evaluate their proposals, and recommend a system contractor

---

* See (17–25) in EDP system bibliography.
** A more detailed discussion of worth is presented later in the chapter.

to the agency administrator (the final source selection authority) for his approval. To do this we need an evaluation procedure to enable us to select a computer which will satisfy the agency that requested it, provide a good balance between cost and performance, and be fair to all vendors submitting proposals.

"I have assigned you to be a member of a computer evaluation project which is being headed up by Tom Jones, my assistant director. John Reynolds and Bill Smith who have had considerable experience in EDP equipment will also be members of the project. I would like you to bring your systems analysis experience to bear on the problem of developing an improved computer selection procedure which we can use to enable us to select a computer best meeting the needs of the using agency. In doing this, I would like you to review the evaluation procedures we are currently using on a present evaluation now underway, as well as those used by industry and the armed forces, and recommend an evaluation procedure for this agency.

"In designing such a procedure I would like you to consider the following relevant factors.

"As you know, it is the job of our new agency to watch government expenditures for computers and, therefore, it is going to be important that the computer selection plan that you gentlemen come up with will show me and the agency administrator that the government will be getting its money's worth for every new computer we buy. Now I realize that it is quite difficult, or even impossible, to forecast accurately what a using agency's computer needs will be five years from now. Hence I don't want the evaluation procedure you design to force an agency to buy a small computer which meets its present needs and only gives them small excess capacity to handle additional computer loads which will inevitably occur. On the other hand, the user should not obtain so large a computer that there will be a large amount of unused computer time on its hands. Similarly, the procedure should not force a user to buy a stripped down computer which lacks some of the newer worthwhile features now becoming available. But it does mean that you will have to find some way of justifying, both to me and to higher level authorities, your recommendations for expenditures on the basis of some balance between computer cost, performance, and what it will do for the using agency.

"Incidentally, as you know, I am a great believer in competition, and we must always make certain there are several vendors bidding on any procurement, since this is the best way to assure that we will get the best computer tailored to our needs at the best over-all price and contract terms. Since government computers are generally large and expensive, I am sure that a number of large computer companies will be interested in entering each announced competition. I am emphasizing this to you so that you can exercise care in requiring a computer feature that only one computer company may supply, ruling out all other companies from the competition; you had better be able to show that there is no other way of getting the required EDP job done without having the feature, and I seriously doubt you can do this. As far as I am concerned, competition is essential.

"Another point I wish to make: In the past, computer suppliers have promised that they would develop certain software features for us by a set delivery date. Unfortunately they often ran into some developmental problems and delivery was delayed. In this agency, I am going to insist that everything we buy will have to be 'off-the-shelf.' You should inform the computer companies that, while they can propose additions such as system expansion capability at a later date to meet an expected increase in EDP jobs, they can only propose hardware and software available and demonstrable at the time we release our specifications.

"Finally, what we will also require for each procurement, in addition to the RFP, is a selection plan that shows the detailed evaluation procedure which you plan to use. This plan should show how you will evaluate the information you are going to receive from the vendors, and how you are going to combine this information to arrive at your selection recommendation. This selection plan will have to be approved by the Source Selection Authority before the RFP is sent to the vendors.

"Letting the vendors know that we have prepared such a plan before any proposals are received will assure them that we plan to conduct a fair evaluation and I think this will encourage them to do a more thorough job on their proposal effort. Incidentally, when considering vendor prices, you should consider two alternatives: outright purchase of the computer or leasing. In making such calculations, you should assume a system operating life of five years."

To obtain a better understanding of the problem, the systems analyst initiated a series of discussions between members of the project team and members of a computer selection team just beginning in the agency. The latter included personnel from the federal agency who will use the computer, as well as evaluation personnel from the data processing administration. The purpose of these discussions was to gain an understanding of the current source selection process being used. This would include an understanding of the user's detailed data needs as the user foresaw them, the technical or performance characteristics which they felt were needed to satisfy these needs, and any thoughts they had regarding the proposal evaluation process they should use to give them the sort of system they required. The following conclusions were reached as a result of these discussions.

OVERVIEW OF THE SELECTION PROCESS

In general, the evaluation and selection process involves three main components as indicated in Figure 14-1:

1. Statement of user needs.
2. Submission of vendor proposed system.
3. Measurement or comparison of the proposal against the stated user needs.

Each of these parts is discussed in more detail and various terms are defined for later reference.

User Needs

The first task to be performed is to determine and make explicit an approved set of user needs which will form the basis of the request for proposal (RFP) to be sent to the vendors and the evaluation and selection procedure. For an EDP system, the user needs can be expressed mainly by the description of the future workload which the user feels he will have to process during the operational life of the EDP system. The main difficulty in expressing

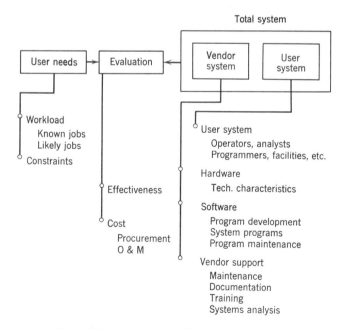

Figure 14-1.   Overview of the selection process.

these needs is that while the user may be able to express accurately his current workload, he is never really sure about the future workload. It is very difficult for the user to predict what he may be asked to process as much as five or more years after the RFP has been issued.

While user needs are expressed mainly in terms of EDP jobs, there are other needs or constraints which relate to the EDP system's ability to perform these jobs. One such constraint is the maximum time allowed to perform any one job or a total set of jobs. For example, the user may feel that:

1. He needs a two-second response time for some task.
2. Some set of jobs must be completed during an eight-hour first shift operation.
3. The monthly workload must be completed in less than, say, 600 hours.
4. The system must be delivered within 90 days after contract award.

It is important to realize that some of these constraints may be quite firm to the user; others may be "open-ended" (i.e., are really "desires"). Several factors complicate the problem of clearly stating user needs. These factors are:

1. The uncertainty of the future workload.
2. The lack of a cost limitation recognized by the user (his main pressure may be to satisfy the future workload he feels he will be called upon to meet independently of cost).
3. The lack of cost-sensitivity information regarding what it costs to meet different combinations of user needs.

Because of these difficulties, some compromises must be made between the user and personnel in the procurement chain such as the Computer Selection Committee, who must approve the selection, to arrive at an approved set of user needs which will be utilized as the basis of the source selection plan for evaluation.

Proposed System

Upon receipt of the RFP which describes the EDP system specifications, the vendor performs various cost-performance tradeoffs to configure an EDP system which in the vendor's opinion will best meet the stated user needs.

This vendor system consists of:

1. Hardware having stated technical characteristics.
2. Software including the various programs required to support operating systems, compilers, etc.

3. Vendor support including required maintenance, documentation, training of user staff, and systems analysis.

However, there are other parts of the total system which the user must provide to make the system function. This user system includes the user's operators, analysts, programmers, and facilities.

Evaluation

The total system corresponding to each vendor's offering must be evaluated against the stated user needs, and the winning vendor selected according to specified criteria. There are two parts to such an evaluation:

1. A system performance evaluation which determines the effectiveness of the system. Here effectiveness is defined * as the degree to which the system will meet the future workload and satisfy the constraints.
2. A cost evaluation which determines the total cost for procurement, operations, and maintenance in performing the future workload over the total required operating life of the system.

Implicit in the evaluation is the need to validate the vendor's proposal. It should be stated that this requirement is common to all evaluation procedures and, from a technical point of view, may represent the most time-consuming part of the evaluation. This area is discussed in more detail later in the chapter.

Selection Objective

To compare alternative selection procedures, an explicit definition of the selection objective is required. The objective selected is as follows:

To select a proposed EDP system which performs a set of future EDP jobs and meets the job constraints at the *lowest total cost* to the government, taking into account job uncertainties and vendor uncertainties.

This objective includes the following three major concepts:

1. All vendors must show that their proposed systems can perform all of the future EDP jobs and meet the job constraints.
2. Lowest total cost to the government should be the selection criterion.
3. Vendor and job uncertainties are the key factors which make the evaluation selection process a difficult one.

* This term is discussed in greater detail later.

Uncertainties

As discussed in the preceding, the selection process is complicated by two classes of uncertainties: vendor uncertainties and job uncertainties. Each of these classes is now discussed in greater detail.

Vendor Uncertainties. The proposal submitted by the vendor, in addition to cost and contractual-type information, will include the following technical information.

1. *Technical Characteristics.* These are the specifications of the components of the proposed computer configuration together with detailed information about the performance of each (e.g., speed, capacity, etc.). Assuming that the equipment will be delivered on time, one can question whether each component will perform at the levels claimed.

2. *Software.* In response to the RFP, the vendor will describe those software packages that he will make available with his equipment. Again, assuming that the packages will be available when needed, one can question what elements and functions are provided, how well each is implemented, and how well each may be used.

3. *System Performance.* Not only must the elements of hardware and software be considered individually, but their interrelationships must also be considered in determining their effects on system performance. For example, a card reader or a printer may not be able to run at rated speeds because of other system requirements, or a software package may degrade system performance because it produces inefficient code, because it may constrain an operating program by reducing the amount of storage space available, or because it is not suitably matched to the available hardware.

4. *Support.* As mentioned earlier, one must always be concerned with the ability of the vendor to deliver his equipment and associated software as scheduled. This is just one example of a number of vendor-dependent activities that can be grouped together under the heading of vendor support. For example, the reliability of the vendor-supplied equipment and programs can very strongly affect estimates of system timing; (i.e., the time required to complete data processing jobs). Also, the user's ability to operate the vendor's equipment will depend on the documentation available and on the professional capability of the analysts and support personnel provided by the vendor. Finally, it must be realized that both equipment and software must be maintained. The vendor's ability to do this efficiently and systematically will also influence the user's ability to attain predicted system performance.

Coping With Vendor Uncertainties. A number of techniques have been developed for dealing with vendor uncertainties. Basically, the requirement that the vendor supply off-the-shelf equipment and undergo a live-test demonstration (to validate the vendor's estimate of the time required to complete certain data processing jobs) removes a large part of the risk associated with making state-of-the-art systems operational. Of course, this procedure has a compensating drawback in that it may prevent the user from acquiring newly developed systems.

The available techniques may be categorized into three major areas: (a) professional personnel, (b) tools, and (c) systematic procedures.

PROFESSIONAL PERSONNEL. The basic ingredient for any evaluation is the availability of competent professional personnel. Such personnel must be carefully trained to stay abreast with the state-of-the-art not just in the equipment alone, but also in the way this equipment may be used, such as through time-sharing or in computer complexes. The ability of these people to interpret and assess vendor claims will be further enhanced through experience. In particular by working with vendors, a better understanding can be acquired of the features of the vendor's equipment and staff as well as of the marketing strategies employed by the various vendors. In addition, through contact with various government user installations, a better understanding can be acquired of the user needs and problems against which the vendor proposals must be assessed.

TOOLS. The problem of validating vendor proposals can be greatly facilitated by having the proper set of tools. Within the inventory of applicable tools, one can identify the following categories:

1. *Simulation programs.* Basically it is desirable to have a program that can take as input a description of the job to be performed together with the specifications of the equipment proposed. The output of the program would be an analysis of system performance (e.g., overall problem timing, buffered times, component times, storage requirements, etc.). Such a program can serve as a check on the vendor's logic, analysis, and calculations. By incorporating into the program an independently derived data base for the equipment specifications, such a program provides a check on the vendor's data accuracy. Simulation techniques are being extended to evaluate some of the dynamic aspects of multiprogramming/multiprocessing systems.

2. *Benchmark programs.* The most satisfactory way to validate a vendor proposal is to run an actual live test. Because of the time and cost involved in testing the whole job, one is led to make use of a program or set of programs that are representative of the jobs to be performed and constitute a predetermined fraction of the total workload. Such programs can be selected to test the performance of individual computer components as well as to measure, through suitable extension factors, the overall system performance. Even though it is difficult to design such benchmark programs to test all of

the significant aspects of the vendor's proposal, nevertheless the use of such programs tends to restrain the vendor and to encourage his use of more defensible estimates. In general, the benchmark programs are selected as portions of the actual expected workload, but programs already developed for other jobs or artificially designed can be used provided they are suitably representative and can be extrapolated to give the information desired.

3. *Software test programs.* An important class of benchmark programs is one especially designed for testing software. Here it is more efficient to design programs to test specific elements or combinations of elements of the software package. However, a job-oriented program may still be useful to test such features as compiler efficiency. There is a modest amount of cooperative effort being expended under the direction of USASI (U.S.A. Standards Institute) to develop such compliance test programs for COBOL.

4. *EDP data base.* Information concerning the availability and performance of EDP equipment can be organized into a data base to facilitate the validation of vendor proposals. Not only does this provide a reservoir of information against which to check figures, but it also serves as a repository for cataloguing acquired experience with vendor equipment and claims. As time allows, one can envisage a sort of *Good Housekeeping* approach to test the performance of computer components and software packages with the results being incorporated into the data base.

5. *Analysis/synthesis.* In support of any validation procedure, there must exist a basic understanding of the performance of the individual computer components and the interrelations that govern how these components work together as part of the overall computer system. For example, the performance of the CPU, storage devices, I/O devices, data channels, file structures, scheduling disciplines, etc., must be analyzed in order to predict the overall system performance or to determine those elements that may be critical to that performance. Such prediction and estimation must take into consideration dynamic conditions that characterize the on-line use of computers today.

SYSTEMATIC PROCEDURES. Because of the large number of parameters that contribute to the overall complexity of validating vendor proposals, systematic procedures must be established to provide an orderly context for assessing the vendor's proposal. Given a competent, professional staff with an appropriate set of tools, it is still necessary to establish an unambiguous set of procedures to assure that the vendor understands the user's requirements, and that the evaluators understand each vendor's proposal. The user's requirements can be formalized into a set of system specifications which can be translated with the cooperation of the evaluation team into the

RFP that is transmitted to the vendor. By carefully establishing the format and contents of the RFP, the vendor will know what to expect and what to look for in the RFP. By establishing lines of communication between the vendor and the user/evaluation team, the vendor can inform the team of critical areas in his proposal and can receive clarification of any questions on the RFP that may arise. By following systematic procedures, one can assure that relevant information is equitably disseminated to all competing vendors. Records can be maintained to determine what information was exchanged in case of misunderstandings that may later arise. By applying established validation procedures and evaluation techniques, one can increase the probability that the vendors will accept the results of the validation and evaluation exercises. In addition, if these procedures were to be used by a professional evaluation organization which would be continually involved in future procurements as new techniques or improvements are developed, they could be more readily incorporated into the established procedures. Finally, by having an established chain of approvals for the selection plan and decisions, there is available a set of checks and balances that will assure the vendor of equitable treatment and avoid the aura of mistrust which might otherwise becloud the vendor/evaluator relationship.

Job Uncertainties. As discussed, a number of factors contribute to the difficulty of explicitly stating the user's needs. For example, given that the user will be asked to perform a certain job in the future, a number of aspects of that job may change in the future and be difficult to predict at the present time. The size of the job may vary due to changes in the lengths of the files to be processed (e.g., the number of fields in a record or the number of records might change). The frequency of running certain jobs may be difficult to predict and consequently the total time demanded for that job becomes uncertain. Complexity of jobs may increase through the incorporation of additional processing steps into the job as experience and requirements evolve, or through the introduction of more refined or sophisticated methodology. Finally, the set of jobs to be performed may change by the addition or substitution of new jobs that were not anticipated when the user originally specified his needs.

The level of credibility of the user's predictions of future workload can be raised by applying more time and resources to the analysis of the user's requirements. For example, if a particular selection involved a specified workload to the exclusion of unanticipated future jobs, then through the use of detailed systems analysis and extensive system simulation, one could very accurately establish the characteristics of even a complex workload. However, in most cases there is insufficient time or manpower or budget to permit the extensive analysis required for accurate workload prediction. In addition, user jobs do change over time to a degree which may be difficult to predict.

Coping With Job Uncertainties. Recognizing that the future workload cannot be considered fixed and completely specified, evaluators have devised a number of techniques to cope with these uncertainties. The most commonly adopted method makes use of a "point-scoring" procedure which establishes a hierarchy of factors or criteria together with appropriate formulas and weights, as described in Chapter 10. Points are then allocated in accordance with how well each vendor has scored on the various factors and upon the relative weights allocated by the evaluation team to these factors. The vendor with the largest total score is then adjudged to be the winner. A detailed presentation of such a procedure for selecting among alternatives has been carried out by Miller.*

This point-scoring procedure is what the agency was currently using in the source selection process. Briefly the procedure used was as follows: The EDP system user asked all units of his agency who currently use the computing center, or might do so in the future, to supply a list of their future possible needs over the next five years. Since there was some uncertainty connected with these jobs, a "best estimate" of the future workload was decided upon and a detailed list of EDP equipment specifications deemed necessary to meet the uncertain workload was formulated. This set of specifications was constructed in the following manner. First, a set of minimum requirements was given, such as the minimum amount of core storage or print characters. It was mandatory that all vendors meet at least these minimum specifications or they would not be considered to be responsive.

However, if a vendor supplied more than the minimum mandatory requirements, this was obviously worth something to the user; but how much? Any feature or specification for which it was desirable that more than the minimum value be supplied was called a "desirable feature," and the point-scoring procedure would provide additional points credit as a function of the excess amount of characteristic provided.

As described in Chapter 10, based on the intuitive judgment of the computer personnel, a maximum number of points were allocated to each evaluation factor and a quantitative evaluation function devised which would convert the numerical value of each performance characteristic (from a minimum to an "overkill" value) to points of credit. Point credits were also assigned to cost or selling price in some inverse relationship. In this fashion,

---

* J. R. Miller, III, "A Systematic Procedure for Assessing the Worth of Complex Alternatives," ESD-TR-67-90, The MITRE Corp., Bedford, Mass., Sept. 1966.

the vendor with the highest point score would be judged the winner (not counting other factors which would be evaluated qualitatively).

When the set of data system requirements, equipments, specifications, and point-scoring procedure for evaluating vendor responses were discussed with the systems analyst, he called the team's attention to the following anticipated problems in applying these procedures:

1. Could the long list of minimum equipment specifications be defended? Some EDP specifications which a user has asked for could be classified as discriminating against some vendors since only one computer company supplied this off-the-shelf feature. When it was indicated to the user personnel that these specifications were discriminatory, they agreed that their data processing jobs could be done to their satisfaction by some alternative method. For example, while they desired multiprogramming, it was agreed that a machine could perform the data processing load if this machine were fast enough. For this reason, these characteristic features were either removed completely, or identified as a desirable, though not mandatory, feature.

Other minimum specifications could lead to higher specifications than are needed because certain *interrelationships* among specifications have not been taken into account by the point-scoring procedure. For example, a vendor could supply the lowest cost computer which had less than the required amount of core storage and which still performed the data processing jobs satisfactorily if his system is sufficiently fast.

2. How should performance and cost be compared? Even if it could be determined that the performance of one system is better than another, it may be difficult to justify its selection if its cost is also higher. While the point system attempts to solve the problem by allocating a number of points for cost, there is always the question of defending the weight allocations among performance and cost.

3. Hesitancy in providing the evaluation procedure to vendors. While the detailed specifications and the qualitative aspects of the evaluation procedure would be provided to each vendor, the current computer selection committee decided that they would not provide either the weights or the evaluation functions to the vendors. They realized that the weights given each factor, and the point evaluation functions to be used were largely based on judgment. Hence there was fear that a vendor would attempt to optimize his system around the evaluation procedure and might win through a loophole. In addition, there was some concern that the losing vendors might protest the decision by disagreeing with some of the evaluation functions used, particularly those based on intuitive judgment.

COST-EFFECTIVENESS ANALYSIS APPROACH

With the previous statement of the problem and the various evaluation difficulties as background, we shall now discuss the approach taken in applying conventional cost-effectiveness analysis to this problem of computer evaluation and selection.

Definitions

We shall start the discussion by defining various terms in a way which permits application to this problem: *

Effectiveness: The degree to which a system will perform the future jobs and satisfy the constraints. Effectiveness is generally considered to consist of the following three main components, as illustrated in Figure 14-2:

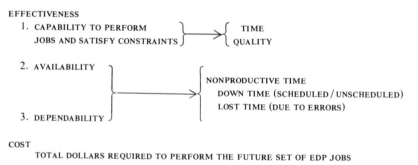

EFFECTIVENESS
   1. CAPABILITY TO PERFORM JOBS AND SATISFY CONSTRAINTS → TIME, QUALITY

   2. AVAILABILITY
   3. DEPENDABILITY → NONPRODUCTIVE TIME, DOWN TIME (SCHEDULED/UNSCHEDULED), LOST TIME (DUE TO ERRORS)

COST
   TOTAL DOLLARS REQUIRED TO PERFORM THE FUTURE SET OF EDP JOBS

Figure 14-2.   Cost-effectiveness definitions.

1. *Capability:* The degree to which a system will perform the future jobs and satisfy the constraints, assuming that the system is always available for operation and will never malfunction. Capability can be measured in various ways, but the two key measures of capability are quality of the work output and time. Quality of the work output is generally multidimensional since it encompasses the many submeasures used to measure the work output. For example, it might include the straightness of a line of print or the maximum number of copies of printout.

---

* These concepts are an application of those used in the evaluation of weapon systems as described in Chapter 4. Reference is made to such reports as "Weapon System Effectiveness Industry Advisory Committee (WSEIAC). Final Summary Report, AFSC-TR-65-6, January 1965."

Given that the quality of the work output can be measured, a second key measure of the EDP system capability is the time taken to perform the future workload, again assuming the system is available for operation and never malfunctions. For example, the measure could be the expected time to perform a given monthly workload.

2. *Availability* may be defined as the probability that the system will be ready for operation when called upon.

3. *Dependability* may be defined as the probability of the system completing the job satisfactorily, given that it was available.

However, when evaluating EDP systems in which the primary measure of capability is the expected time to perform a given workload (given that the quality of the system meets a certain level of acceptability), both availability and dependability can correspondingly be measured in units of nonproductive time expected during the performance of a given (monthly) workload. This nonproductive time consists of two sources: (a) down time: both scheduled maintenance to prevent malfunctions and unscheduled maintenance to detect and correct malfunction; and (b) lost time: nonproductive run time requiring rerun because of suspected or detected errors.

Availability also includes how well the vendors can meet the desired delivery date, since this has an impact on the non-productive time of the system.

Cost: The total dollars required to procure, operate, and maintain the system to perform the future set of EDP jobs. As indicated in Figure 14-1, all costs are to be included in making this computation for both the vendor and the user.*

The Selection Problem

Given that the effectiveness and cost of a particular system can each be measured separately, the evaluation team will still be faced with the problem of how to combine these two factors to reach a final selection. For example, as illustrated in Figure 14-3, we might have a situation where System B provides a higher level of effectiveness than System A but costs more.** The source selection problem concerns which is the better system to buy. This could be restated as, "Is the additional amount of effectiveness worth the added amount of cost?" It is impossible to answer this question, except on a purely intuitive basis (which may be wrong or difficult to defend), without resorting to either of the following two source selection criteria used in a cost-effectiveness analysis:

* In some cases, certain nondifferentiating costs may be omitted.
** Note that the decision is straightforward if there is a dominant case where one system provides more effectiveness at a lower cost.

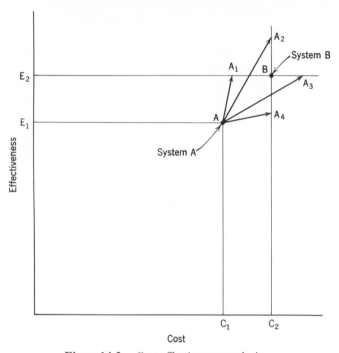

Figure 14-3.   Cost-effectiveness analysis.

1. Specify a level of effectiveness which all systems must meet, and select that system which meets this level at lowest total cost. This criterion is called "pivoting on constant effectiveness." Thus, if E_2 is chosen as the comparison level of effectiveness and the effectiveness of System A is increased accordingly, its new operating point on Figure 14-3 might be either at A_1 (lower cost than B and hence selected), or A_3 (higher cost than B and hence rejected).

2. Specify a level of cost which all systems must not exceed, and select that system which provides the highest level of effectiveness. This is called "pivoting on constant cost." Thus, if C_2 is chosen as the comparison level of cost, and the cost of System A is increased accordingly, its new "operating point" in Figure 14-3 might be either at A_2 (higher effectiveness than B and hence selected) or A_4 (lower effectiveness than B and hence rejected).

The first selection criterion is used for illustration for the remainder of this chapter, since it is more generally applicable.

Proposed Selection Process

We shall now discuss in greater detail the three categories which we have used to describe user needs, a method for quantitatively communicating these needs to the vendors, and a procedure for evaluating how well each vendor's proposed system meets each of these three categories of needs.

Classification of User Needs. Taking into account the many jobs which could make up a future workload and the various uncertainties associated with each job and with the vendors' proposals, efforts were made to apply cost-effectiveness analysis methods and techniques to the source selection process of Figure 14-1. The analytical approach used concentrated on making explicit all of the characteristics of the possible jobs and on quantifying the uncertainties associated with each job.

It soon became apparent that this might not be practical to do on every source selection since some might require an excessive expenditure of time and user/analyst manpower. Hence it was decided to restructure the user needs portion of the selection process as shown in Figure 14-4. User needs can be considered to be made up of three primary parts. The first part describes the representative workload. The second and third parts which were formerly encompassed by the term "constraints" are more explicitly defined here as mandatory requirements and desirable features.

1. *Representative workload.* Out of the many jobs which the user predicts may make up the futue workload, only the most important or a selected subset would be used to represent this future workload by adjusting job

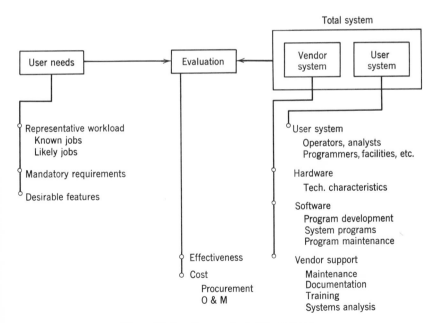

Figure 14-4. Proposed selection process.

types and their frequencies of operation. Jobs again consist of "known jobs," for which the user has a high degree of confidence in their occurring and in their characteristics, such as size and frequency of operation, and "likely jobs," for which the degree of confidence is lower. An analytical technique for expressing this uncertainty by using quantitative probabilistic estimates is described later in the chapter.

2. *Mandatory requirements.* These are absolute requirements which must be met if a system is to be considered for further evaluation. A system that does not satisfy one or more of these requirements is considered nonresponsive and is essentially disqualified. Since this is such a strong constraint, every effort must be made to keep these requirements to a minimum.

3. *Desirable features.* This category is used to express user needs for the following reasons:

As indicated previously, the representative workload only approximates the actual expected workload. Since there may be other elements of the workload that will not be measured in the system timing determination, inclusion of selected desirable features allows the user to account for the additional capability provided by these features to satisfy the preceding additional requirements. In a sense, these features offer the user a "hedge" against uncertainty in his statement of the expected workload.

Since all measurement techniques have some uncertainty connected with them, the inclusion of a list of desirable features also serves as a hedge against this uncertainty in measuring a vendor's ability to meet the future workload.

Many times a user need which is called mandatory is really only a desire. For example, is a two-second response time for a task absolutely mandatory, or would 2.1 seconds be permissible for a system which is 20% less expensive? It is better to list nondiscriminatory design goals to the vendor in this category rather than in mandatory requirements, since elimination of a vendor for slight nonresponsiveness (at large decreases in cost) may not really be defensible.

As long as the production of computing equipment continues to be competitive with the introduction by different vendors of improved design features, the user is forced to consider the benefits to be obtained from these features. If the availability of these features will be of benefit to the user and if those benefits are not adequately incorporated in the system timing determination, some means should be provided in the selection process to account for them.

The Representative Workload. We shall now discuss a method for explicitly describing the representative workload. This description will be used by the vendors in preparing their system proposals and will be used as a basis for evaluating the performance of each vendor.

DESCRIPTION. In explicitly describing the representative workload, the analyst must deal with the uncertainties that may exist for each point of time in the future. This can be treated as two basic problems: first, there is the estimation of the user's workload for a particular future time including a quantification of the uncertainties in this workload expressed in probabilistic form; secondly, there is the estimation of how the workload may change with time in the future which may be based on extrapolations of current and past workload data. Note that we are using the same techniques described in Chapter 11 for modeling an uncertain system demand function.

Probabilistic workload description. To express this uncertainty analytically, the user is asked to define for a given point in time a set of the different workload elements which are apt to exist at the point in time, a reference workload, and to provide a quantitative estimate of the probability P that the actual workload occurring at this time may exceed the specified reference. This can be expressed by selecting certain multiples of the reference workload and having the user specify the probability that the actual workload at the selected time will be equal to or exceed each multiple of the reference workload. For example, in Figure 14-5 the user has stated that for a particular time there is a probability P_1 equal to 1.0 that the future workload will exceed the reference workload, and a probability P_2 equal to 0.80 that it will exceed 1.1 times the reference workload, etc. Several comments can be made about such an estimate:

1. These are the user's subjective estimates and may have to be justified to approving authorities such as the computer selection committee.

2. While theoretically there may be some small probability that the ac-

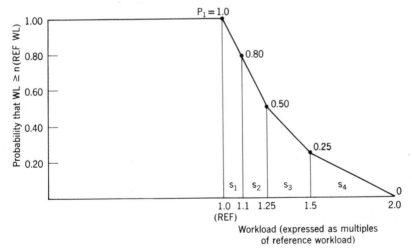

Figure 14-5.   Probabilistic workload description for a given time.

tual workload will exceed the upper bound shown (e.g., 2.0 times the reference workload), the probability can be taken as zero if it can be mutually agreed that this will be taken as the practical upper design limit for the EDP system.

3. We have described a situation where a workload may vary in size. The actual workload may also vary in complexity or in any fashion which the user chooses to make explicit and include in his estimate. The probabilistic description presented above can be used for each element of such a workload specification. The workload discussed below will consist, where appropriate, of combinations of such elements.

4. The units for expressing workload will depend upon the particular situation in question. In general, some common measure such as number of equivalent EDP problems will be used.

5. Since the user has provided the analyst in the example illustrated in Figure 14-5 with five estimates of workload levels, the entire workload range may be divided into four segments of interest (S_1, S_2, S_3, S_4) as shown in the figure. To obtain estimates for intermediate workload levels, the original four estimates have been connected by straight lines. If greater estimation accuracy is desired, additional estimates must be provided by the user.

Workload growth with time. As indicated previously, workloads change and generally grow with time. Hence a probabilistic estimate of each of the representative workload elements is needed for various periods of time. The summation of these data may be structured as a function of operational year, as shown in Figure 14-6. In the example shown, an estimate of the average yearly workload for each of the five operational years is shown in probabilistic form.

In Figure 14-6 each year contains five workload levels corresponding to the five probability levels selected in Figure 14-5. For example, the lowest line in any year represents that workload which the user has indicated to have 100 % probability of being experienced or exceeded. In other words, the user has specified complete certainty that his workload will be at least as great as the amount shown by this lowest line.

The lowest line for the first year in Figure 14-6 has also been labeled as the "reference workload." The user's specification of his reference workload is actually independent of the probability level he attaches to that level. In other words, the user will first specify his reference workload by whatever predictive tools he has available. Then, through an independent operation, he will assign a probabilistic level to that workload.

One final analytical factor should be made explicit. For the work element being described, the user states that the probability distributions of this workload element for the different years are independent, and this simpli-

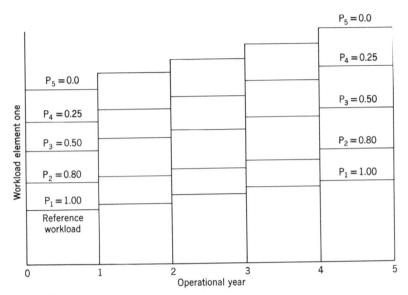

Figure 14-6. Probabilistic workload description averaged per year.

fies the analysis which follows. If the user felt that the distributions were dependent, probabilities could be assigned to different "paths" through time and the subsequent analysis would reflect the different paths available.

CALCULATION OF TOTAL EXPECTED SYSTEM COST. By constructing an explicit demand function (i.e., the probabilistic workload), the user has stated the range of possible workloads for each workload element which concerns him. The constant level of effectiveness on which we are pivoting is the ability to satisfactorily meet every one of these workloads. Hence every vendor must show that his EDP system can meet each of the workload elements up to the indicated maximum which the user may require. These means may include equipment expansion or replacement at a later time. It may also include the use of service bureau leasing or any other type of operation acceptable to the user.

Since the vendor will be provided with the user's estimates of the predicted workload, he may now perform various cost-performance trade off analyses, resulting in a proposal of an initial system installation together with system growth when and if the actual total workload reaches certain levels. The vendor costs for the proposed initial system and its growth will also be provided in the proposal.

Based on this information, the expected total cost of meeting the probabilistic workload can be calculated. The example shown in Figure 14-7 il-

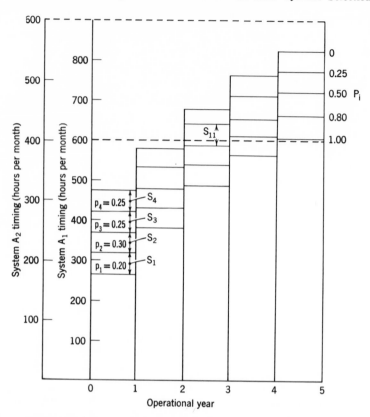

Figure 14-7.   Vendor response to perform representative workload.

lustrates the hypothetical response of Vendor A to the workload described in Figures 14-5 and 14-6. This vendor has proposed to install initially a System A_1 which can perform the monthly workload for the first two years of operation within a stated mandatory requirement of less than 600 hours per month (allowing the remaining time of 120 hours for scheduled and unscheduled maintenance as well as any necessary reruns of errors). However, during subsequent years, the probable increase in workload may exceed the 600 hours per month available on System A_1. In fact, based on the stated workload, there is a 100% chance that this will occur in year five, if it did not occur sooner. To cope with this increase, the vendor has proposed that his initial System A_1 be altered through addition or replacement to a new system (System A_2) which can perform the increased workload within the 600-hour limitation and which will be available when the user so directs. The validated timings for each system to perform the different workload levels are shown in Figure 13-7. The vendor also provides cost information indicating his proposed costs for all elements of each system (i.e., A_1 and A_2) as a function of system running time. Such cost information includes shift costs, if relevant, as well as lease versus buy information.

To these costs which the vendor provides, the cost analyst adds the costs which the user would incur in operating the system over the total operational life. Based on these total cost data, the total expected cost $\bar{C}_T$ for operating each vendor's proposed system can be calculated for each year from the formula:

$$\bar{C}_T = p_1 C_1 + p_2 C_2 + \ldots + P_n C_n,$$

where P_i is the probability that the actual workload will be contained in the segment S_i and incur a total cost C_i; and n is the total number of segments used to represent the workload range for that year. These segments can be determined by the analyst based on the user's description of each of his workload elements and the probability that any combination of workload elements will occur.

In the example we have selected, the probability that the actual workload will fall within any one segment S_i is found simply by taking the difference between the two cumulative probabilities that bound that segment. For example, the probability that the workload will fall within the segment S_1 is the difference between the probability that the workload will equal or exceed the reference workload (1.0) and the probability that the workload will equal or exceed 1.1 times the reference workload. Referring to Figure 14-5, the probability that the workload will fall within segment S_1 is $p_1 = 0.20$ $(1.00 - 0.80)$. Similarly, we could determine the probability p_i that the workload will fall within each of the other segments S_i.

These probabilities p $(i = 1, \ldots, 4$ in this example) can then be used in the preceding equation to determine the total expected cost for that operational year. The determination of the cost C_1 to be applied to each segment will depend upon the amount of information available to the analyst and the accuracy that the analyst requires in his calculation. For example, referring to Figure 14-7 for vendor System A_1, there is a probability of 0.20 that the actual workload will fall within segment S_1 (i.e., between 260 hours and 320 hours). If we can assume that the probability is distributed uniformly within this range of workload and that the cost is proportional to the workload, then we can determine C_1 as the cost for System A_1 to perform an average workload of 290 hours.

Similarly, it can be seen from Figure 14-7 that there is a probability $p_2 = 0.30$ that the actual workload will fall within segment S_2 (between 320 hours and 370 hours) for vendor System A_1. Again, assuming it is permissible to use the average workload, cost C_2 will be determined for vendor System A_1 to perform the average workload of 345 hours. In a similar fashion, probabilities p_3 and p_4 together with costs C_3 and C_4 can be determined. The preceding equation can now be used to determine the expected cost of operation for the first year $(p_1C_1 + p_2C_2 + p_3C_3 + p_4C_4)$. The same procedure could then be applied to each operational year to determine the expected cost for that year.

Several comments regarding this method should be made:

1. System discontinuities may occur inside a segment. For example, the vendor may indicate a shift from System A_1 to A_2 in S_{11}, the third segment of the third operational year. Hence to calculate properly the expected cost of the third year's operation, a separate calculation for each of the two subsegments must be made. In the above example, the probability that the workload will lie within each subsegment is determined from a further analysis of the cumulative distribution function of Figure 14-5, reproduced again in Figure 14-8. If an assumption is made that the cumulative distribution function of Figure 14-8 consists of straight line segments which connect the estimates provided by the user, then linear interpolation may be employed. For example, the probability that the actual workload is between 1.0 and 1.05 times the representative workload is 0.10. A similar breakpoint or subsegment can be used to represent other discontinuities that may occur in system costs such as might accompany shift changes. Again, linear interpolation can be used to determine the probability associated with each of the resultant subsegments.

2. Nonproductive time, previously discussed under system availability

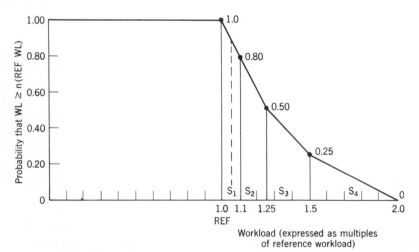

Figure 14-8.   Probabilistic workload showing linear interpolation.

and dependability, can be handled in several ways. The method previously described assumed that the reliability of all vendor systems cannot be differentiated from one another, since they will all meet certain minimum standards. Hence an arbitrary maximum productive time (e.g., 600 hours) may be chosen for all vendors, and the vendor system expansions can all be based on this upper limit. On the other hand, the evaluators can permit each vendor to calculate his maximum productive time, and use this figure for expansion design purposes. Obviously, this method will require validation of vendor claims, but this can be done by using vendor-claimed reliability as part of the contract, if the vendor agrees to do so. The burden of proof for such validation is still on the vendor.

3. Note that total system running time was used to describe each workload for which the corresponding cost element C_1 was determined. In an actual case, the time corresponding to the utilization of each equipment component in the proposed configuration would be determined depending upon the vendor's cost elements. Again, depending upon the information available and the accuracy desired, the analyst could introduce simplifying assumptions to keep the calculations tractable and commensurate with the evaluation model selected. Actually the expected cost calculation could be readily programmed for computer calculation.

4. After the expected cost for performing each year's workload is calculated, we must still combine this cost stream over time into one total expected cost. This can be done using the standard approach of obtaining the total present worth by reflecting each of these costs back to time zero and using an appropriate interest rate. In the same fashion, the lease versus buy calculation may be performed to determine which present worth is the lower cost. Thus in the case of leasing, the calculation is straightforward since the cost stream would consist of the series of expected leasing and operating costs, each at its particular point in time. In the case of purchase, we similarly have a cost stream of expected operating and maintenance costs, but, in addition, have expected purchase costs of the initially installed system and any expansion capabilities which we expect to add at later times. For example, it can be seen from Figure 14-7 that we would install System A_1 at time zero (with a probability of 1.0). However, System A_2 could be installed either at year 3, 4, or 5, with different probabilities for each. These probabilities may be obtained in the manner described in Chapter 11, where we calculated the probability of purchasing a bridge expansion in any given year; that is, the probability of purchasing the new system is equal to the product of the probabilities of needing the new system and not having purchased it in previous years.

BENEFITS. We shall now indicate some of the benefits of the proposed method of specifying the workload in probabilistic form as contrasted with a deterministic method of specifying workload. The deterministic procedure requires that the user provide one estimate of the representative workload for any given time rather than a "band" of estimates, as in the probabilistic approach. Thus the uncertainty is hidden rather than explicit. Under that procedure, a user is forced to insert some factor of safety in making his estimate (which may be unduly high) since there are pressures on him to provide service to his users.

Providing only one estimate of workload to the vendor in the RFP does not permit him to perform suitable cost-performance tradeoff analyses, since the vendor is not given any information indicating the worth of excess system capacity to the user. Providing the vendor with a range of values permits him to see the upper limit that has been set, as well as the estimated likelihood of reaching different workload levels. It thus permits the vendor to design a system capable of expanding to meet possible future growth requirements and to determine the worth of such an evolutionary system design in terms of the costs and expectation of using these growth increments. In this way the vendor can more effectively evaluate his alternative system configurations prior to submitting his proposal. This may reduce the number of alternative proposals which a vendor submits.

Using the proposed approach the source selection team can evaluate the vendor proposal in terms of its total expected cost. By including considerations of growth and determining their cost implications rather than asking the vendor if growth is available but not costing it, a more accurate estimate of the total cost of each vendor's proposed system can be obtained.

Mandatory Requirements. As discussed in reference to Figure 14-4, a second part of the selection process is the satisfying of the mandatory requirements. Each system can be readily evaluated against the mandatory requirements since, by definition, all systems must meet these or the vendor is considered nonresponsive. For this reason, when the source selection plan is constructed, the list of mandatory requirements should be limited to those characteristics which can be firmly defended on a "go/no/go" basis. Any feature which the user desires, but cannot firmly defend, should be categorized as a desirable feature.

Desirable Features. As discussed before, the evaluation team must also consider a set of desirable features as a source of additional vendor capability that was not adequately covered in the system timing; this will act as a hedge against uncertainty in the user's statement of his expected workload and as a hedge against the evaluator's uncertainty in measuring the vendor's capabilities.

PROBLEM STATEMENT. There are several reasons why the problem of evaluating desirable features is a much more difficult one than handling the first two elements of user needs (i.e., representative workload and mandatory requirements). First, it is almost certain that each of the vendors will submit a different "mix" of desirable features, ranging from none at all to all of the features requested. However, even though two vendors submit the same feature, each may have a different level associated with it, such as one billion versus two billion characters of IAS (Intermediate Access Storage). Thus, the first problem concerns how to quantitatively measure the effectiveness of the combination of desirable features which each vendor offers. The evaluator can do this by attempting to determine the benefits to the user jobs which each feature contributes and then taking into account the interrelationship of several features as they contribute jointly to the accomplishment of user jobs. In developing a method for evaluating a particular feature, a way must be found to relate the characteristics of that feature to the jobs whose performance will be benefited by it. In general the direct effect of a system feature is felt in the time (system and/or staff) or quality of performing the jobs. If it can be determined that the effects of a feature have been adequately covered in the estimates of system timing previously calculated, then that feature need not be considered separately in the list of desirables. If this is not the case, then specific steps must be established for evaluating the feature.

Even if the evaluator could solve the first problem of evaluating the benefits contributed by each desirable feature, he still faces the problem of determining if the difference in effectiveness among vendors is worth the difference in cost. Figure 14-9 illustrates in simplified form this problem of source

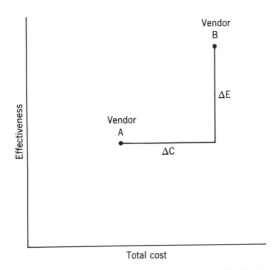

Figure 14-9. The problem of evaluating desirable features.

selection with respect to desirable features. Consider two vendors, each of whom performs the future workload at the same expected cost. Assume that vendor A has proposed a minimal EDP system containing none of the desirable features listed in the RFP, but that vendor B has provided one of the desirable features as part of his proposal at a cost ΔC greater than vendor A's. Assuming that the performance of both machines is identical in all respects except for this desirable feature, we could state qualitatively that the effectiveness of vendor B's system is greater than vendor A's. However, is the increased effectiveness worth the additional cost of ΔC?

As indicated previously, the fundamental principle employed in the evaluation is to pivot on some constant level of effectiveness and to choose the vendor who provides this level of effectiveness at lowest cost. The inclusion of desirable features would appear to make it difficult to define a constant level of effectiveness. But this is not so since the user has stated that, in addition to accomplishing a certain representative workload, he *desires* that additional features also be provided. The solution to this problem can be found in the realization that if the user desires these features for meaningful reasons, then he must expect to have certain jobs which will benefit from these features. However, since the user has not made the provision of these features a mandatory requirement, he is implying that there must be alternative ways of accomplishing these jobs if the desirable feature is not available. This information enables the analyst to choose the proper level of effectiveness for selection purposes. This will be the level of satisfactorily performing the *entire approved set of user jobs* in accordance with approved standards of performance. Thus it should be emphasized that while each desirable feature may be optional, it is mandatory that the entire approved set of user jobs be done (with or without the desirable feature), while satisfying all the previously defined constraints. Since each desirable feature contributes to some of these jobs, there now arises a question of comparing the proposed cost of any desirable feature against the cost of other alternatives which can be used to do the same job(s). This approach to system selection translates the task of evaluating desirable features to one of cost analysis, and leads to the concept of the worth of a desirable feature in doing a job.

The term "worth" has been subjected to diverse economic interpretation. For our purposes we will define the worth of a feature in doing a job as the lowest incremental cost to do the same job if the feature is not available. If the vendor's cost is less than the user's worth, then that feature will be acquired from the vendor and the cost will be added to the total system cost. If the vendor's cost exceeds the user's worth or if the vendor does not provide the feature, then the user will make use of an alternative and add the corresponding cost to the vendor's total system cost. If the vendor's cost for the feature is not separately identifiable, then there is no way to determine if his cost exceeds the user's worth and his proposed cost will not be changed.

This process for evaluating desirable features will be illustrated by an example later. A key element in this process is the determination of the worth of a desirable feature.

METHODS OF EVALUATING WORTH. We can distinguish between two basic ways for determining the worth of the desirable feature. One method makes use of analysis; the other makes use of comparative ranking. Each of these methods is now to be examined.

Analytical Determination of Worth. Since the worth of a feature in doing a job has been defined as the lowest incremental cost to do the same job if the feature is not available, then to determine the worth, one must first identify the alternative ways acceptable to the user of doing the job if the feature were not available. As indicated in Figure 14-10, there may be several ways of doing the job without using the feature. The cost of each alternative should be determined; then, making use of the probability that the associated job will be performed, the total expected cost over the operational life

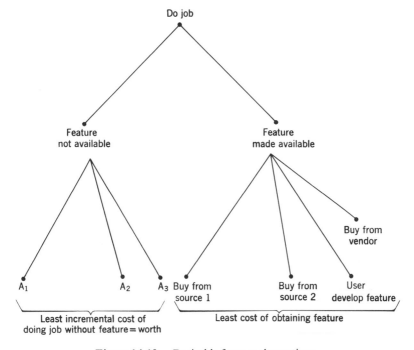

Figure 14-10. Desirable feature alternatives.

of the system should be determined for each of these alternatives. The worth of the feature will be the least of these costs.

In deciding on whether to utilize a particular desirable feature, the evaluator must also consider the cost of alternative ways acceptable to the user of obtaining the feature and doing the job using the feature. For example, he might buy the feature directly from another source. Alternatively, as in the case of a software feature, the user might develop the feature using his own resources (in-house). The least cost of obtaining the feature can also be found. If possible the evaluator should make his decision based on doing the job(s) for which the desired feature is intended at lowest total cost. This might be called his "efficient solution."

Thus the efficient solution may be chosen by determining the lowest cost method of obtaining the feature (and doing the job), comparing this cost against the worth (cost of doing the job without the feature), and choosing the lowest cost alternative. This process is illustrated in Figure 14-11. This figure shows the cost-effectiveness of two proposals, both of which perform the same basic workload and satisfy the mandatory requirements. These two proposals are assumed to be identical in all respects, differing only in that one provides a desirable feature F at a total system cost of C, whereas the

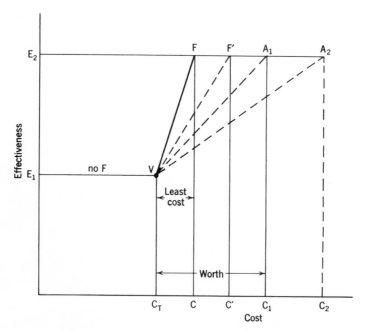

Figure 14-11. Selection of least cost alternative.

other does not. It is immaterial to the present discussion whether these proposals come from the same or different vendors.

While the cost axis of Figure 14-11 can be quantified in a unidimensional scale, the effectiveness axis may involve a number of dimensions to represent the elements of effectiveness. However, since we are pivoting on the constant level of effectiveness specified by the analyst/user, we can represent this level diagrammatically. This means that we are insisting that those user requirements which would benefit from the availability of the feature, be satisfied by some other means if the vendor does not provide feature F. Analyzing the alternatives available to the user results in alternatives A_1 and A_2 with their respective costs as shown. Thus, by our definition, the worth of the feature is the incremental cost of providing A_1 (i.e., $C_1 - C_T$).

However, it should be noted that once the worth of a feature is determined by analyzing the cost of a number of different ways of getting the job done if the feature is not used at all, this worth must be compared with the cost of all alternative ways of obtaining the feature and doing the job using this feature. Assume in the example that the same feature is available from one other source, labeled F', and that the job could be done with this feature at a total system cost of C . The proposed strategy is to choose the lowest method of getting the job done, whether by acquiring the feature or using a lower cost alternative. Hence, in this case, feature F would be purchased.

Determination of Worth by Comparative Ranking. Sometimes, because of time and manpower limitations, it may not be possible to determine the worth of all desirable features by analysis and considered judgment. In such cases, intuitive judgment can be used as a part of the quantitative evaluation. Such an approach would be implemented as follows:

1. Rank all of the desirable features in order of importance. The ranking should be supported by deliberation utilizing whatever quantitative analysis and considered judgment may be available.

2. Allocate points to each feature, establishing its relative worth. Such relative worth will be based on the rationale developed.

3. Translate points to dollars of worth. This is accomplished by calibrating one or more of the features through determining its worth on an analytical basis.

4. Review the results obtained for intuitive soundness, which is the only real test of this procedure. If the final results of dollar worth of each feature do not agree with the intuitive feelings of the evaluators or source selection plan reviewers, an iteration of the previous three steps should be performed, focusing on the following two potential sources of error. First, should the relative worth (i.e., points assigned) be changed? Second, are there ways of obtaining the features in question at a lower cost than the worth assigned to

the feature? Obviously the more features that are calibrated by analysis, the more accurate will be this procedure.

This determination of worth by comparative ranking is really a modification of the point-scoring approach, and the reader may wonder why it is permissible to use this here and not for the entire evaluation. Here, the analyst is confronted with the practical tradeoff between accuracy of results and analytical resources available to do the job of evaluation. On the one hand, the EDP specialists and the project leader feel that the desirable features should be included in the evaluation, but they cannot supply sufficient resources (of time and manpower) to perform the suggested analysis. Since evaluation of desirable features generally plays a relatively small role in the total evaluation (e.g., EDP specialists using the point-scoring system generally allocate only 10% to 15% of the total number of points to desirable features), it was felt that the evaluation procedure is more rational and defensible with the features included (even at the introduction of a relatively small error), than when they are not considered at all.

OTHER EVALUATION ALTERNATIVES. The most credible way to evaluate a desirable feature is to design a live-test demonstration or a simulation test that will include the effects of that feature in accomplishing certain jobs. In this way, the results of the test will provide an explicit quantitative measure of the benefits of that feature. If these results can be incorporated into the overall system timing, then the particular feature need not be given any further separate consideration. If this is not the case, then the results of the test may be used in an evaluation by worth.

One of the practical constraints in an actual source selection, however, is the cost and effort expended in the live-test demonstrations. This means that the number of tests must be kept at a minimum with each test designed to serve as many testing functions as possible.

Under special circumstances, the following two evaluation alternatives may be justified:

1. *Establish design goals.* The user may wish to establish a certain level of hardware or software performance that is characteristic of the present generation of equipment. If it is difficult to express the system requirements or to design the live-test demonstration in such a way as to rule out the proposal of equipment considered by the user to be substandard, then it may be desirable and justifiable to specify the feature as a nondiscriminating standard or design goal which all qualified vendors can be expected to meet. For example, specifying the level of performance of a card reader, card punch, or printer might be justified in this way. If this requirement is discriminatory

among the competing vendors, it would be necessary to support it more carefully in terms of system requirements.

2. *Qualitative evaluation.* If the worth of a desirable feature cannot be evaluated quantitatively by any of the above techniques, a qualitative evaluation should be made and documented for consideration by the source selection authority. Such qualitative factors would only be considered and used as "tie-breakers" if several vendors were sufficiently close, based on the quantitative evaluation.

EXAMPLES OF EVALUATING DESIRABLE FEATURES. The following examples are offered to illustrate how desirable features might be evaluated. It should be emphasized that these examples are only representative. Actually such features must be considered in the context of the user's system requirements.

Example 1: Additional Core Storage. The user may feel that additional jobs not included in the representative workload may occur which will require additional core storage over and above that provided by the vendor in meeting the basic workload. While the best way of measuring this feature would be to include a job requiring large amounts of storage as one of the workloads, it may not always be possible to do this. Hence, the evaluator must explore ways of performing the job if the additional core storage were not available. One way of doing this would be to segment the job into smaller parts and determine what this will do to system costs. First, programmer time will increase due to the additional programming load. Second, the system running time will increase due to the lower efficiency of the operation. Both of these will lead to increased machine and staff time. The size of the increase will depend on the complexity of the job and the frequency of its operation. If no other alternatives were available, the cost of segmentation would be the worth of this desirable feature.

The evaluator must also consider alternative ways of obtaining the feature. For example, it may be possible to contract the jobs requiring additional core storage to a service bureau or some facility equipped to handle it, if this is satisfactory to the user. In this case, the resulting costs would be estimated and the least cost of all alternatives would be determined.

Example 2: Software Feature. If the job needing the feature has been included in the live-test demonstrations, evaluation of the feature is implicitly included in the system timing obtained and another evaluation is not needed. If the feature is not included in system timing, nonavailability of the feature will most likely affect the programmer hours required to develop and maintain the system's programs. Programmer hours would be affected since the programmer would now have to do additional programming to compensate for not having the software feature at all. One way to handle this would be for the analyst to determine the total number of programming hours which would be required to do the programming if the software feature were available at some standard reference level. Based on this reference level, the term "programmer performance factor" can be defined in the following fashion:

$$\text{Programmer Performance Factor} = \frac{\text{Reference Time}}{\text{Proposed Time}},$$

where the reference time is the total estimated programmer time (in man-hours) taken to program a particular job (using this reference level of software capability for the feature in question) and proposed time is the total estimated programmer time taken using the vendor proposed level of that feature. Using either live-test demonstration or simulation results, or the judgment of the evaluators in estimating the capability of a software feature to perform a certain class or classes of jobs, the evaluation function illustrated in Figure 14-12 can be constructed. Making use of this evaluation function and the total programmer hours estimated by the analyst to be required if the software feature were available at the reference level, the programmer hours required for the proposed level can be determined. For ex-

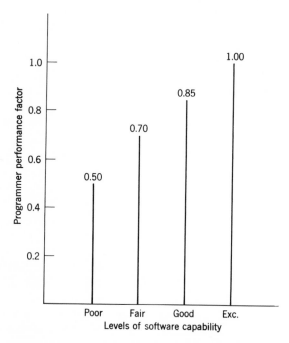

Figure 14-12. Software feature evaluation function.

ample, if the analyst estimates that for the reference level of the software feature, the programmer time would be 3000 hours, and if he determines that the proposed feature has an efficiency of 85%, then the estimated programmer hours required by the vendor's system is given by:

$$\text{Estimated Programmer Hours} = \frac{\text{Reference Programmer Hours}}{\text{Programmer Performance Factor}}$$

$$= \frac{3000 \text{ hours}}{0.85} = 3450 \text{ hours}.$$

By subtracting the cost of the estimated programmer hours from the least costly method of programming the job(s) if the desirable feature were not available, the worth of the feature can be determined.

Alternative ways of obtaining the software feature must also be considered. For example, it may be possible for the user to develop the feature in-house, requiring additional programmer and machine time. Alternatively, he may turn to other software sources and purchase the feature as a package. The least cost of these alternatives would then be used along with the worth in the selection process.

Example 3: Documentation. Again the question can be asked, "How much does it cost the user if the user is forced to use inadequate documentation as opposed to excellent documentation?" The added cost might be the extra time required for readers of the documentation as they struggle to understand what the author had in mind. Thus, documentation may be evaluated using the same concept of efficiency as described previously, and calculating the larger number of staff hours required, based on the lower efficiency factor. If possible, the analyst might include in this determination of worth some measure of the costs incurred due to system malfunction that might occur because of the inadequate documentation. As before, the analyst would also want to consider alternative ways of obtaining excellent documentation. The least cost of these alternatives would then be used along with the worth in the selection process.

Note that for this example the analyst might prefer for various reasons (such as economy) to use an alternative evaluation procedure. He could handle documentation as a design goal by requiring in the RFP that certain standards of documentation be satisfied. In this way the feature becomes a mandatory requirement. Alternatively, the analyst might choose to process vendor differences in documentation qualitatively by noting the differences and including the relevant information for consideration only as part of a tie-breaking procedure.

REMARKS. The determination of the worth of a desirable feature by constructing an evaluation function which relates the performance benefits of the feature with cost is the most satisfactory way of evaluating a desirable feature if its effects cannot be directly included in system timing. It should be emphasized that the judgment of experienced personnel will be required to construct the evaluation functions. In fact, the accuracy of the analysis is only as good as the experienced judgment and substantiating data available. Undoubtedly, the most reliable substantiation would come from benchmark tests. While errors in judgment are never completely avoidable, there are two compensating features to the above approach:

1. This type of analysis forces the user and evaluator to think through and develop the rationale for the need of desirable features; hence, it is superior to a purely intuitive judgment approach since it is more defensible.

2. The rationale and evaluation functions developed are made explicit and are thereby subject to review by the source selection authority. Thus they can be changed if additional information is available.

Selection Approach

We have indicated how to evaluate vendor proposals with respect to workload, mandatory requirements, and desirable features. We shall now expand upon the steps to be followed in evaluating the vendor proposals for their desirable features and in making a final selection using all of the data gathered. To illustrate the approach, a simplified example is used.

Determining the Worth of Desirable Features. The source selection plan approved by the Source Selection Authority will include a list of all desirable

Cost Elements							System Cost		
							C_A	C_B	C_C
1. Total proposed vendor cost							300K	310K	330K
2. Expected cost to do representative workload C_T							300K	305K	310K
3. Cost of additional job benefits:									
	Desirable Feature	User Worth	Least Cost	Vender Cost					
				C_A	C_B	C_C			
	F_1	10K	15K	—	incl.	15K	10K	—	10K
	F_2	25K	20K	—	—	5K	20K	20K	5K
	F_3	10K	20K	—	5K	incl.	10K	5K	—
4. Total expected cost to do user jobs							340K	330K	325K*

* Vendor C selected—lowest total cost.

Figure 14-13.   Evaluator's worksheet.

features to be quantitatively evaluated, the dollar worth (or evaluation function which describes such worth) for each feature, as well as the lowest known cost of obtaining the feature separately. An example of the worksheet to be used in the evaluation (which can be constructed as an appendix to the selection plan) is shown in Figure 14-13. This figure corresponds to an example in which there are three desirable features to be considered (F_1, F_2, and F_3) whose user worths and least costs are indicated. In the example shown, each feature is either provided completely or not at all. If various levels of a feature could be provided, the evaluation function showing worth as a function of level provided would be used instead of the single number.

Vendor Submits Proposed Costs. The proposal submitted by each vendor provides the following information to the evaluators:

1. Total proposed cost for entire system.
2. Cost of each system and expansion capability required to meet the probabilistic workload.
3. Sufficient information to calculate the total expected cost of performing the probabilistic workload.
4. Cost of each separate desirable feature not included as part of the basic system.

Calculating Cost of Representative Workload. Utilizing the vendor-supplied information, the evaluator calculates the total expected cost of each vendor's system to perform the total representative workload. These results are then entered into the evaluator's worksheet as shown in Figure 14-13.

Validation of Mandatory Requirements. Based on the vendor-supplied information, the evaluator must validate that the mandatory requirements have been satisfied.

Calculating Cost of Additional Job Benefits. The evaluator inserts into the evaluator's worksheet all of the desirable features which each vendor has proposed and the incremental costs associated with each of these options. Note that vendor A does not provide any of the three features, whereas the cost of F_1 and F_3 are included in the costs of vendor B and vendor C, respectively. Based on the cost information of Figure 14-13, the evaluator can determine for each vendor the least costly of the three alternative ways of receiving the benefits provided by each of the desirable features. These three alternatives are:

1. Buying the desirable feature from the vendor (at the vendor's proposed cost).

2. Obtaining the desirable feature from another source (at the least cost of feature if obtained separately).

3. Not buying the feature, but using the least costly alternative way to provide the benefits (at a cost equal to user worth).

The lowest additional user cost for obtaining the desirable feature (or its equivalent) is shown in Figure 14-13 as system cost. Note in the example that the user has stated that the worth of F_1 is $10K (i.e., he can perform the jobs associated with F_1 at an expected cost of $10K). Since vendor A does not provide this feature, the user will be forced to spend $10K in addition to vendor A costs to meet those jobs associated with F_1. Vendor B includes this feature as part of his basic system and has stated that it cannot be removed or priced separately. Hence, the user will not have to spend the $10K when using vendor B's system. Vendor C can provide F_1 at a cost of $15K. Hence, the evaluator decides to eliminate this optional feature from vendor C's proposal since its cost is higher than its worth to the user (i.e., the cost of an alternative method for the user to perform the related jobs). This same approach is followed in determining which of the other desirable features are to be included in the evaluation.

Calculating Total Expected Cost. The total expected cost to the user is then calculated by adding the cost of each desirable feature (or user cost equivalent) to the expected cost of performing the probabilistic representative workload. This total cost, shown in Figure 14-13, completes the cost calculation.

Several significant observations can be made from analyzing this illustration:

1. Vendor A had the lowest proposed cost (since he provided no desirable features) as well as the lowest cost of performing the representative workload. On the other hand, winning vendor C had the highest proposed cost (since he had proposed all three desirable features) and the highest cost of performing the representative workload. However, neither of these costs is the proper measure for selection. If one believes that the user really does have need for the additional capability represented by the list of desirable features, and that he will have to spend additional funds (i.e., the worth) if a desirable feature is not provided, the true criterion of choice must be based on the total system costs. There were two reasons why vendor C had the lowest total cost in spite of his other higher costs. First, he included F_3 at no additional cost, and this was worth $10K. Second, he provided F_3 for $5K and the evaluators estimated its worth to be $20K.

2. With this approach, there are definite advantages to the vendor to separate as many desirable features as possible from the basic system and pro-

vide these as optional cost features at a stated price for each feature. The reason for this is that if the calculated worth of each feature is not stated to the vendors (and it should not be, since this information may affect the vendor's price), the vendor has no logical way of determining whether to propose a desirable feature or not. Hence, he is forced to hedge his bets by submitting alternative proposals which may increase the vendor's proposal costs and the evaluator's selection costs. With the proposed procedure, however, the vendor knows that the evaluator will choose only those desirable features which have value to the user and reject those whose costs are too high. Hence, the vendor will feel free to offer a "shopping list" of optional desirable features, each at a separate price, as part of his proposal, knowing that he cannot be penalized by this strategy.

In the preceding illustration, if vendor C had included the high cost of F_1 as part of his basic system, his total cost would have been higher and he would have tied with vendor B.

CONCLUSIONS

The method described provides not only a means of evaluating an EDP system, but also a means of describing the system specifications in such a way that each vendor will know precisely what type of system is desired and can design his system in accordance with the objective function provided. The method offers the following implications:

User Implications

This evaluation procedure will permit the user to acquire a cost-effective EDP system. It should be emphasized, however, that additional analysis and data will be required from the user, relating system specifications to its expected use, if this procedure is to be implemented. Such data will consist of a representative workload expressed in probabilistic form (as agreed upon by the user and the Source Selection Authority), a set of defensible mandatory requirements, and a set of desirable features together with the worth to the user for having each feature, as determined by the user and the evaluator. By stating his data system needs as an explicit system demand function (including his forecasting uncertainties), the user can better defend the characteristics of the computer which he feels he needs.

Vendor Implications

The system specifications provide the vendors with a more useful statement of user needs, and the selection criteria to be used allow the vendors to construct a better system design by performing more and better cost-performance tradeoff analyses, based on better information. In addition, by permitting the vendor to propose optional features, some of which will be selected by the evaluators on the basis of the cost being less than the worth, the vendor may not have to propose separate, alternative proposals, each containing different combinations of desirable features.

Evaluation Implications

From the evaluator's point of view, the proposed approach is for several reasons more defensible than other approaches examined. First, it is operationally oriented; hence it is more rational and should be more understandable to reviewers. Second, it avoids combining cost and performance factors, which is always difficult to justify, in favor of choosing that system which will satisfy approved user needs at lowest total cost to the government purchaser.

Since it is more explicit and rational than other procedures examined, it offers a means of resolving differences of opinion regarding the worth of system features. It should be stressed that the overall evaluation framework that has been developed does not eliminate the need for vendor validation.

Approval Authority Implications

By providing the approval authority with the system demand function in explicit operational terms, he is now in a better position to validate the system requirement and alter it if he disagrees with it. A second reason for altering the specifications is operative when a preliminary analysis of these specifications indicates that the expected cost of the system will far exceed the procurement funds available, and thus some of the jobs must be removed from consideration.

The proposed approach is consistent with procurement practices which compare cost and benefits received. In addition, by quantifying the uncertainty in workload, the major factor in the evaluation, rather than using a single deterministic estimate, increased confidence in the final selection is obtained.

Applicability to Other Source Selections

One last point should be made to those readers who have an interest in source selection of systems other than EDP systems. It has been shown that the general principles of cost-effectiveness analysis which have been applied so often to the concept formulation or systems planning phase of the systems acquisition process can also be applied to the source selection process, specifically of EDP systems. The same approach can also be applied to other type systems. In fact, it may be easier to apply this approach to other areas where the measure of effectiveness is more easily defined. Another example of a systems evaluation procedure which also contains the element of development uncertainty is contained in Chapter 15.

15

Dealing with Development
Uncertainties

In Chapter 13 we discussed methods for designing and evaluating a system
or subsystem based on its set of estimated performance characteristics and
costs, as well as system delivery time. Presumably the system designer will
not permit consideration of a system alternative which cannot be delivered
on or before some desired delivery date. The system description is then forwarded to the contracting agency in the form of a proposal, where it will be
evaluated along with all other proposals received by an evaluation team.
While it is possible to check each vendor's calculations of performance
characteristics, system effectiveness and cost, there is always the question
"Do you, the evaluator, believe the vendor's claims?" or "What is your confidence in his predictions of performance, cost, and delivery date?" particularly when there is any system development involved. While Chapter 14 described an evaluation method for off-the-shelf equipment, we shall now
consider the problems of including developmental uncertainties in the systems evaluation as in the evaluation of a system development or even manufacturing proposal. This factor is particularly important from the contracting agency's point of view because, while the contractor may not receive full
payment if he fails to deliver a system which meets the contract specifications on the delivery date, the user would not have the expected system and
there may be great penalties or risk connected with this.

DEPARTMENT OF DEFENSE EVALUATION AND
SOURCE SELECTION

It should be emphasized that the main reason there is uncertainty in what
the contractor will deliver is that he still may have to develop, design,
and manufacture the equipment. Hence one approach which is sometimes
used to greatly reduce this uncertainty is to insist on off-the-shelf equipment

only, such as in the procurement of electronic data processing equipment for most government installations. Since the system components must be available prior to the procurement announcement, they can be combined into a system and its operation tested on some representative jobs which may be typical of those encountered in future operating years.

Of course, sometimes such restrictions to off-the-shelf equipment cannot be made, particularly in defense systems acquisitions. Hence, a second approach is to divide the entire procurement problem into a series of sequential procurement decision problems, as was done in the DOD system acquisition process described in Chapter 3. These separate phases are as follows.

Concept Formulation (the systems planning phase)

This is the initial step in showing feasibility, expected system performance and effectiveness, development and operational schedule, and the cost-effectiveness of alternatives. Here alternative system concepts may be compared by configuring all systems for equal effectiveness and then calculating the costs and delivery dates of the different system alternatives. Since no particular contractor is involved in providing equipments, the best estimates of performance, cost, and delivery date are inserted, based on judgment and contacts with appropriate vendors in the different areas. Sensitivity analyses are conducted to cope with the various uncertainties involved.

Contract Definition

This is also an analytical phase in which two or more competing vendors perform a more detailed analysis of the system. Uncertainties are coped with in two fashions. First, this phase can begin only if the technological feasibility has already been established; this helps reduce but not eliminate the uncertainties in system performance. Second, the output of the phase is a proposal for a fixed price, or fixed price incentive, engineering development program. This helps to make the cost estimates more realistic. While this phase does provide more detailed information than the systems planning phase, the evaluator still has the problem of evaluating and selecting one of the vendors involved, and hence must evaluate the credibility of each of the proposals. A fixed price contract may guarantee costs but does not guarantee performance characteristics or delivery date.

Engineering Development

The same comments made for contract definition apply here, only more so, since even more information is available regarding the risks involved. Having divided the entire program into a series of phases, the decision-maker is now in a better position to evaluate whether sufficient information is available to predict the risks involved in continuing the program into the next phase. While the costs increase with each succeeding phase, the risks should not proportionately increase since more information is available with each completed phase; hence there is less uncertainty about final system performance, delivery date, and cost.

Source Selection Criteria

Now given a proposal for any one of the systems engineering phases, the vendor's ability to deliver what he has proposed must be evaluated. In DOD source selection, evaluation teams are directed to consider not only the factors of performance and cost, but also many other factors which are addressed to evaluating the degree of confidence in each contractor's proposal. For example, Air Force Manual No. 70-10, "Procurement, Source Selection Board Procedures," contains a list of criteria to be employed in a system evaluation. This list is reproduced in Appendix II as a useful checklist for the evaluator. Some of the factors which the evaluation team is directed to consider include:

1. *Producibility*. If certain components of the systems are not now in production (or even designed for production as yet), there exists some element of uncertainty in the contractor's ability to deliver final hardware having the proposed performance characteristics.

2. *Maintainability and supportability*. Will the various elements of the logistics support system provide the performance characteristics proposed?

3. *Management capability*. Will the contractor exert sufficient management control to provide the proposed performance, particularly when technological or other problems arise?

4. *Cost realism*. Are the proposed costs credible enough to provide sufficient confidence that the system can be delivered at the proposed cost?

5. *Systems analysis and integration*. Does the contractor have sufficient understanding of the problems involved, and does it appear that the approach he has indicated can solve these problems and produce the proposed performance characteristics?

While the preceding discussion is directed toward the ability of the contractor to deliver the proposed performance characteristics, we emphasize that implicit in this are the other two constraints (i.e., by the proposed delivery date, and at the proposed cost). Obviously the chances of meeting the performance increase as the cost and delivery date limits are extended.

One procedure for combining these criteria in the evaluation of a contractor is through the point system evaluation procedure, in which relative weights based on subjective estimates are assigned to each of the criteria to be considered. And while it is difficult to defend a point-scoring procedure which combines performance characteristics and cost, it is even more difficult to defend the use of points for evaluating all of these other factors which are treated independently of one another and quite often evaluated by independent groups. For example, the technical evaluation group may give points to a vendor for recognizing a problem area. However, this evaluation group may not see the management plan or cost proposal, each of which may indicate how much resources the vendor is planning to assign to this technical problem area. On the other hand, it is important that these factors be included in the evaluation; hence a framework must be developed which will permit their inclusion on some rational, defensible basis.

THE USE OF THE OPERATIONAL FLOW MODEL IN SOURCE SELECTION

We shall now consider an alternative evaluation method in which the key measures in evaluating a proposal are:

1. System effectiveness as obtained from the set of performance characteristics provided by the vendor, and the given environmental characteristics.
2. System delivery date.
3. System cost.

The role of the systems analyst prior to the evaluation should be to try to establish relationships which indicate the effect of various factors being considered on these three key measures. The analyst may be able to do this using available data (such as cost overruns by a vendor on past related projects). If the data are not available, subjective estimates of the measure should be made, accompanied by a statement of the logic used by the evaluator. In this case the evaluation uncertainties may be reduced by using a comparative method of evaluation (i.e., having the same evaluator provide an evaluation of all vendors for a given factor or a relative basis).

Potential problem areas which may cause one of the three evaluation measures to be lower than that proposed should be examined by asking:

1. What is to be delivered (i.e., performance characteristics)?
2. When will it be delivered?
3. What are the problem areas the contractor has recognized?
4. How does he propose to take care of these areas?
5. What resources (manpower, time, etc.) is he scheduling in these problem areas?
6. Is there much slack time or resources available in these problem areas?
7. What has been the contractor's past performance in these areas?

In any one technical area, all of these factors should be examined and evaluated by the same technical experts so that the interrelated factors described above can be simultaneously considered.

We shall now discuss how this information might be used under different contracting procedures available, so that the appropriate one can be chosen for the particular procurement involved.

Cost Plus Fixed Fee (CPFF) Contract

Under this method of procurement, there is a high likelihood of obtaining the proposed performance and delivery date, but at the expense of additional cost which would be used to increase the resources used in the problem areas as they arise. This method of procurement is almost "open-ended" in the cost dimension, since the only efficiency incentives to the contractor are that low performance on one contract may be taken into account on future contract awards, and since the fee is fixed, the percent return on investment may be reduced as additional time is expended. Even this may not be a satisfactory cost incentive in the case of a contractor who is short of other work and may wish to keep his development force on "applied time" instead of "unapplied time" as they finish the job, thereby reducing profits even further. Hence, the performance characteristics, delivery date, and costs expected could be evaluated as follows:

1. Assume a fixed delivery date (unless there is reason to believe that a particular vendor may "stretch out" the delivery date).

2. Determine a set of system performance characteristics which the technical evaluators feel the vendor will achieve by this delivery date. Each performance characteristic would be either equal to that proposed or lower if the technical evaluator feels that there are problems that the vendor would probably not overcome by the given delivery date.

3. Estimate the resulting costs to achieve this given level of performance. Again this total cost would be equal to or greater than that proposed, depending on the problem areas involved.

After this is evaluated for each vendor, the analyst could combine these expected values of system performance and cost in the operational flow model, to determine the total expected cost of performing the mission (i.e., achieving a fixed level of effectiveness) and recommend a selection based on lowest total cost (assuming the same delivery date for each vendor). If the expected delivery dates among the various vendors are found to differ, the source selection authority will have to decide intuitively whether a delivery slippage is warranted by either the improved performance or cost savings expected.

Fixed Price Contract

While this contracting approach may be considered an improvement over the CPFF approach since it contains a guarantee that the total cost will not be exceeded, there is still no real guarantee on what performance characteristics or delivery date will be achieved. If the contractor runs into technological or other problems, he is reluctant to increase total resources very much over those originally proposed, since the costs involved will come out of his profits. Many times a time lag occurs before a problem area which will affect the schedule is detected. Since the contractor does not wish to spend the premium costs associated with overtime, a schedule slippage may occur. When the situation is detected by the contracting agency's program manager, he is just about confronted with a *fait accompli* that is difficult to handle. What are his choices? He does have legal powers on his side; hence he need not accept the system if it does not meet the contract specifications, and the contractor will not receive the full remaining costs, over and above any partial payments already received, until he finishes the work according to specification. On the other hand, the user will not have his system on the predicted delivery date and this can result in certain losses from the other complementary elements, such as idle personnel. Perhaps the most important loss is the reduced mission effectiveness over that which was planned. Hence there is great pressure on the contracting agency's program manager to agree to a contract negotiation to include a reduction in certain performance characteristics (and cost) to minimize the delivery lateness.

How can such a situation be predicted and evaluated properly during the initial proposal evaluation? Under a fixed price contracting arrangement, the performance characteristics, delivery date, and costs expected can be evaluated as follows:

1. Total cost can be considered fixed (at the vendor's proposed price) because of the strong pressures not to overrun cost.

2. A fixed delivery date can be assumed due to the user pressures.

3. Now the evaluator must estimate each of the performance characteristics he feels the given vendor will attain under the above constraints of time and cost. In this case, each performance characteristic predicted will be equal to or less than that proposed, depending on the technical specialist's appraisal of the vendor's ability to cope with technical problems within the time and resource constraint. While the technical evaluator looks for many different kinds of information of the form already mentioned (see also Appendix II) to aid him in his estimation of the final performance characteristics predicted, the burden of proof is on the vendor to show how he intends

to solve the problems expected. This problem of estimating the expected performance by attempting to validate each vendor's proposal claims is not unlike that currently used in the technical proposal evaluation process. For example, consider the case of a vendor who claims his radar system will detect targets at a range of 200 miles and supports this with range equation calculations. If an error is found in the arithmetic calculation, such that the range is now 180 miles, this is used as the performance characteristic. If technical negotiation is permitted, this new information may be reported to the vendor who then has a choice of improving the design to achieve the 200 mile range, or leaving it at 180 miles. His decision would be based on his cost-performance tradeoffs.

In a different case, the vendor proposes the use of a new antenna design which he claims will achieve an antenna efficiency of 80%, but the evaluator does not believe it will be this high when the antenna is produced, again resulting in a possible decrease in radar detection range. Here, the evaluator can request more information from the vendor, such as experimental test data, prototype production models constructed and tested, etc. In the final consideration, the evaluator would have to make a judgment of the performance expected based on the information provided him.

When evaluating a technical area where some engineering development or manufacturing difficulties have been anticipated by the vendor, the evaluator can inspect the people and facilities available for the job, as well as the past performance of the vendor in this area. However, even here there is some uncertainty in the estimation, since the vendor can always later change the allocation of resources from that proposed. Also while past performance in a given area is a good indication of the future, the vendor may change his method of doing business in many ways (for better or for worse). Hence, some evidence is needed to support vendor claims.

Finally, after the evaluators' estimates of all system performance characteristics are completed * (based on holding the submitted cost and delivery schedule constant), these characteristics are inserted and combined in the operational flow model to determine the level of system effectiveness expected from each vendor and each proposal is adjusted so that it pivots on a constant level of effectiveness in doing the total job, in order to arrive at a common basis of comparison. Technical negotiations with the vendors are helpful in making such readjustments. Finally the source selection can be made on the basis of lowest total cost.

Fixed Price Incentive Contract

This type of contract is like the fixed price contract expect that, in addition, various financial incentives are offered to the contractor to improve differ-

---

* Group discussion and the Delphi Method can be used to obtain the best estimate of the evaluators.

ent elements of his performance (as well as penalties for performance degradation). Each incentive may be viewed as a price bonus (or loss) function to be added to (or subtracted from) some base price. Figure 15-1a, for example, might represent a cost incentive transfer function in which the amount of profit awarded the contractor is a function of his cost. Thus he has an incentive to increase his operating efficiency and reduce cost. The same type of transfer function would be applicable as an incentive for reducing warhead dispersion (i.e., CEP as measured in feet), or delivery time. Figure 15-1b is an example of an increasing type transfer function which might represent the incentive for increasing system reliability, or speed of an aircraft.

Several comments should be made regarding assumptions and pitfalls inherent in the applications of incentive systems. All of these are based on an assumption that the agency program manager knows the vendor's objectives and that the vendor will tailor his actions on the program to meet these objectives and optimize his total return. A cost incentive scheme provides an

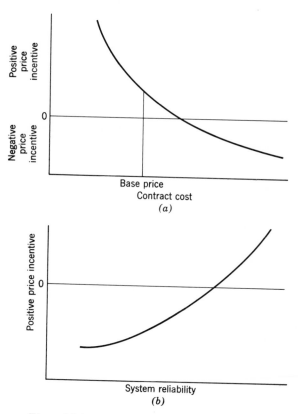

Figure 15-1.   Cost incentive evaluation function.

opportunity for the contractor to optimize his program profit (or fee) by finding ways of reducing his total program costs. However, the contractor is really concerned with increasing the profit of the total organization. These objectives may not be compatible, such as in the example of the contractor not having other work and willing to take a reduced profit or loss by stretching out the program. This increases his recoverable, applied costs rather than stopping the program earlier and thus having to place resources on unapplied time, reducing company profits.

Even if the contractor is interested in increasing his program profit, however, there are still some problems to be considered in designing the incentive functions which involve the interrelationship of system performance characteristics. First, there must be an objective way of measuring a system performance characteristic (which is common to the problem of properly preparing specifications). For example, suppose an aircraft speed is specified, or an incentive given for increased speed. At what altitude must this speed be attained, and with what payload? There are many such opportunities for misunderstanding to arise. In addition, the incentive transfer functions for performance characteristics should, in general, be designed on a multi-dimensional basis, rather than using a single dimension as in Figure 15-1. If they do not, the vendor will place his effort on those areas where his marginal effectiveness (i.e., increase in incentive to increase in resources required) is greatest, which may not result in the best mix of system performance characteristics to the user. Fortunately, we have already examined ways of finding the best mix of performance characteristics which resulted in the construction of the operational flow model. Hence this model can be used to derive the incentive functions in the same way that it was used for the cost-effectiveness evaluation.

Concluding Remarks

In evaluating different vendor proposals, the system evaluator must also consider the ability of each vendor to deliver the predicted performance characteristics on the desired delivery date and at the predicted cost. The uncertainties involved in this ability may be handled as described in Chapter 8, where pertinent data are collected and applied (including the use of intuitive judgment) as subjective estimates of these performance, time, and cost characteristics. All expected performance characteristics are then combined in the operational flow model to obtain each vendor's system effectiveness in performing the total job. A recommended selection is then made by pivoting on equal effectiveness and delivery time (or equal cost and delivery time).

VII

CONCLUSIONS

16

Concluding Remarks

In this book we have tried to show that the systems planning process can be made explicit and to show the role which systems analysis plays in this process through case examples illustrating the key principles involved. We emphasized that, while systems analysis may be only an aid to a decision-maker, it is intimately related to the management process, as illustrated by the six key management questions to which systems analysis is directed. In this chapter we shall make some concluding remarks, particularly regarding the application of these systems analysis principles to practical problems which the reader may face.

DEALING WITH TIME AND RESOURCE CONSTRAINTS

One of the biggest problems an analyst has is in performing a suitable analysis when there is a shortage of time and resources to obtain the data and to structure information which the analyst feels is needed. Invariably there is such a constraint. However, there is also a systematic approach for meeting the request for systems analysis assistance within the constraints of available time and manpower. Many times a systems analyst is asked to contribute to a systems planning effort in which an analytical solution is required, and is asked to do so in what may be an unreasonably short amount of time or with only a small amount of manpower resources. Analysts respond to such a request differently, some rejecting it unless sufficient resources are provided. However, this approach provides no assistance at all to a genuine need for information. Hence there is a need for the analyst to work in concert with the decision-maker by tailoring the analysis to the need for information as well as to the time and resource constraints. One way of doing this and satisfying both requirements is by conducting the analysis in an iterative fashion, as depicted in Figure 16-1. This figure is a flow diagram indicating phases of the work activity, starting from a request for system analysis support by the client (a systems planning project leader) to the final

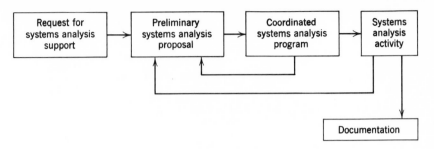

Figure 16-1.   Program planning process.

documentation of the systems analysis activity. In general, the systems analysis activity is initiated by a request for systems analysis support. This request may take one of the following forms.

1. Evaluate the effectiveness of this system in performing a given mission.

2. Determine one of the effectiveness components of a system. For example, how survivable is this airborne command post when operating in a tactical war environment?

3. Develop a system to meet the following need: . . .

Sometimes the request consists of a vague discussion of some "felt need" or the problem as given. This felt need is obtained after verbal discussions with the system planners regarding their intuitive opinions about the problem or some solution they are considering.

After some preliminary examination of the problem as given, a structure of the problem can be formulated. This results in a preliminary systems analysis proposal indicating the many ramifications of the problem (problem as understood). After the systems planning project leader has viewed this preliminary proposal, he is in a better position to determine the degree of depth in which he wishes systems analysis support to pursue this problem by evaluating different work options with the manpower required for each. Presumably, he has a better understanding of the problem than he originally did in the unstructured discussions. It is possible that several iterations with the project leader may be needed to result in a coordinated systems analysis program which can be supported at an agreed-on level. This coordinated program is the basis for continuation and implementation of systems analysis activity. This activity culminates in documents consisting of periodic and final reports, and may also result in additional proposals being made to the systems planning project leader for additional activities which have resulted from the systems analysis effort.

With this in mind, each of the eight tasks of the systems evaluation process, shown in Figure 4-7b, can be used as a structured data base for program review and program planning. Such a data base will aid not only the analysis but also the communications process among the participants in the project.

Data Base Maintained During Systems Analysis Activity

The following data base can be generated, bearing in mind that a systems analysis program is always dynamic and hence subject to change at any point in time:

Program Plan. This contains the following information:

1. Program objective(s). In addition to a statement of program objectives, this may also include a list of questions which the program will seek to answer.

2. Program products. This is the type of information which is expected as an end product of the analysis.

Structure of the Work Elements. For each designated major program, a systems analysis data base will be structured for review purposes in accordance with the eight task approaches as described in the WSEIAC reports, and illustrated in Figure 4-7b.* It is intended that the data base be dynamic so that it currently indicates the status of each of these eight tasks, modified as described below:

Task 1: Mission Definition. The missions definition is a precise statement of the intended purposes of the system. (The environmental conditions, including system operation scenarios, are described under Task 4.)

Task 2: System Description. This description should contain not only a description of each system alternative under study, but also a block diagram showing its relationship to any other competing system which can perform the same mission. Thus the system description will also include the hierarchy of systems leading up to the performance of a mission (or missions for a multipurpose system).

Task 3: Measures. Given the hierarchy of systems, a hierarchy of objectives will be constructed, indicating various system performance and effectiveness measures which may be used within the hierarchy.

Task 4: Identification of Accountable Factors. This should include not only all of the assumptions and boundary conditions made, but also the assumptions of the external environment, particularly the scenario or description of the various operational conditions under which the system is expected to operate. If there is more than one scenario to be considered, these should be listed accordingly.

* Reference is made to WSEIAC report AFSC-TR-65-2, Vol. 2, "Prediction–Measurement" (Concepts, Task Analysis, Principles of Model Construction)," pp. 5–16.

TASK 5: MODEL CONSTRUCTION. An operational flow model indicating the flow of information and activities for each of the parts of the model should be constructed. This flow model should indicate system performance/effectiveness measures as contained in Task 4, as well as the set of mathematical and/or logical equations (transfer functions) to be utilized. Connections of submodels should be indicated.

TASK 6: DATA ACQUISITION. This file indicates the data to be acquired, its location, and its difficulty in being obtained.

TASK 7: ESTIMATE OF MODEL PARAMETERS. This file indicates how the data will be extrapolated to estimate model parameters to be used.

TASK 8: MODEL EXERCISE. The different approaches for model exercise which have been considered should be indicated prior to model exercise. These should include analytical approaches as well as Monte Carlo simulation.

Application of Data Base to New Program Planning

We shall now construct a program plan indicating the work efforts to be undertaken in each of the eight tasks. It is realized that such a program plan is a dynamic one; hence it is subject to modification at any time when additional data are obtained during the project upon concurrence with the systems analysis management and the systems planning project. It will be constructed as follows:

Preliminary Systems Analysis Proposal. This should contain the preliminary analysis of the client's problem, including the broader aspects of his problem as obtained from the preliminary discussions with the client and others. To accomplish this, an initial approach to performing each of the eight tasks already indicated should be made. It is important that this be made in an explicit fashion, showing the many aspects of the problem, including various scenarios which might be considered, and modeling the problem up to a mission effectiveness level. This is useful not only in indicating to the client how his felt need relates to mission effectiveness, but also to other related aspects of the problem.

To assist the client in better understanding the relationship between the type of analysis he may obtain and its cost in time and manpower resources, a program plan could then be constructed, indicating different options of analysis available in terms of degree of depth and breadth of study as a function of required manpower resources. A ranking of importance of the different options as evaluated by the systems analyst would be provided. This would include a ranking of the various scenarios to be considered as well as the amount of data which could be obtained and the models which might be constructed and exercised.

In this fashion the analyst can identify for the client that program which requires the least resources and consists of perhaps only one (the most important) scenario, and data which could be rapidly assembled. However, even this preliminary effort which is actually problem formulation can be done fairly rapidly, and is generally the most important effort since it focuses on the hierarchy of objectives and related systems. The other analytical options available (and their estimated costs) will then enable the client to find that option which meets his informational accuracy needs as well as resource constraints. Many times this "shopping list" strategy results in the client choosing the minimal effort to meet a time constraint and then continuing the more detailed analysis as a second iteration after the initial request has been met and he sees the value of performing the subsequent studies. This is why analyses should be designed to be performed on an iterative basis.

Of course, sometimes the client may lack the resources for anything but a superficial study, but even this may be better than nothing, as long as the necessary assumptions made and approach used are clearly identified and emphasized.

Coordinated Systems Analysis Program. Presentation of the preliminary systems analysis proposal to the systems planner can result in obtaining his feedback regarding the proposal and his ranking of the various options at his disposal. These are all compared with the manpower which he wishes to expend. This program and manpower allocation thus form the basis for the systems analysis effort.

Application to Management Review of Programs

Program reviews of all designated major programs should take place periodically. The main advantage of such a review procedure is that the analyst knows in advance the questions which his manager is seeking and will ask. The review can be structured in accordance with the eight tasks already outlined. A review is particularly in order upon completion of Task 5 and prior to entering Task 8. Unless there is a general agreement in the model to be used and the data inputs to the model, the results of model exercise will not be valid.

SUBSYSTEM OPTIMIZATION

Ideally the objective of any systems planning effort is to configure a system which will provide a given level of effectiveness at lowest total cost to the highest echelon of the organization. Many times there are organizational constraints which prevent optimization at the highest level, and lower-level optimization (suboptimization) is used. In either case, an operational flow model and a cost model must be constructed to determine the mission effectiveness a system will provide and the costs to be incurred.

The most common failure in conducting a subsystem planning effort is the failure to relate subsystem performance and cost to the mission to be performed. A decision-maker can justify spending money on a new system only if one of two conditions can be shown: First, that these expenditures will save other greater expenditures elsewhere in the system, keeping the level of mission effectiveness constant; or second, that a higher level of mission effectiveness is required and the proposed system addition is the least cost alternative of providing this higher level of effectiveness. This is the strength of the mission-oriented operational flow model which relates all of the functional performance characteristics of all interacting subsystems to mission effectiveness.

ANALYZING MULTIPURPOSE SYSTEMS

We have concentrated in this book on methods for analyzing systems which perform a single mission. When several missions are simultaneously in operation, the systems analysis becomes more difficult to perform because of the interactions involved. In this case, the analyst must generate a scenario which relates and combines the various missions. This is what was done for the strategic planning case when both offensive and defensive missions were involved, using a higher-level objective of survival following offensive and defensive actions. A multipurpose computer or communication system offers the same challenge at a lower level. Here the analyst must examine the system effectiveness for each mission or determine how the missions may interrelate (or demand service from the system at the same time since such parallel demands may result in performance degradation).

LIMITATIONS OF SYSTEMS ANALYSIS

Systems analysis can provide an answer to the question, "What would happen if . . . ?" Its main advantage is that the answer provided is based on very explicit logic and data inputs. For this reason the analysis is subject to verification and reproducibility of results by others.

One of the main sources of error in a systems analysis is the error of omission. For example, if a competitor adopts a system or strategy which we neglected to consider, the results may be greatly different than had been anticipated. Similarly, if the logic of our models or the data used were incorrect, the results will also be incorrect; since the objective of an analysis is to make all of these factors explicit, there is high likelihood that such error will be detected. Furthermore, all other methods involved in decision-making also suffer from the same "deficiency."

Another limitation of systems analysis is its inability to cope quantitatively with certain higher-level factors. What is the worth of a human life? How do we measure the morale or prestige of a nation? How are many performance factors combined into national security or happiness? We have tried to show in the case situations how quantitative measures can be used. Invariably there will be factors to be considered which cannot be quantified; hence, they cannot be combined with the other key factors. Here a qualitative discussion of the impact which the factor can have on each system alternative (or vice versa) should accompany the quantitative analysis. It is then up to the decision-maker to consider this information along with the quantitative work, and combine both types of information in an intuitive manner. However, any other method used in the decision-making process has the same limitations and deficiencies just cited. Still, decisions must be made and alternatives will be chosen at appropriate times. Hence in the final analysis, the decision-maker must compare the information he can obtain from the systems analysis approach with that which he obtains from any other. It is hoped that many of the principles explained and applied in this book will be found to be meaningful and applicable to the problems with which you will be faced. This is, perhaps, the real test of this book.

Appendix 1: Notes on Probability and Statistics*

Probability and statistics, as we are applying them to problems involving random processes, include the tasks of gathering data which pertain to the set of past outcomes of the process and beliefs concerning what is involved in the process, structuring this data, and then interpreting the resulting information so that a prediction can be made regarding future occurrences of the process.

PROBABILITY DISTRIBUTIONS OF A DISCRETE VARIABLE

Consider as a simple example the process of launching an artillery shell whose function is to destroy a target. Assume that 1000 test firings of this type of shell have occurred, and 700 of these firings have been successful. Further assume that all of the shells are identical to one another, the firing process is the same, and that the success (or failure) results appear to be in a random fashion. Based on this information of past firings, what can be said about the results of firing the 1001st shell? Here we are dealing with a discrete variable, since the results of each missile firing have only two discrete states: success or failure. If an individual were to make a wager on this occurrence and felt that this particular shell was identical to the other shells previously fired, and that the operational conditions were the same as for the test data, he might feel there is a 70% chance of the shell being successful, since 70% of the previous shells were successful. Thus he might choose to use this as his estimate of the probability of success of this process. This is defined as the "most likely estimate." In the preceding case, the analyst's estimate is based on a frequency estimate since 70% was the frequency of success

---

* For a fuller treatment of the subject see Parzen (1960).

of the past occurrences. He "feels safe" in this estimate, based on two key factors:

1. A large number of trials (1000) have been run.
2. He feels that the set of conditions which now confronts him is the same as for the previous data samples.

Thus, this estimate of the probability of success depends on both the relative frequency of past successes and personal belief in the similarity of conditions.

Assuming a probability of success of 0.7, the probability of failure is 0.3 (i.e., $q = 1 - p_s = 0.3$). A probability function can be constructed, as shown in Figure A-1, to illustrate the uncertainty associated with this random process. This is called a probability distribution since it indicates how a unit of probability is distributed (spread out) over the set of all possible values of the outcome (zero or one missile success). The probability function is also called a frequency function since it is a prediction of the frequency which each of the possible discrete states can take. This particular function is further called a binomial distribution.

A second question which could be examined, based on the past sample data, concerns what might occur if two shells were fired (each at a different target), and what the chances associated with each possibility would be. Since there are two possibilities, success (S) or failure (F), associated with each shell, there are four specific possibilities associated with the two shells, as shown in Figure A-2. If the physical phenomenon is such that we can assume each trial to be independent of the others, and the same probability of success to be constant for all occurrences,* the total probability of success for each of the four possibilities is the product of the individual probabilities. These four possibilities may be translated into another variable outcome (the number of successful shells which yields the number of targets destroyed) and there would then be three possible outcomes in terms of this

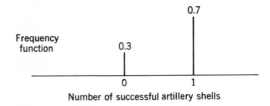

Number of successful artillery shells

Figure A-1. Probability distribution for the launching of one artillery shell at one target.

---

* A process which meets these conditions is called a "Bernoulli process with known parameters." See R. Schlaifer (1959), Chapter 10.

Specific Outcomes	Probability of Event	Number of Shell Successes
Shell 1 successful, Shell 2 successful	$p_1p_2 = 0.49$	2 Successes
Shell 1 successful, Shell 2 failure	$p_1q_2 = 0.21$	1 Success
Shell 1 failure, Shell 2 successful	$q_1p_2 = 0.21$	1 Success
Shell 1 failure, Shell 2 failure	$q_1q_2 = 0.09$	0 Successes

Figure A-2. Logic table of missile outcomes.

variable (i.e., 0, 1, or 2 shell successes). The probability of each of these three possible outcomes can be obtained by taking the sum of the probabilities of each of the same outcomes. This frequency distribution function can be plotted as shown in Figure A-3. Notice that the sum of the probabilities of all possible events of both Figures A-2 and A-3 equal unity.

Another important characteristic which can be derived from the frequency distribution is the "expected value" which is the mean or average value expected of all possible results X_i, when the probability of each result is factored in. Mathematically the expected value $\overline{X}$, is defined as follows:

$$\overline{X} = p_1X_1 + p_2X_2 + \ldots \ldots + p_nX_n.$$

Thus from the frequency function of Figure A-2, it can be calculated that if two missiles are launched, each at one different target, using the frequency distribution of Figure A-3, TD, the expected number of targets destroyed, is:

$$\overline{TD} = 0(.09) + 1(.42) + 2(.49) = 1.40.$$

It can be shown that for a binomial distribution the expected value can also be found by taking the product of the number of trials and the probability of success of each trial. Thus,

$$\overline{TD} = nP = 2(0.7) = 1.4.$$

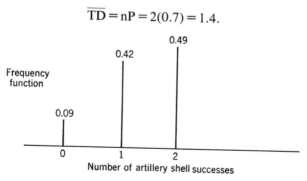

Figure A-3. Probability distribution for the launching of two artillery shells at two targets (assuming a probability of success of 0.7 for each shell).

From this frequency distribution function, the analyst can also determine the probability of achieving one or more successes, since this is equal to the sum of the probabilities of one or two successes. Thus,

$$P_{1or2} = P_1 + P_2 = 0.91,$$

where P_1 = probability of 1 success   = 0.42,
P_2 = probability of 2 successes = 0.49.

In fact, the distribution function for the probability of n or more successes (or n or less successes) can be determined and plotted as shown in Figure A-4a and b, for n = 0, 1, or 2 successes. These distribution functions are called "cumulative distribution functions." As seen, the cumulative distribution may be obtained from the frequency function by "accumulating" (i.e., addition). Conversely, the frequency function may be obtained from the cumulative distribution by the process of subtraction.

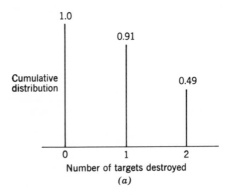

(a)

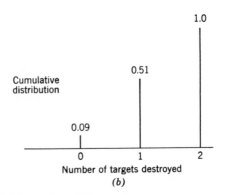

(b)

Figure A-4.   (a) Probability of destroying n or more targets (where p = 0.7); (b) probability of destroying n or less targets (where p = 0.7).

Another question arises from this same problem. We shall now suppose that the single-shot kill probability of $P = 0.7$ was not considered high enough, and it was decided to launch a salvo of two shells simultaneously at each target. Now what is the likelihood of destroying each target?

Since the target will be destroyed if either one or both shells are successful, the analyst can compute the new frequency distribution for the salvo attack by adding the probabilities of each of the two means of success from Figure A-2, as shown in Figure A-5. Thus, it can be seen from this distribution function that launching a salvo of two missiles increases the total (salvo) kill probability to 0.91.

A further question which could be examined would be to predict the probability of the various possible results in launching a series of m shells, each at one target. This result can be found by determining the binomial probability function. Here the frequency function showing the probability of exactly n successes (targets destroyed) out of m trials is:

$$f(n) = C_n^m p^n q^{m-n},$$

where $C_n^m =$ the combination of m things taken n at a time,

$$C_n^m = \frac{m!}{n!(m-n)!},$$

$p =$ the probability of success of each trial,
$q =$ the probability of failure of each trial
$= 1 - p$.

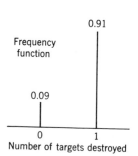

Figure A-5. Probability distribution for the launching of two missiles at one target (where $p = 0.7$).

Numerical values for this and the other probability functions presented here may be found in most texts on probability theory.

As an example, the frequency distribution for the original problem using three shells fired at three targets is:

$$f(0) = 1(0.7)^0(0.3)^3 = 0.027,$$
$$f(1) = 3(0.7)^1(0.3)^2 = 0.189,$$
$$f(2) = 3(0.7)^2(0.3)^1 = 0.441,$$
$$f(3) = 1(0.7)^3(0.3)^0 = 0.343.$$

This frequency function is illustrated in Figure A-6.

Some Important Probability Functions of a Discrete Variable

There are several types of probability functions which approximate physical occurrences and of which the systems planner should be aware. These

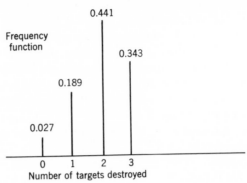

Figure A-6. Probability distribution of launching three artillery shells at three targets (where p = 0.7).

will be described qualitatively. For a more quantitative description, the reader is referred to an introductory text on probability.

The first function is the binomial distribution, previously discussed and illustrated in Figure A-1. Here there are only two primary possibilities (or states) which the random variable may take: "Go" or "No Go," (or success or failure, or "Works" or "Doesn't Work"), and the frequency function provides the probability associated with each possibility. When n trials are taken from the same distribution, a new binomial distribution is created, showing the resulting probabilities for the n trials, as shown in Figure A-6 for n = 3.

The binomial probability function is actually a special case of the multinomial distribution, which is concerned with n possible states of the random variable. A special example of this frequency distribution is the tossing of a die in which there are six possible outcomes, each having a probability of ⅙.

The Poisson distribution illustrated in Figure A-7 is another discrete distribution given by:

$$f(x) = \frac{e^{-m}m^x}{x!} \quad (x = 0,1,2 \ . \ . \ .),$$

where m is the expected value (mean) of x.

Note that the only parameter is m, the mean of x. An example using this probability distribution is in traffic problems, such as in determining the probability that x cars will pass a given point in a unit of time (e.g., one hour) when the average number of cars passing in this time is found to be m. A detailed application of this distribution is discussed in Chapter 11 which deals with the failure rates of equipment components.

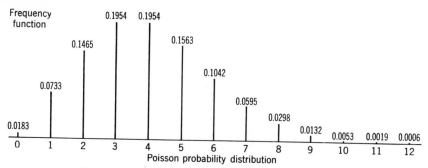

Figure A-7. Poisson probability distribution for m = 4.

Variance and Standard Deviation

To summarize the discussion thus far, the frequency function completely describes all of the uncertainties involved in a random process since it indicates all of the possible values that the outcome may take, and the numerical value of probability associated with each possible outcome. As previously defined, the expected value of a probability distribution which can be obtained from the frequency functions gives only partial information about the possible outcomes. It tells only what mean value would be expected if a large number of trials were made, but tells nothing about the other values which may occur. For example, consider the possible results of each of two different gambling propositions, A and B, whose multinomial frequency functions are illustrated in Figures A-8 and A-9. Note that in each case there is an equal chance of winning or losing a given amount of money. Hence the expected gain from each proposition is zero. However, proposition B has a greater possible loss (and gain) association with it. Proposition A has a range of values which encompasses $2000 while B has a range of values of $20,000. However, notice that there is only a very small chance of winning or losing more than $500 in proposition A or $5,000 in proposition B. Hence the total range may not be a good measure of the spread of each distribution around the mean. A better measure is the variance or the standard distribution of the probability distribution. The variance σ_2 is defined as follows:

$$\sigma_x^2 = (x_0 - \bar{x})^2 f(x_0) + (x_1 - \bar{x})^2 f(x_1) + \ldots ,$$

where $\bar{x}$ is the expected value of the distribution.

Notice that the variance is the weighted mean of the squares of the variation from the expected value. The standard deviation σ is then defined as the positive square root of the variance.

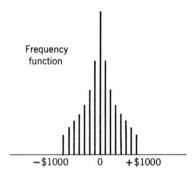

Figure A-8.   Probabilistic results of gambling opportunity A.

PROBABILITY DISTRIBUTIONS OF A CONTINUOUS VARIABLE

We have discussed several probability functions used to describe the uncertainties which are associated with some random process involving a discrete variable. If, on the other hand, the random variable is continuous, the probability distribution which would be analogous to the frequency function for the discrete variable is called a "probability density function." In addition, the same factors as presented for the discrete variable apply.

An example of this continuous function is the time taken to repair a piece of equipment, as shown in Figure A-10a. This function can be used to determine the probability that the event will occur (i.e., the equipment will be repaired) within a period of time Δt between t_1 and t_2. This probability is obtained by integrating to find the area of the curve between the limits of t_1 and t_2. By definition of the probability density function, the total area under probability density function from $x = -\infty$ to $x = +\infty$, is always equal to unity.

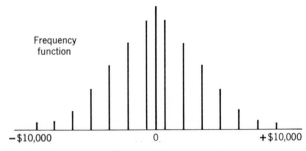

Figure A-9.   Probabilistic results of gambling opportunity B.

The function can also be used to determine the expected value of the variable, which, as in the case of the frequency function for a discrete variable, is equal to the mean or average value of the variable. This would be the "mean time to repair" (MTTR) in this example. However, for the continuous variable the summation of the products of each value of the variable times its probability of occurrence must be obtained by integration. The variance and standard deviation of the distribution could also be found by the process of integration.

As in the analogous case of the discrete variable, the density function can be used to determine the probability that an event will occur by a certain time (or less). For example, using Figure A-10a we can determine the probability that the repair will be complete by time t_3 (or less). This is obtained by constructing the cumulative distribution, which is the integral of the density function as shown in Figure A-10b. Thus, either of these two functions provides the probability distribution of the phenomenon, since given the cumulative distribution, the density function can be obtained by differentiation.

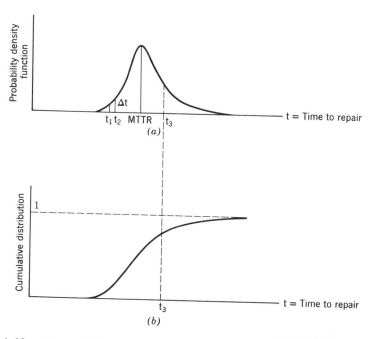

Figure A-10. (*a*) Probability that equipment will be repaired during time Δt: (*b*) probability that equipment will be repaired by time t.

Some Important Probability Functions of a Continuous Variable

There are a number of distribution functions of a continuous variable which are mathematically described in various texts on probability. Three with which the reader should be particularly acquainted are the normal (or Gaussian) distribution, the exponential distribution, and the uniform distribution, since these are widely used in systems analysis.

The normal distribution, illustrated by the two functions of Figure A-11, is expressed mathematically by:

$$f(x) = \frac{1}{\sigma\sqrt{2\pi}} e^{-(x-\bar{x})^2/2\sigma^2} \quad (-\infty < x < +\infty),$$

where $\bar{x}$ = expected value of x,

σ^2 = variance of x.

Note that the normal distribution is a function of two parameters. Thus, as shown in Figure A-11, of the two normal distributions shown, both have the same expected value, but distribution B has the larger standard deviation. Any text on probability contains tables for obtaining the values of both the density and cumulative distributions as a function of the two parameters, making use of the mathematical formula unnecessary.

UNCERTAINTY IN ESTIMATING NUMERICAL VALUES OF PROBABILITY

We shall now return to the original problem which began the discussion (i.e., assuming 1000 test firings of which there have been 700 past successes, what is the likelihood of the next artillery shell being successful?) Here the analyst must make an estimate of the probability of success. Or

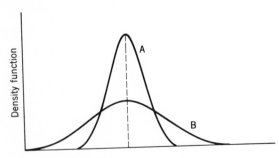

Figure A-11. Normal probability distributions with same mean but different standard deviations.

consider a second case in which only 100 data samples (occurrences) have been accumulated, seventy of which were successful, or a third case of ten samples, seven of which were successful. In all of these cases a layman might estimate that the probability of success is 0.7 but he should feel less certain about the outcome when the amount of available data is small. For example, in the extreme case of only two trials, one of which is successful, the estimator could not be at all confident that the probability of success is 0.5.

We shall now examine the following sample problem: Assume that eight out of ten test firings have been successful; how well can the analyst estimate the probability of success for additional firings, based solely on the preceding data? Here the analyst is confronted with the following analogous situation. Consider a large number of balls contained in an urn, some proportion of which are red and the rest are blue. Draw out a ball, record whether it is red or blue, and replace the ball in the urn. Do this ten times. Now, assuming that exactly eight red balls have been drawn, how accurate an estimate can be made about the percentage of red balls in the total (i.e., what is the probability of drawing a red ball on any one trial)?

If a layman were asked for one number which would estimate the total probability of success (or proportion of red balls) for the above examples, he would probably estimate this probability (proportion) as follows:

$$P \approx \frac{\text{total number of missile successes}}{\text{total number of missile firings}}$$

or

$$P \approx 8/10 = 0.80.$$

How accurate an estimate is this?

The statistician might approach this problem as follows: If it can be assumed that each of the trials was independent of any other, a binomial probability distribution can be used. Thus the probability of exactly eight successes (or drawing red balls) out of ten trials is given by the frequency function:

$$F(8) = C_8^{10} p^8 (1-p)^2.$$

Thus the problem becomes, "For what value of p will the probability $F(8)$ be a maximum?" This value, called the "maximum likelihood estimate of p," would be found to be the value:

$$p = x/n = 8/10 = 0.8.$$

Thus the mathematics verifies our intuitive feelings.

Confidence Limits *

Unfortunately, the maximum likelihood estimate for a small sampling of data will almost certainly differ from the true value of the quantity being estimated, particularly if only a small sample of data is used in making the estimate. Hence the analyst should provide some indication of the accuracy of his estimate. This is the concept of a "confidence interval." To develop this concept, consider the following example.

Assume that the true proportion of red balls in the previous example is really 0.7 and, again, eight red balls were drawn out of ten trials. From this information the frequency function can be constructed of the probability of drawing x red balls, using a binomial distribution as shown in Figure A-12. Each of these values of x, if drawn, would provide a most likely estimate (x/n) of the true probability; this value of x/n is also shown in Figure A-12. Thus since ten trials were taken, we have the same probability of obtaining this estimate of p as we have of drawing x balls. This figure permits us to calculate the probability of the true probability, lying within a given interval of the most likely value of p obtained. For example, note that there is approximately a 99.9% chance that the true value of p (i.e., 0.7) lies between $p = 0.3$ and 1.0. Similarly, there is approximately:

a 99% chance that p lies between 0.4 and 1.0 (an interval
of ± 0.3 away from p),

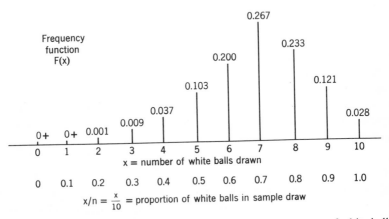

Figure A-12. Binomial distribution of ten trials where the percentage of white balls is 0.7.

---

* For a more detailed treatment of this subject the reader is referred to R. Schlaifer (1959), Chapter 42, and F. Mosteller, et al. (1961), Chapter 8.

a 93% chance that p lies between 0.5 and 0.9 (an interval of ± 0.2 away from p),

a 70% chance that p lies between 0.6 and 0.8 (an interval of ± 0.1 away from p).

This indicates that if this series of ten draws were repeated a large number of times, the true value of p (0.7) would be within 0.2 of the most likely value calculated, 93% of the time. This percentage of the time (93%) is called the confidence coefficient. The interval (±0.2) is called the confidence interval or limits. As the confidence coefficient increases, the confidence interval also increases.

Confidence intervals have been calculated for the entire set of probabilities. For example, Clopper and Pearson * have published the chart shown in Figure A-13 which provides confidence limits for a confidence coefficient of 0.95 and a binomial probability distribution. Thus for any value of x/n (i.e., the most likely estimate of the probability), the confidence limits may be obtained. If, as in the previous example, eight successes are obtained out of ten trials, the analyst can estimate that the true probability will lie between 0.43 and 0.98, with 95% correctness of this estimate (i.e., over the long run of such estimates, the estimator using this rule would be correct 95% of the time).

Suppose more information were available. For instance, if 16 successes are obtained from 20 trials, the true probability can be estimated to lie between 0.56 and 0.95. If 80 successes are obtained from 100 trials, the true probability can be estimated to lie between 0.70 and 0.88. Again, we must emphasize that we can never *guarantee* any of these statements to be true, only that 95% of them are true.

Subjective Estimates

Sometimes the analyst may have no data or previous trials at all but is still forced to make an estimate of the probabilities involved. For example, consider the analyst visiting a gambling casino and playing with a new set of dice. Before the first play he could estimate the probability of rolling a particular number, assuming each die to be perfectly balanced. He could examine the dice to see if they are loaded. He could factor in other information such as his "feel" for the honesty of the gambling establishment. Of course he could change his estimate of a particular probability, based on new information obtained. But, if one die comes up "one" four times out of ten, is this because it is "loaded" or because ten trials are quite limited data? The major

* From C. J. Clopper and E. S. Pearson, "The use of confidence or fiducial limits illustrated in the case of the binomial," *Biometrika,* Vol. 26 (1934), p. 410.

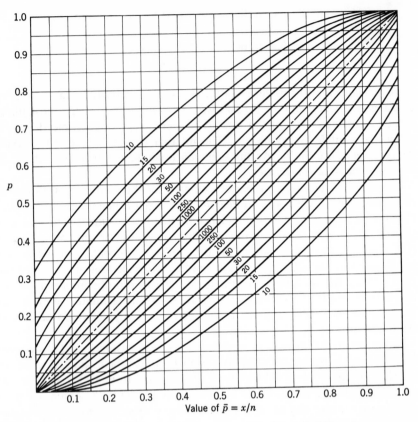

Figure A-13. Chart for 95% confidence limits on p, the probability of success on a single binomial trial, assuming confidence coefficient $= 0.95$. To obtain confidence limits for p enter the horizontal axis at the observed value of p. Read the vertical axis at the two points where the two curves for n cross the vertical line erected from p. [By permission of the Biometrika Trustees this chart has been reproduced from C. J. Clopper and E. S. Pearson, "The use of confidence or fiducial limits illustrated in the case of the binomial," *Biometrika,* Vol. 26 (1934), p. 410.]

point is that the estimate of probability can include any explicit reasoning (considered judgment) or intuitive feelings (intuitive judgment). Both of these are based on past, but limited, data (i.e., experience).

Conclusions

In concluding this topic of estimating probabilities (or other performance characteristics), we shall emphasize the following general principles:

1. Accumulate whatever data is available within the time and resource constraints.

2. Apply appropriate techniques (statistical, subjective, etc.) to reduce data to desired parameter.

3. If subjective opinions are used, document the reasons for the opinion and the spread of the uncertainties involved.

Appendix 2: Source Selection: Evaluation Criteria to be Employed*

INTRODUCTION—EVALUATING AREAS OF CAPABILITY

Evaluating an entire system requires the input of the most highly qualified representatives from the Air Force Systems Command, Air Force Logistics Command and (using command). Many of the factors to be evaluated are of major concern, and require the simultaneous and integrated application of the specialized and technical knowledge available in each of these three commands. For greater ease in managing this evaluation, the factors to be evaluated are organized into these three areas of capability:

1. Operational area.
2. Management, production, scientific, technical, and cost area.
3. Logistics (supply, maintenance, and transportation) area. The command representative who has the principal responsibility for the evaluation must insure that all of the inputs from each of the commands are thoroughly considered, so that the numerical rating for each factor represents a coordinated position.

GENERAL CONSIDERATIONS—CONTRACTOR'S CORRECTION POTENTIAL AND PAST PERFORMANCE

Throughout the evaluation, the Air Force must consider:
1. *Correction Potential.* The contractor's present competence and the

* Excerpts from Air Force Manual AFM 70-10, "System Source Selection Board Procedures, 18 January 1963."

452

probability of his successfully overcoming outstanding problems in meeting his proposed time schedule. An item in one proposal that is deficient, but which can be easily and readily rectified, shall lower the contractor's overall rating less than an item in another proposal with a similar deficiency, that would require a major reorientation of the proposed design, management, production, or logistics concept. This consideration is termed the contractor's "Correction Potential."

2. *Contractor Past Performance.* The contractor's past performance must also be weighed and evaluated independently; this evaluation must specifically cover the contractor's ability in past performance to (a) meet Air Force technical requirements, (b) build a quality product, (c) make timely delivery, (d) control costs, (e) produce without undue government assistance, and (f) correct past performance deficiencies.

These considerations may include others deemed pertinent to past performance. The weight assigned to this factor must be a substantial, rather than an insignificant, one. The consideration of the contractor's past performance is made a part of source selection records.

OPERATIONAL AREA

In the operational area, general consideration is given to the operational utility of the system that will result from the contractor's proposed program, and to the practicability of maintenance and over-all utility of the system, from the viewpoint of the using command. The specific criteria are:

1. *Mission Suitability.* This refers to the over-all adaptability of the system to the planned missions; it must also demonstrate suitability for the primary mission.

2. *Configuration Suitability.* This refers to the operational and functional suitability of the provisions required for the effective accomplishments of the ease of convertibility and flexibility.

3. *Maintainability and Supportability.* This refers to the personnel requirements, spares support planning, support equipment, maintenance, storage and checkout facilities, subsystem maintainability, simplicity, reliability, turn-around time, ruggedness, and adaptability to environment (including noise level tolerance and impact on logistic support requirements).

4. *Impact on Using Command Operations and Organization.* This involves the changes, if any, in operational philosophy and organization that might be required to optimize the system.

5. *Methodology.* This requires consideration of the procedures, programs, etc., required to achieve the optimum use of the system.

6. *Operator and Technician Training.* This requires consideration of the

time and complexity of training (*i.e.*, skill) required to operate the system, and the selection and availability of the required skills.

MANAGEMENT AND PRODUCTION AREA

In evaluating the criteria described below, general consideration is given to the realism of data in contractor's presentation, and confidence in his proposed performance. Evidence from the proposal is of primary concern in analyzing these general considerations; however, past experience with the contractor, and any other pertinent facts needed to reach a sound conclusion, should be considered in the evaluation process. The specific criteria are:

1. *Management and Organization.* Study the corporate organizational structure to determine position of _____(*name*)_____project in its organization, and look at the stature of the management personnel who will be assigned to it; there must be assurance of integrated effort by the various working units.

(a) Make a thorough analysis of the procedures, policy, and organization for subcontracting (including the contractor's selection procedures); purchasing (including his system for determining financial and technical subcontractor capability, and the adequacy of his purchasing system); reliability and maintainability; supportability; plant loading; implementing engineering changes; quality control; value engineering; and project planning and control.

(b) Evaluate the contractor's system of top management control of middle and lower echelon managers, to insure adequate attention to design simplicity and cost performance tradeoffs.

(c) Examine whether or not management has shown evidence that the purchasing system has been approved by the government; if it has not been approved, state the nature and status of each of the deficiencies.

2. *Master Plan.* This includes all major elements and the time-phased relationship necessary to plan the development, production, and support of this system.

(a) Analyze and evaluate the completeness and realism of the basic elements that are used to integrate the major elements, in terms of timing and use of resources by this and other programs.

(b) Evaluate the realism of the proposed schedules, considering the time required for development, testing, procurement, manufacturing, tooling, installation, and testing to meet operational requirements.

3. *Production.* Evaluate the production plans to determine the adequacy of existing facilities, labor, tooling, production methods, materials; the an-

ticipation of requirements in each of these areas; and the action planned to overcome any resulting deficiencies in these areas. Evaluate the controls and procedures for "make-or-buy," standardization, subcontracting and integration, past record in meeting production schedules, utilization of readily available components, and the contractor's control over subcontractors to insure that there has been adequate attention to design simplicity and cost performance tradeoffs.

4. *Facilities.* Evaluate the availability of engineering, test, and production facilities as required by time scheduling, the contractor's funding commitments, and the requirements for government financing.

5. *Manpower.* Evaluate the quantity and skill level of the manpower that will be required, and its availability.

6. *Financial Capability.* Analyze the over-all financial capability of the contractor to handle this program, considering his other workloads, financial commitments, and credit availability.

7. *Quality Record.* Evaluate the contractor's past quality record, considering the quality control problems he has experienced in the past, and his present production programs.

8. *Accounting Policies.* Examine the contractor's accounting system to determine if it is capable of supporting and justifying budgetary data, financial plans and reports, "make-or-buy" decisions, and segregation apportionment of contract costs. Have his "Contractor Estimating Methods" been reviewed and found adequate by the government? If inadequate, report the nature and status of each of the deficiencies.

SCIENTIFIC AND TECHNICAL AREA

These criteria are to be used in assessing the contractor's over-all program and design; they fit the requirements given in the work statement forwarded to the contractor by letter dated. . . . The proposal is the basic and primary source of data for evaluation purposes but not the exclusive source; other pertinent factual data must be considered in the evaluation process. On all items, the following factors must be considered:

1. *System Analysis and Integration.* Examine the contractor's understanding of, and approach to, the problem; the compatibility to all elements of the system; the use of system analysis and simulation techniques in the proposed development approach; how producibility, simplicity, maintainability, reliability, and operational considerations are integrated into the proposed design approach; the degree to which the technical features and objectives of this proposal satisfy program requirements; and the realism of the proposed time schedule for research, development, and test.

2. *Design Simplicity and Cost-Performance Tradeoff.* Evaluate the contractor's approach to shortening development lead-time and reducing development cost, both without affecting the required performance of the system.

3. *Major Elements of the System.* (That is, elements such as aerodynamics, propulsion, guidance, or breakdowns for a propulsion system, such as propellant, tankage, controls, etc.). For each element, evaluate the contractor's demonstrated technical ability, general and specialized experience background, quality and availability of engineering talent, his compliance with the requirements of specifications or exhibits, the soundness of the proposed design, and back-up programs.

4. *Test Program.* Evaluate the completeness and the adequacy of proposed testing methods, the experience in conducting test programs similar to the one required, and whether the timing of test efforts is appropriate.

5. *Aerospace Ground Equipment.* Evaluate the adequacy of the proposal, the understanding of requirements, and the time-phasing of the development.

6. *Growth Potential.* Evaluate the adaptability and capability of the proposed system to incorporate developments which would yield increases in performance without redesigning the basic system.

7. *Reliability Program.* Evaluate the contractor's awareness, organization, plans, procedures, and his effort to achieve the maximum reliability of the complete system; in particular, evaluate his methods for (a) collecting reliability information, and (b) incorporating the relevant data into the design.

8. *Value Engineering Program.* Evaluate the contractor's awareness, his organization plans, and his procedures and methods that will allow him to reach a balance between low cost and best performance; in particular, evaluate the outline of his training program and his organized approach to value engineering (particularly in such areas as specification reviews, design reviews, materials, production processes, inspection, testing, maintenance, etc.).

COST AREA

1. *Cost Realism.* Evaluate the completeness of coverage, realism, and validity of the cost estimates for all segments of the contractor's program, including the effectiveness and applicability of the techniques and the methods by which they were derived; this includes: (a) the identification and description of major cost deficiencies and their causes, (b) planned expenditure rates for separate program segments to be related time-wise and cost-wise, to determine efficient dollar utilization, and, (c) past overrun experiences, as well as effective use of funds as proposed by the contractor.

2. *Program Costs.* Evaluate the total program costs broken down by: development, test, production, GFAE (Government Furnished Aircraft Equipment), support, AGE (Aerospace Ground Equipment), and facilities, etc.; also, analyze both commitments and expenditures by fiscal year, as well as the profit pattern proposed by the contractor.

LOGISTICS, (SUPPLY, MAINTENANCE, AND TRANSPORTATION) AREA

In evaluating the proposed logistical support of the system, examine the adequacy of planning for supply, standardization, calibration, maintenance, transportation and handling of the operational vehicle; also, its support equipment. Analyze the organization and experience needed to support and maintain complex equipment over an extended period of time. Another important item to evaluate is the contractor's plan for handbook and technical manual development.

HOW TO ASSIGN A SCORE IN AN EVALUATION

1. In evaluation scoring, a numerical rating is assigned to each item in the "Criteria To Be Employed." These are:

> 10—Excellent
> 8—Very Good
> 6—Good
> 4—Fair
> 2—Poor
> 0—Unacceptable

2. Each item to be rated will cover a broad area; to arrive at the numerical rating, consider all factors that should influence the rating of that item. First, you must identify all the factors that affect the numerical rating for an item; each of these factors is in turn evaluated as follows:

> † Above Normal
> / Normal
> — Below Normal
> O Unacceptable
> N Insufficient information for evaluation

(*Note:* Normal is defined as "that quality of design, approach to the problem, planning, contractor capability or economy that meets the minimum USAF requirements.")

3. Some of the factors rated on an item are more important than others and, therefore, deserve more weight. In such circumstances, a simple arithmetic calculation, adding up the number of †, /, —, O, and N ratings, will not suffice; each must be weighed in accordance with its importance.

(a) The sum of the weights, multiplied by the sum of the maximum evaluation score possible for all items, must equal 1,000.

(b) These weights are established by a select few members of the working group who are also members of the SSSB. These weights are never made known to the evaluation group, and the members assigning these weights must not be rating members of the evaluation group.

Bibliography

GENERAL REFERENCES AND FOOTNOTES

Brown, B. and O. Helmer, *Improving the Reliability of Estimate Obtained from a Consensus of Experts,* The RAND Corp., P-2986, September 1964.

Campbell, R., "A Methodological Study of the Utilization of Experts in Business Forecasting," unpublished Ph.D. dissertation, UCLA, 1966.

Chestnut, H., *Systems Engineering Methods,* New York: John Wiley and Sons, 1967.

Clopper, C. J., and E. S. Pearson, "The Use of Confidence or Fiducial Limits Illustrated In the Case of the Binomial," *Biometrika,* Vol. 26, 1934, p. 410.

Dalkey, N. C., "Delphi," The RAND Corp. Paper presented to the Second Symposium on Long Range Forecasting and Planning, Almagordo, N.M., October 11–12, 1967.

Dalkey, N., and O. Helmer, "An Experimental Application of the Delphi Method to the Use of Experts," *Management Science,* 9, 1963, 458–467.

Davis, D. J., "An Analysis of Some Failure Data," *J. Amer. Statistical Assoc.,* June 1952.

Department of the Air Force, *Source Selection Procedures,* AFM 70-10, January 1968.

Enke, S. (ed.), *Defense Management,* Englewood Cliffs: Prentice-Hall, 1967.

Feller, W., *An Introduction to Probability Theory and its Application,* New York: John Wiley and Sons, 1957.

Forrester, J. W., *Industrial Dynamics,* Boston: MIT Press, 1961.

Gordon, T. J., and O. Helmer, *Report on a Long Range Forecasting Study,* The RAND Corporation, P-2982, September 1964.

Grant, E. L., *Principles of Engineering Economy,* New York: Ronald Press, 1950.

Hall, A. D., *A Methodology for Systems Engineering,* Princeton: D. Van Nostrand, 1962.

Hitch, C. J., *Decision-Making for Defense,* Berkeley: University of California Press, 1965.

Hitch, C. J., and R. N. McKean, *The Economics of Defense in the Nuclear Age,* Cambridge: Harvard University Press, 1960.

Kahn, H., and I. Mann, *Techniques of Systems Analysis,* The RAND Corp., RM-1829-1, 1957.

Kaufmann, W. W., *The McNamara Strategy,* New York: Harper and Row, 1964.

Maier, N. R. F., "Assets and Liabilities in Group Problem Solving: The Need for an Integrative Function," *Psych. Rev.,* Vol. 74, No. 4, July 1967, pp. 239–249.

Mosteller, F., R. Rourke, and G. Thomas, Jr., *Probability and Statistics,* Reading, Mass.: Addison-Wesley, 1961.

459

Novick, D. (ed.), *Program Budgeting,* The RAND Corp., 1964.
Parzen, E., *Modern Probability Theory and its Application,* New York: John Wiley and Sons, 1960.
Porter, J. D., and B. H. Rudwick, "Application of Cost-Effectiveness Analysis to EDP System Selection," MITRE Technical Report ESD-TR-67-412, March 1968.
Quade, E. S., (ed.) *Analysis for Military Decisions,* The RAND Corp., 1964.
Quade, E. S., and W. I. Boucher. *The Role of Analysis in Defense Planning,* The RAND Corp., R-439-PR, 1965.
Schlaifer, R., *Probability and Statistics for Business Decisions,* New York: McGraw-Hill, 1959.
Weibull, W., "A Statistical Distribution Function of Wide Applicability," *J. Appl. Mechanics,* Trans. ASME, Vol. 18, 1951, pp. 293–297.

GENERAL REFERENCES—COST ANALYSIS

Cost Concepts and Principles

Abert, J. G., *Some Problems in Cost Analysis,* Institute for Defense Analyses, Research and Engineering Support Division, June 1965.
Air Force Systems Command, U.S. Air Force, *Cost Estimating Procedures,* Andrews Air Force Base, AFSCM 173-1, November 1967.
Curry, D. A., *Costing Concepts for the Defense Programs Management System,* Stanford Research Institute Memo Report No. SD-84, March 1962.
Executive Office of the President, Bureau of the Budget, Bulletin No. 66-3 *Planning-Programming-Budgeting,* October 12, 1965, Supplement to Bulletin No. 66-3, February 21, 1966.
Fisher, G. H., "Costing Methods" in E. S. Quade (ed.), *Analysis for Military Decisions,* The RAND Corporation, R-387 (DDC No. AD 453887), November 1964. (Also published by Rand McNally, Chicago, 1964, Chapter 15.)
Fisher, G. H., *The Role of Cost-Utility Analysis in Program Budgeting,* The RAND Corp., RM-4279-RC (DDC No. AD 608055), September 1964.
Hoch, S., *Cost Criteria in Weapon Systems Analysis and Force Studies,* Office Assistant Secretary of Defense (Comptroller), August 16, 1965.
Grosse, R. N., and A. Proschan, "Military Cost Analysis," *Amer. Econ. Rev.,* Vol. LV, No. 2, May 1965, pp. 427–433.
Johnson, L. L., *Joint Cost and Price Discrimination in the Case of Communications Satellites,* The RAND Corp., 1963.
Jones, M. V., *An Approach to Command and Control Systems Costing,* The MITRE Corp., September 15, 1967.
Jones, M. V., *System Cost Analysis: A Management Tool for Decision-Making,* The MITRE Corp., TM-4063, 1964.
Large, J. P. (ed.), *Concepts and Procedures of Cost Analysis,* The RAND Corp., RM-3589-PR (DDC No. AD 411554), June 1963 (for official use only).
McCullough, J. D., *Cost Analysis for Planning-Programming Budgeting Cost-Benefit Studies.* The RAND Corp., November 1966.
Noah, J. W., *Concepts and Techniques for Summarizing Defense System Costs,* Center for Naval Analyses, Systems Evaluation Group, Research Contribution No. 1, September 24, 1965.
Noah, J. W., *Defense Systems Cost Analysis,* Center for Naval Analyses, Cost Analysis Group, April 1965.

Novick, D., *Costing Tomorrow's Weapon Systems*, The RAND Corp., RM-3170-PR (DDC No. AD 287997), June 1962. (Also published in *The Quarterly Review of Economics and Business*, University of Illinois, Spring 1963.)

Novick, D., et al., *Program Budgeting: Program Analysis and the Federal Budget*, The RAND Corp., 1965. (Published by the Government Printing Office, Washington, D.C., 1965; and by the Harvard University Press, Cambridge, 1965.)

Cost Estimating Relationships and Techniques

Bradley, B. D., *Building a New Force Structure Cost Analysis Model*, The RAND Corp., RM-4764-PR, October 1965.

Early, L. B., S. M. Barro, and M. A. Margolis, *Procedures for Estimating Electronic Equipment Costs*, The RAND Corporation, May 1963.

Fisher, G. H., *Derivation of Estimating Relationships: An Illustrative Example*, The RAND Corp., RM-3366-PR (DDC No. AD 290951), November 1962. [Also published in J. P. Large (ed.), *Concepts and Procedures of Cost Analysis*, The RAND Corp., RM-3589-PR (DDC No. AD 411554), June 1963 (for official use only), Chapter V.]

Fleishman, T., *Current Results from the Analysis of Cost Data for Computer Programming*, The System Development Corp., TM-3026/000/01, July 26, 1966.

Fowlkes, T. F., *Aircraft Cost Curves*, General Dynamics, August 1963.

General Dynamics, *Cost Model, Launch Vehicle Systems*. Prepared for the Marshall Space Flight Center, NASA, June 15, 1965.

Gradwohl, A. J., G. S. Beckwith, S. H. Wong, and W. O. Wootan, *Phase II Final Report on Use of Air Force ADP Experience to Assist Air Force ADP Management*, Vol. I, Summary, Conclusions, and Recommendations, December 1966.

Grosse, R. N., *Army Cost Model*, The RAND Corp., RM-3446-ASDC (DDC No. AD 293801), December 1962 (for official use only).

Jannsen, T. J., and H. Glazer, *Electronic System Cost Model*, The MITRE Corp., TM-3364, August 8, 1962.

Jones, M. V., *A Generalized Cost Structure for Electronic Systems*, The MITRE Corp., TM-3299, May 1962.

Jones, M. V., *Cost Factors as a Tool in Military System Cost Analysis*, The MITRE Corp., TM-3172, November 1961.

Jones, M. V., *Estimating Methods and Data Sources used in Costing Military Systems*, The MITRE Corp., June 21, 1965.

Leach, R., *Development of a Price Index for Aircraft and Missile Manufacturing*, Defense Research Corp., March 1966.

Noah, J. W., *Concepts and Techniques for Summarizing Defense System Costs*, Center for Naval Analyses, Systems Evaluation Group, September 24, 1965.

Noah, J. W., *Identifying and Estimating R & D Costs*. The RAND Corp., RM-3067-PR (DDC No. AD 283794), May 1962. [Also published in J. P. Large (ed.), *Concepts and Procedures of Cost Analysis*, The RAND Corp., RM-3589-PR (DDC No. AD 411554), June 1963 (for official use only) Chapter VII.]

Petruschell, R. L., *The Derivation and Use of Estimating Relationships*, The RAND Corp., RM-3215-PR (DDC No. AD 276673), June 1962 (for official use only).

Planning Research Corporation, *Methods of Estimating Fixed-Wing Airframe Costs*, PRC-547, February 1965.

Slivinski, S. C., *The RAND Cost Analysis Department Data Bank*, The RAND Corp., P-2985 (DDC No. AD 606581), September 1964.

Teng, C., *An Estimating Relationship for Fighter/Interceptor Avionic System Procurement Cost*, The RAND Corp., May 1966.

Tenzer, A. J., O. Hansen, and E. M. Roque, *Relationships for Estimating USAF Administrative and Support Manpower Requirements*, The RAND Corp., RM-4366-PR (DDC No. AD 611587), January 1965.

Walters, A. A., "Production and Cost Functions: An Economic Survey," *Econometrica*, Vol. 31, Nos. 1–2, January–April 1963, pp. 1–66.

Yaross, A. D., *Cost Prediction Based on CER Utilization*, USAF, Air Force Systems Command, Aeronautical Systems Division, Cost Analysis Information Report 62-6, May 1962.

Young, S. L., *Misapplications of the Learning Curve Concept*, Performance Technology Corp., Waltham, Massachusetts.

Techniques for Handling Uncertainty in Cost Estimates

Air Force Systems Command, United States Air Force, *Cost Estimating Procedures*. (Attachment 1, "AFSC Form 27—Cost Estimate Confidence Rating.") Andrews Air Force Base, AFSCM 173-1, November 1967.

Fisher, G. H., *A Discussion of Uncertainty in Cost Analysis*, The RAND Corp., RM-4071-PR (DDC No. AD 279936), April 1962. Also published in J. P. Large (ed.), *Concepts and Procedures of Cost Analysis*, The RAND Corp., RM-3589-PR (DDC No. AD 411554), June 1963 (for official use only), Chapter VI.

Pardee, F. S., *Weapon System Cost Sensitivity Analysis as an Aid in Determining Economic Resource Impact*, The RAND Corp., P-2021 (DDC No. AD 224289), June 1960.

Sobel, S. A., *A Computerized Technique to Express Uncertainty in Advanced Systems Cost Estimates*, The MITRE Corp., TM-3728, September 1962.

Summers, R., *Cost Estimates as Predictors of Actual Weapon Costs: A Study of Major Hardware Articles*, The RAND Corp., RM-3061-PR (DDC No. AD 329265), March 1965.

Tenzer, A. J., *Cost Sensitivity Analysis*, The RAND Corp., P-3097 (DDC No. AD 620836), March 1965.

Yates, E. H., H. M. Stanfield, and D. K. Nance, *A Method for Deriving Confidence Estimates in Cost Analysis*, Defense Research Corp. Technical Memorandum 231, March 1966.

GENERAL REFERENCES—EDP
EQUIPMENT EVALUATION—CHAPTER 14

[1] O. Dopping, "Test Problems Used for Evaluation of Computers," *BIT 2*, 1962, pp. 197–202.

[2] J. A. Gosden, "Estimating Computer Performance," *The Computer Journal*, 1962, pp. 276–283.

[3] J. A. Gosden and R. L. Sisson, "Standardized Comparison of Computer Performance," Information Processing, *Proceedings of the IFIP Congress 1962*, pp. 57–61.

[4] J. R. Hillegasse, A. C. Nester, J. A. Gosden, and R. L. Sisson, "Generalized Measures of Computer Systems Performance," *1962 ACM National Conference*, pp. 120–121.

[5] R. W. Rector, "Measuring the Capability of Computing Equipment," SDC internal working paper N-19243, November 1, 1962.

[6] C. C. Hendrie and R. W. Sonnenfeldt, "Evaluating Control Computers," *ISA Journal*, August 1963, pp. 73–78.

[7] P. W. Abrahams, M. F. Lipp, and J. Harlow," Quantitative Methods of Information Processing System Evaluation," ESD-TDR-63-670, October 1963, published by the Directorate of Computers, ESD (USAF), available through DDC and OTS.

[8] N. Statland, "Methods of Evaluating Computer Systems Performance," *Computers and Automation*, February 1964, pp. 18–23.

[9] L. Fein, "Assessing Computing Systems," *Data Processing for Management*, February 1964, pp. 19–21.

[10] E. Joslin, "Application Benchmark—The Key to Meaningful Computer Evaluations," *Proceedings of ACM Conference—1965*.

[11] G. F. Weinwurm, "The Description of Computer System Performance," *Proceedings of the Symposium on Economics of Automatic Data Processing*, Rome, October 19–22, 1965.

[12] R. A. Arbuckle, "Computer Analysis and Thruput Evaluation," *Computers and Automation*, January 1966, pp. 12–15, 19.

[13] J. R. Hillegasse, "Standardized Benchmark Problems Measure Computer Performance," *Computers and Automation*, January 1966, pp. 16–19.

[14] Martin B. Solomon, Jr., "Economies of Scale and the IBM System/360," *Comm. ACM 10*, June 1966, pp. 435–440.

[15] P. Calingaert, "System Performance Evaluation: Survey and Appraisal," *Comm. ACM 10*, Jan. 1967, pp. 12–18.

[16] M. Schatzoff, R. Tsao, and R. Wiig, "An Experimental Comparison of Time Sharing and Batch Processing," *Comm. ACM 10*, May 1967, pp. 261–265.

[17] M. B. Solo, "Selecting Electronic Data Processing," *Datamation*, Nov.–Dec. 1958, pp. 28–32.

[18] J. DeParis, "Evaluating Available Equipment," *Data Processing*, September 1960, pp. 36–37.

[19] Castillo-Fernandez, "Technique for Evaluating Electronic Computers," *Data Processing*, September 1962, pp. 32–34.

[20] C. R. Pack, and J. L. Hancock, "When a Commander Picks a Computer . . . ," *Armed Forces Management*, March 1963, pp. 38–42.

[21] Q. N. Williams, R. S. Perrot, J. Weitzman, and J. A. Murray, "A Methodology for Computer Selection Studies," *Computers and Automation*, May 1963, pp. 18–22.

[22] E. Joslin, "Cost-Value Technique for Evaluation of Computer System Proposals," 1964 Spring Joint Computer Conference.

[23] S. Rosenthal, "Analytical Technique for Automatic Data Processing Equipment Acquisition," 1964 Spring Joint Computer Conference.

[24] C. Koenig, and K. Bruck, "EDP Equipment Selection," *Management Services*, Sep.–Oct. 1964, pp. 37–43.

[25] E. O. Joslin, and J. J. Aiken, "The Validity of Basing Computer Selections on Benchmark Results," *Computers and Automation*, January 1966, pp. 22, 23.

[26] L. R. Huesmann, and R. P. Goldberg, "Evaluating Computer Systems Through Simulation," *Computer Journal*, Aug. 1967, pp. 150–155.

[27] N. R. Nielsen, "Computer Simulation of Computer System Performance," *Pro-*

ceedings of 22nd National Conference, A.C.M. Publication P-67, Thompson Book Company, Washington, D.C., 1967, pp. 581–590.

[28] A. J. Dowkont, W. A. Morris, and T. D. Buettell, "A Methodology for Comparison of Generalized Data Management Systems: PEGS," DDC AD-811682L, Informatics Inc., Sherman Oaks, Calif., March 1967.

[29] J. R. Miller, III, "A Systematic Procedure for Assessing the Worth of Complex Alternatives," ESD-TR-67-90, The MITRE Corp., Bedford, Mass., Sept. 1966.

[30] Weapon System Effectiveness Industry Advisory Committee (WSEIAC), Final Summary Report, AFSC-TR-65-6, January 1965.

[31] C. J. Hitch and R. N. McKean, *The Economics of Defense in the Nuclear Age,* Harvard University Press, Cambridge, 1960.

[32] R. Schlaifer, *Probability and Statistics for Business Decisions,* McGraw-Hill, New York, 1959.

[33] D. W. Fife, "Alternatives in Evaluation of Computer Systems," MTR-413, The MITRE Corp., Bedford, Mass., April 1967.

Index

465